Essentials of Biotechnology for Students

Essentials of Biotechnology for Students

Edited by

Satya N Das PhD, MAMS
Additional Professor
Department of Biotechnology
All India Institute of Medical Sciences
New Delhi 110029

PEEPEE
PUBLISHERS and DISTRIBUTORS (P) LTD.

Essentials of Biotechnology for Students

Published by
Pawaninder P. Vij
Peepee Publishers and Distributors (P) Ltd.

7/31, Ansari Road, Daryaganj,
New Delhi 110002 (India)

Ph: 55195868, 9811156083

e-mail: peepee160@yahoo.co.in

e-mail: peepee160@rediffmail.com

First Edition: **2005**

ISBN: 81-88867-18-7

Cover design by Prashant Das

Printed at
Lordson, C-5/19, Rana Pratap Bagh, Delhi-110 007

To my father
Late Shri TN Das (1922-2002)
who was a popular teachter,
caring father and a decent human being.
He always stood by me
whenever I needed him.

Preface

Biotecnology is one of the powerful technology of 21st Century which has revolutionized the Biological Sciences and is growing as a separate discipline encompassing multi-disciplinary subjects like cell biology, immunology, molecular biology, recombinant DNA technology and bioinformatics. It has started showing its importance in varied areas such as biology, medicine, agriculture and environment. Human genome project, animal and plant genome studies are the latest contribution of biotechnology that has already started showing their worth.

Why this book was written? Recently, Biotechnology has been added to the school and university curriculum as a separate discipline. Unfortunately, standard Text Books are not available on this subject. Although the market is flooded with many excellent and nicely illustrated books on the various topics related to biotechnology, most fail in their purpose due to their exhaustive texts that are often choking to students. During my twelve years of teaching career, I have also noticed that students have little knowledge of immunology or they are simply not interest in it. **Essentials of Biotechnology for Students** was therefore, conceived to provide students with bare essentials of biotechnology with little extended but defined topics on immunology in a palatable form. Each chapter of the book starts with a short introduction to give readers a background of the topic, followed by the main topic described under suitable headings and sub-headings explained through figures and illustrations. Excessive details and discussions are avoided. Study questions are added at the end of each chapter for revision and learning. New concepts and terms are highlighted in italics and bold face for easy recognition and recall. Appropriate experimental techniques are described wherever necessary. I hope that this book will help students to not only increase awareness and understanding of the subject but to sufficiently pass their course with a high score.

Satya N. Das

Acknowledgements

I acknowledge unconditional support from Vinita, Prashant and Priyanka who relieved me from my domestic duties; Mary and Shalini for their help during preparation of the manuscript.

Contributors

Shivani Arora MSc
Research Fellow
Department of Biochemistry
All India Institute of Medical Sciences
Ansari Nagar, New Delhi

Radhika Bakshi PhD
Lecturer
Department of Biochemistry
Venkateswara College
New Delhi

Shyam S Chauhan PhD
Additional Professor
Department of Biochemistry
All India Institute of Medical Sciences
Ansari Nagar, New Delhi

Satya N Das PhD, MAMS
Additional Professor
Department of Biotechnology
All India Institute of Medical Sciences
Ansari Nagar, New Delhi

Anushree Gupta PhD
Assistant Professor
Department of Biotechnology
All India Institute of Medical Sciences
Ansari Nagar, New Delhi

Rashmi Jalah PhD
Research fellow
Department of Biotechnology
All India Institute of Medical Sciences
Ansari Nagar, New Delhi

Gulshan Wadhwa PhD
Scientist and Programme Officer
Biotechnology Patent Facilitating Cell
Department of Biotechnology
Ministry of Science and Technology,
New Delhi

Contents

PART I: BASIC CELL BIOLOGY AND BIOCHEMISTRY

1. **Cell Structure, Function and Sub-cellular Compartments** — 3
   Rashmi Jalah
2. **Basic Principles of Biochemical Reactions** — 17
   Rashmi Jalah
3. **Enzymes and Mechanisms of Enzyme Action** — 26
   Rashmi Jalah and Satya N. Das
4. **Molecular Composition of Cells** — 42
   Rashmi Jalah
5. **Nucleic Acids** — 67
   Rashmi Jalah and Satya N. Das

PART II: IMMUNOLOGY

6. **Immune System: An Overview** — 85
   Satya N. Das
7. **Innate Immunity** — 93
   Satya N. Das
8. **Antigens Immunization Strategies** — 105
   Satya N. Das
9. **MHC Structure, Genomic Organization, Antigen Processing and Presentation** — 114
   Satya N. Das
10. **Antibody Structure, Function, Genetics and Antibody Engineering** — 128
    Satya N. Das
11. **Antigen-Antibody Interactions and its Applications in Immunodiagnosis** — 151
    Satya N. Das
12. **Biology of B-lymphocytes and Generation of Humoral Immunity** — 172
    Satya N. Das
13. **Biology of T-lymphocytes and Generation of Cellular Immunity** — 180
    Satya N. Das
14. **Cytokines and Cytokine Receptors** — 195
    Satya N. Das

15. Chemokines, Chemokine Receptors, Cell Migration and Homing 205
Satya N. Das
16. Autoimmunity 215
Satya N. Das
17. Transplantation Immunity 223
Satya N. Das
18. Tumour Immunity 235
Satya N. Das
19. Infection Immunity and Strategies for Vaccine Development 251
Satya N. Das

PART III: MOLECULAR BIOLOGY AND RECOMBINANT DNA TECHNOLOGY

20. DNA, Replication and Repair 273
Anushree Gupta
21. Transcription 308
Anushree Gupta
22. Translation 335
Anushree Gupta
23. Regulation of Gene Expression 363
Satya N. Das
24. Recombinant DNA Technology 379
Shivani Arora, Radhika Bakshi and Shyam S. Chouhan

PART IV: APPLIED BIOTECHNOLOGY

25. Good Laboratory Practice 393
Satya N. Das
26. Applications of Biotechnology 408
Anushree Gupta
27. Basic Tools in Bioinformatics 424
Satya N. Das
28. Intellectual Property Rights 436
Gulshan Wadhwa

Index 459

PART I:

BASIC CELL BIOLOGY AND BIOCHEMISTRY

•

Cell Structure, Function and Subcellular Compartments *3*

•

Basic Principles of Biochemical Reactions *17*

•

Enzymes and Mechanisms of Enzyme Action *26*

•

Molecular Composition of Cells *42*

•

Nucleic Acids *67*

1

Cell Structure, Function and Subcellular Compartments

I. INTRODUCT`ION

All living things are composed of cells or cell products. Organisms may be single cells (**unicellular**) or collections of many small cells (**multicellular**). Either way, everything the organism is composed of is cells and everything it does and is capable of doing is directly determined by the chemistry that occurs within its cells. This chemistry is controlled by enzymes, which are themselves controlled by genes.

A. Cells: A Brief History

- **Robert Hooke** built the first compound microscope and in 1665 used the term "cell" to refer to the compartments in a slice of cork.
- Dutch amateur scientist **Antoni van Leeuwenhoek** discovered microscopic animals in water.
- **Brown** named the nucleolus, which he observed in all plant cells.
- German scientist **Matthias J. Schleiden** in 1838 concluded that all plant tissues are composed of cells. In the same year, another German scientist **Theodor S. Schwann** noted cells in animal tissues. Together they proposed that *all organisms are made of one or more cells*.
- German biologist **Virchow** in 1858 stated that all cells come from the *division of pre-existing cells*.

Hence the **cell theory** or **cell principle** had been proposed as follows:

1. All living organisms are made up of one or more cells.
2. The cell is the basic unit of life.
3. All cells come from the division of pre-existing cells.

B.Cell Shape

The cell shape may be:

a. *Variable,* i.e. constantly changing (e.g. *Amoeba* and leucocytes) or

b. *Fixed* – Here, the cells may be:

i. **Flattened**: e.g. squamous epithelium, endothelium and the upper layers of the epidermis.
ii. **Cuboidal**: e.g. in thyroid gland follicles.
iii. **Columnar**: e.g. the cells lining the intestine.
iv. **Discoidal**: e.g. erythrocytes.
v. **Spherical**: e.g. the eggs of many animals.
vi. **Spindle-shaped**: e.g. smooth muscle fibres.
vii. **Elongated**: e.g. nerve cells.
viii. **Branched**: e.g. pigment cells of the skin.

C. Cell Size and the Reason for Multicellularity

The sizes of the cells vary from the very small cells of bacteria (0.2-10 mm) to the very large egg of the ostrich (6 inches). In fact, most eggs are relatively large cells as a considerable part of the volume is made up of yolk, which is not protoplasm. Some cells such as nerve cells may be a meter or more in length. Eukaryotic cells are about 10-100 mm. Several constraints place an upper limit on the size of cells. The important forces limiting cell size are as follows:

1. **The nucleo-cytoplasmic ratio** or the ratio between the volume of the nucleus and the cytoplasm – The cell size is limited by the ability of one set of chromosomes to provide mRNA to the cell's cytoplasm. A cell can get larger by being metabolically inactive (e.g. yolky eggs) or by having extra nuclei (skeletal muscles).
2. **The ratio of the cell surface to the cell volume** (S/V) - All objects have a SURFACE on the outside and a VOLUME on the inside. This is of interest to us because the metabolic processes and reactions that are essential to the cell mostly occur in the volume whereas the supplies and wastes for these reactions cross the surface to enter or leave the cell. There must be sufficient surface area to supply the cell with its needs and remove its wastes.
   So why are cells small? As we move from small size to large the surface increases and the volume increases. So far this is intuitive. The problem is that as size increases the volume increases FASTER than the surface area (as for a spherical cell, **Area = $4\pi r^2$ and Volume = $4/3\pi r^3$**). The consequence is that large objects have RELATIVELY LESS surface area than do small objects. For cells, a large surface to volume ratio is required which only small cells can achieve. Nerve cells are large because they are long and narrow and thus have greater surface area.
3. **Why are large organisms not composed of a single cell or a few very large cells?** Building a body of numerous small cells allows for **specialization** in multicellular organisms. Various cells can differentiate or specialize in particular functions such as locomotion, reproduction, secretion, etc. In unicellular organisms, each cell must accomplish all necessary functions and consequently unicells are far more complex than any given cell of a multicellular organism.

II. THE ORIGIN AND EVOLUTION OF CELLS

Cells are defined into two main classes, initially defined by whether they contain nucleus. **PROKARYOTIC cells** (bacteria) lack a nuclear envelope whereas **EUKARYOTIC cells** have a nucleus in which the genetic material is separated from the cytoplasm. Prokaryotic cells are comparatively smaller and simpler; their genomes are less complex and they do not contain cytoplasmic organelles or a cytoskeleton (Table 1. 1).

Table 1.1: Differences in prokaryotic and eukaryotic cells

S.No.	*Feature*	*Prokaryotic cell*	*Eukaryotic cell*
1.	**DNA**	Present in a **nucleoid** in direct contact with the cytoplasm. enclosed **nucleus**	Located within double-membrane
2.	**Size of cell**	Smaller (~1 mm).	Larger (10-100 mm).
3.	**Type of reproduction**	**Asexual** reproduction. **sexual** reproduction	**Both asexual and** occurs.
4.	**Mode of cell division**	By **DNA replication and binary fission** (mitosis and meiosis does not occur).	Nuclear division during normal cell division (**mitosis**) and **meiosis** (forming sex cells) occurs.
5.	**Chromosomes**	**Single circular chromosome**, **haploid**, generally **no introns**, **histone proteins are absent.** However, there are proteins associated with prokaryotic chromosomes but they are easily separated.	**>1 (multiple) linear chromosomes**, generally **diploid**, **histonene proteins are present**, genes contain intervening non-coding seq. i.e. **introns are present.**
6.	**DNA content (base pairs)**	1×10^6 to 5×10^6	1.5×10^7 to 5×10^9
7.	**Cell membrane composition**	Cell membranes **do not contain sterols** (except *Mycoplasma*). cholestrol.	Cell membranes **contain sterols** like
8.	**Cell wall composition**	Cell walls **contain peptidoglycan** (except for *Archaea*).	Cell walls **do not contain** Peptidoglycan.
9.	**Cytoplasmic organelles and cytoskeleton**	**Membrane-bound organelles do not occur.** Any internal membranes are elaborations of the plasma membrane and are known as **mesosomes**. For example, photosynthetic pigments must be on membranes and in prokaryotes are on mesosomes.	Outside the nucleus is the **cytoplasm** consisting of a semifluid, i.e. the **cytosol** in which are embedded a variety of **membrane bound organelles** such as the **mitochondria, chloroplasts, endoplasmic reticulum, etc.** These organelles are intracellular structures that carry our specific functions.
10.	**Existence**	Found in bacteria and blue green bacteria (i.e. the Kingdom Monera)	Found in the remaining four kingdoms, i.e. the Protista, Fungi, Plantae, and Animalia.

In spite of these differences, the same basic molecular mechanisms govern the lives of both prokaryotes and eukaryotes, indicating that all present day cells are descended from a single primordial ancestor.

A. The First Cell

It appears that life first emerged at least 3.8 billion years ago, approximately 750 million years after earth was formed. The origin and evolution of cells have been predicted to follow the following sequence of events:

1. It was first suggested in the 1920's that **simple organic molecules could form and spontaneously polymerize into macromolecules under the conditions thought to exist in primitive Earth's atmosphere,** i.e. the presence of principally CO_2 and N_2 in addition to smaller amounts of gases such as H_2, H_2S and CO_2 and no or little O_2. Such an atmosphere provides reducing conditions in which organic molecules, given a source of energy such as sunlight or electrical discharge, can form spontaneously. The spontaneous formation of organic molecules was first demonstrated experimentally in the 1950s, when **Stanley Miller** showed the discharge of electric sparks into a mixture of H_2, CH_4 and NH_3, in the presence of water, led to the formation of a variety of organic molecules, including several amino acids (Fig. 1.1). Although Miller's experiments did not precisely

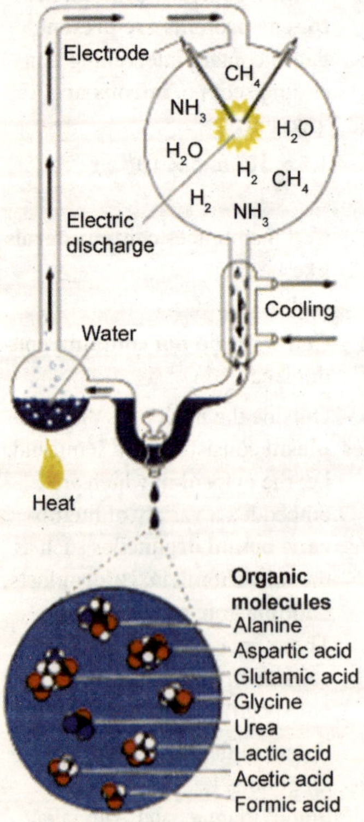

Fig.1.1: Spontaneous formation of organic molecules. Water vapour was refluxed through an atmosphere consisting of H_2, CH_4 and NH_3, into which electric sparks were discharged. Analysis of the reaction products revealed the formation of a variety of organic molecules, including the amino acids alanine, aspartic acid, glutamic acid and glycine

reproduce the conditions of primitive Earth, they clearly demonstrated the plausibility of the spontaneous synthesis of organic molecules, providing the basic materials from which the first living organisms arose.

2. The next step in evolution was the **formation of macromolecules**. The monomeric building blocks of macromolecules have been demonstrated to polymerize spontaneously under plausible prebiotic conditions, e.g. heating dry mixtures of amino acids, results in their polymerization to form polypeptides. However, only nucleic acids (and not proteins) are capable of directing their own self-replication, i.e. they can serve as templates for their own synthesis as a result of specific base pairing between complementary nucleotides. A critical step in understanding molecular evolution was thus reached in the early 1980s, when it was discovered by **Sid Altman and Tom Cech** that RNA is capable of catalyzing a number of chemical reactions, including the polymerization of nucleic acids. RNA is thus uniquely able both to serve as a template for and to catalyze its own replication. Consequently, **RNA is generally believed to have been the initial genetic system**, and an early stage of chemical evolution is thought to have been based on self-replicating RNA molecules – a period of evolution known as the **RNA world. Ordered interactions between RNA and amino acids then evolved into the present-day genetic code, and DNA eventually replaced RNA as the genetic material.**

3. Further, **the first cell** is presumed to have arisen by the **enclosure of self-replicating RNA in a membrane composed of phospholipids**, which are the basic component of the present-day biological membranes, including the plasma membrane of both prokaryotes and eukaryotic cells. They are **amphipathic** molecules having long water-insoluble (**hydrophobic**) hydrocarbon chains joined to water-soluble (**hydrophilic**) head groups that contain phosphate. When placed in water, phospholipids spontaneously aggregate into **bilayer** with their phosphate-containing head groups on the outside in contact with water and their hydrocarbon tails in the interior in contact with each other. Such a phospholipid bilayer forms a stable barrier between two aqueous compartments, e.g. separating the interior of the cell from its external environment.

   The enclosure of self-replicating RNA and associated molecules in a phospholipid membrane would thus have maintained them as a unit, capable of self-reproduction and further evolution. RNA directed protein synthesis may already have evolved by this time, in which case **the first cell would have consisted of self-replicated RNA and its encoded proteins.**

4. **The evolution of metabolism.** Since cells originated in a sea of organic molecules, they were able to obtain food and energy directly from the environment. But such a situation is self-limiting, so cells needed to evolve their own mechanisms for generating energy and synthesizing the molecules necessary for their replication. The generation and controlled utilization of

metabolic energy is central to cell activities, and the principal pathways of energy metabolism are highly conserved in present-day cells. All cells use adenosine 5'-triphosphate (ATP) as their source of metabolic energy. The mechanisms used by cells for ATP generation are thought to have evolved in three stages, corresponding to the evolution of **glycolysis, photosynthesis and then oxidative metabolism** (Fig. 1.2). The development of these metabolic pathways changed Earth's atmosphere, thereby altering the course of further evolution.

1. Glycolysis

$C_6H_{12}O_6 \rightarrow 2\ C_3H_6O_6$ **(Generates 2 ATP)**
Glucose Lactic acid

2. Photosynthesis

$6\ CO_2 + 6\ H_2O \xrightarrow{\text{Sunlight}} C_6H_{12}O_6 + 6\ O_2$
Glucose

3. Oxidative metabolism

$C_6H_{12}O_6 + 6\ O_2 \rightarrow 6\ CO_2 + 6\ H_2O$ **(Generates 36-38 ATP)**
Glucose

Fig.1.2: Generation of metabolic energy. Glycolysis is the anaerobic breakdown of glucose to lactic acid. **Photosynthesis** utilizes energy from sunlight to drive the synthesis of glucose from CO_2 and H_2O, with release of O_2 as the byproduct. The O_2 released by photosynthesis is used in **oxidative metabolism**, in which glucose is broken down to CO_2 and H_2O, releasing much more energy than is obtained from glycolysis.

In the initially anaerobic (without oxygen) atmosphere of Earth, the anaerobic breakdown of glucose to lactic acid, with the net energy gain of two ATP molecules, i.e. **glycolysis** must have occurred. The development of **photosynthesis** is thought to have been the next evolutionary step, which allows the cells to harness energy from sunlight and provided independence from the utilization of preformed organic molecules. The first photosynthetic bacteria, which evolved more than 3 billion years ago, probably utilized H_2S to convert CO_2 to organic molecules–a pathway of photosynthesis still used by some bacteria. The use of H_2O a donor of electrons and hydrogen for the conversion of CO_2 to organic compounds evolved later and is thought to have been responsible for making O_2 abundant in Earth's atmosphere. This led to the development of **oxidative metabolism**. Alternatively, oxidative metabolism may have evolved before photosynthesis, with the increase in atmospheric O_2 then providing a strong selective advantage for organisms capable of using O_2 in energy-producing reactions. In either case, O_2 is a highly reactive molecule and oxidative metabolism, utilizing this reactivity, has provided a mechanism for generating energy from organic molecules that is much more efficient than anaerobic glycolysis. Hence, with few exceptions, present-day cells use oxidative reactions as their principal source of energy.

III. THE EUKARYOTIC CELL STRUCTURE AND CELL COMPONENTS

Structure and components of the cell have been shown in Figure1.3.Eukaryotic cells are organized into different compartments. Cellular compartments are

A. Animal cells

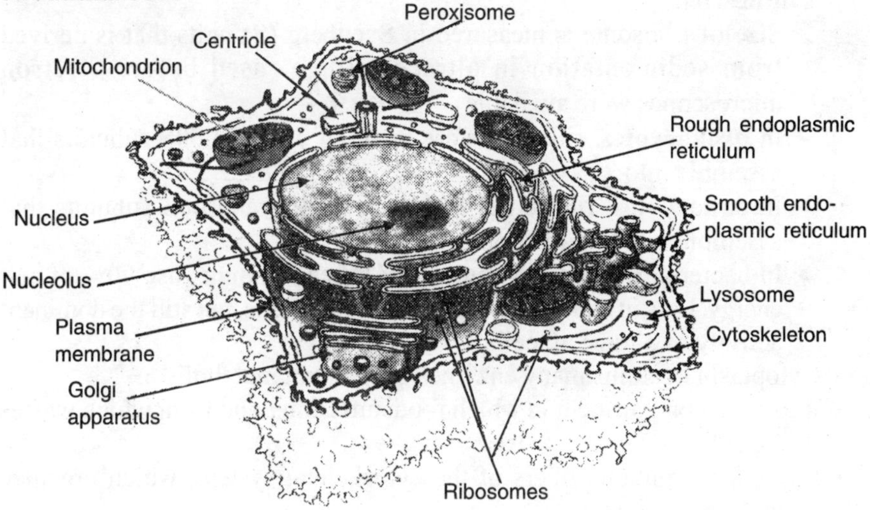

B. Plant cells

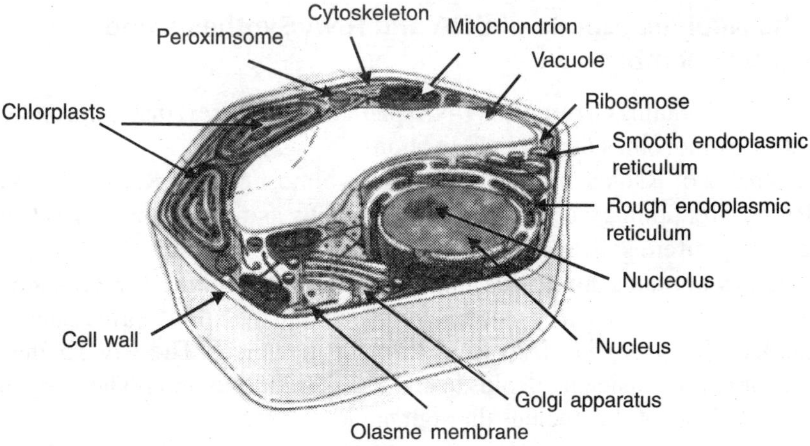

Figs 1.3A and B: Structures of (a) animal (b) plant cells. Both animal and plant cells are surrounded by a plasma membrane and contain a nucleus, a cystoskeleton and many cytoplasmic organelles in common. Plant cells are also surrounded by a cell wall and contain chloroplasts and large vacuoles

bound by membranes. Cytoplasm constitutes a central metabolic compartment, bounded by cell membrane. Other compartments inside cytoplasm are called organelles. Compartmentation allows specialized functions to be carried out in different locations.

A.The Cytoplasm: Site of Protein Synthesis and Many Metabolic Events

- Cytoplasm contains many **ribosomes (particles on which proteins are synthesized).**
  - ➢ Size of ribosome is measured in Svedberg (S) units that is derived from sedimentation in ultracentrifuge (used before electron microscopes were available).
  - ➢ In **prokaryotes**, ribosomes are made of **30S** and **50S** subunits that assemble into **70S** ribosome.
  - ➢ In **eukaryotes**, ribosomes are made of **40S** and **60S** subunits that assemble into **80S** ribosome.
  - ➢ In bacteria, ribosomes occupy 25% of cell volume, use 90% of cell energy. Less in many specialized eukaryotic cells, but still the dominant activity of almost all cells.
- Cytoplasm contains many enzymes for general metabolism.
- It forms a compartment in which foodstuffs enter and from which wastes leave cell.
- Cytoplasm contains fibers of the cytoskeletal system, which organize cytoplasmic structure.
- It also contains many different organelles.

B. The Nucleus: Locus of DNA and RNA Synthesis and Protein Assembly

- Nucleus contains **chromatin (DNA-protein complexes)** that can condense into chromosomes during cell division.
- It is a site of RNA synthesis and contains almost 80% of RNA (ribosomal RNA). Remaining 20% RNA leaves nucleus as t-RNA and m-RNA and directs **protein synthesis.**
- Nucleus contains **nucleolus** that is an **assembly plant for ribosomes**. Ribosomal proteins are made in cytoplasm, must be transported back into nucleus. Ribosomal RNA (r RNA) is made in nucleus. These two elements are integrated inside nucleolus to create ribosomal subunits. These are then exported out of the nucleus through nuclear pores.
- Nucleus is bounded by **nuclear membrane** that is a **double-layered structure**. It contains many **nuclear pores** that allow material to move in and out of nucleus.
- Nuclear pores have octagonal "doors" made of protein; open and close on either side depending on specific signals. Pore has diameter of about

10 nm, smaller than diameter of a complete ribosome. Pore can open up to as much as 26 nm in response to certain signals. Some signals allow motion in but not out, other signals control reverse transport.

C. The Endomembrane System: Moving Materials into Different Compartments

Endomembrane system is a set of interconnected compartments: endoplasmic reticulum (ER), Golgi body, lysosomes, cell membrane.

1. Endoplasmic reticulum

- **Rough ER** (r ER): Synthesizes proteins for export or movement to different cell compartments (but **not** to cytoplasm).
- **Signal hypothesis**: Certain mRNAs encode proteins designated for export. These carry a peptide signal at growing end, causes growing protein to move to ER ("docking"), insert peptide into membrane, translocate growing polypeptide chain across ER membrane. When protein synthesis is complete, polypeptide folds up inside ER, not in cytoplasm.
- **Smooth ER** (sER): Synthesizes lipids, detoxifies drugs and poisons (in liver).

2. Golgi body

- Functions as intracellular "postoffice" for sorting new proteins made on rER.
- Vesicles containing protein pinch off from ER, fuse with *cis* face of Golgi. Inside Golgi, oligosaccharide chains on proteins are modified. Vesicles pinch off from *trans* face of Golgi, carry proteins to several possible destinations such as lysosomes, peroxisomes, cell membrane, etc.and export it outside the cell.

3. Lysosomes

- Compartments to breakdown old proteins, foreign materials, and many wastes.
- Contain ~40 hydrolytic enzymes: lipases, proteases, nucleases, etc. Break-down organic polymers of all types.
- "Suicide bags" if opened up causes **apoptosis** of cell itself.
- Lysosomes are used in phagocytosis, a process in which foreign materials are brought into the cell and "chewed up".

4. Cell membrane (or plasma membrane).

5. Vacuoles

- Large membrane compartments (contrasted with small membrane bags called vesicles).
- Plant cells have especially large vacuole called the **central vacuole**, can occupy most of the volume of a plant cell. Stores pigments, wastes, water, poisons, and more.

D.Organelles Involved in Energy Transformations are Separate from the Endomembrane System

"Energy organelles" have unique properties:

- They are enclosed by **double membrane** system.
- They contain DNA and ribosomes (70S, not 80S like cytoplasmic ribosomes).
- They make some of their own proteins from their own genes.
- They divide by binary fission (but not autonomous, cannot grow or sustain life outside of cell.

1. **Mitochondria (centers for respiratory catabolism)**. Oxygen combined with chemicals to breakdown foods, generate cell energy. Contain outer and inner compartments, with many membranous **cristae** that "criss-cross" the internal space. Found in virtually every eukaryotic cell. They form small structures similar to the size of bacteria in some cases.
2. **Chloroplasts (centers for photosynthetic anabolism)**. Chloroplast belongs to group of plant organelles called plastids. It include chloroplasts (photosynthesis), amyloplasts (store starch), chromoplasts (store pigments). It traps light and converts energy to sugars (+ CO_2, water). Chloroplasts contain stacks of **thylakoids**, where green-pigmented chlorophyll is embedded in membrane to trap light.
3. **The endosymbiont theory:**

   All organelles seem to share many properties with bacteria: contain 70S ribosomes (whereas rest of eukaryote cells contain 80S ribosomes), divide by binary fission, contain circular DNA without nucleus, etc. **Lynn Margulis** proposed **endosymbiont hypothesis**: that organelles derived from ancient colonization of large bacteria (became the eukaryotic cell) by smaller bacteria (became the mitochondria, chloroplast, etc.) Symbiosis = "living together". Eventually, organelles lost ability to exist as separate organisms, cannot be separated from cell. Recent evolutionary taxonomy by comparing ribosomal RNA shows that this idea has lots of merit. Mitochondrial and plastid ribosomes are very similar to current bacteria, very different from eukaryotes (Fig. 1.4).

E. Cytoskeletal System Provides Internal Fibrous Structure to Cells

Cell is not "just a bag in a bubble". Lots of internal fibers forms internal **skeleton**. Cytoskeleton is **not** rigid like bone but it is capable of being assembled and broken down in minutes. It allows cell movement, cell division, internal motion of compartments. Cytoskeletal system consists of three main components:

1. **Microtubules**
   - Largest diameter fiber. Found in cytoplasm of all eukaryotes.
   - Involved in many structures: cilia, flagella (9+2 arrangement); spindle fibers that polymerize from centrioles during mitosis/meiosis.

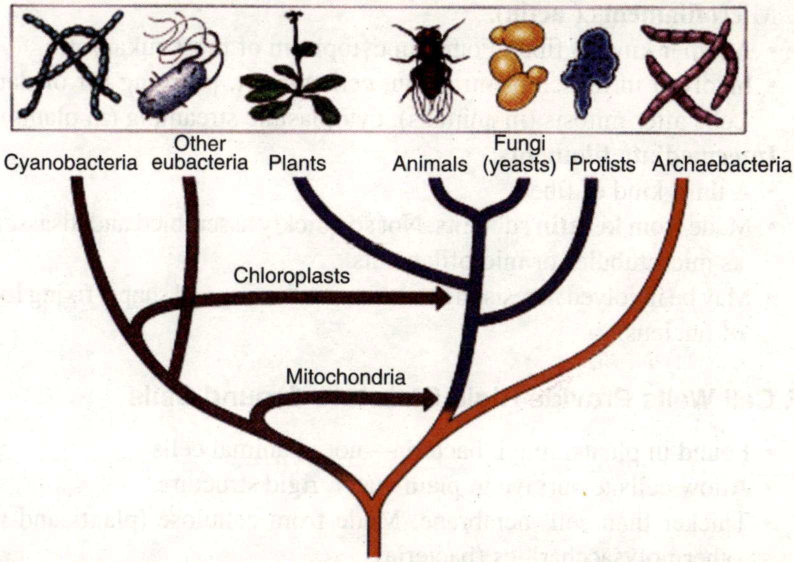

Fig.1.4: Evolution of cells. Present-day cells evolved from a common prokaryotic ancestor along three lines of descent, giving rise to archaebacteria, eubacteria and eukaryotes. Mitochondria and chloroplasts originated from the endosymbiotic association of aerobic bacteria and cyanobacteria, respectively, with the ancestors of eukaryotes

- Made of tubulin protein; polymerizes into hollow tubules 25 nm diameter.
- View cell treated with anti-tubulin fluorescent antibody
- **Cilia and flagella**: organelles of locomotion. Contain 9 double rings of microtubules, 2 central microtubules.
- Two motor proteins allow motion along microtubules
  i. Motor protein 1—Dynein
    - Side arms allow one tubule to "walk up" along neighboring tubule, cause bending (powered by ATP).
    - Causes motion from positive (+) end of the microtubule (where new tubulin is added to the microtubule) toward the minus (-) end of the microtubule.
    - When coordinated, this causes rotating or whip-like motion of entire cilium or flagellum ⟶ motion.
  ii. Motor protein 2 — Kinesin
    - Also powered by ATP, also allows protein to move along microtubule
    - Causes motion from negative (-) end of the microtubule toward the positive (+) end of the microtubule (where new tubulin is added to the microtubule).
    - Pulls things toward outer reaches of cell
    - Example: in nerve cells, kinesin pulls vesicles away from center towards nerve endings.

2. Microfilaments (actin).

- Another kind of fiber. Found in cytoplasm of most eukaryotes.
- Involved in muscle contraction, cell support, pinching off of daughter cells after mitosis (in animals), cytoplasmic streaming (in plants).

3. Intermediate filaments.

- A third kind of fiber.
- Made from **keratin** subunits. Not so quickly assembled and disassembled as microtubules or microfilaments.
- May be involved in resisting tension, reinforcing cell shape, fixing location of nucleus.

F. Cell Walls Provide Rigid Structure Around Cells

- Found in plants, fungi, bacteria—not in animal cells
- Allow cells to survive in plain water, rigid structure.
- Thicker than cell membrane. Made from cellulose (plants and fungi), other polysaccharides (bacteria).
- Cells maintain contact by **plasmodesmata**—thin cytoplasmic connections, lined by membrane, that pass across cell wall junctions.

G. Cells Do not End at their Outer Membrane; They Possess an Extracellular Matrix (ECM)

- Animal cells do not have walls, but do have ECM (meshwork of macromolecules outside plasma membrane). Consists mainly of glycoproteins (proteins with oligosaccharide chains), especially collagen.
- Some cells attached directly to ECM by bonding to collagen or fibronectin.

H. Cells are Joined by a Variety of Intracellular Junctions

- In multicellular organisms, adjacent cells are held together by several types of specialized junctions:
  - **Tight junctions** (found in animals): specialized "belts" that bind two cells tightly to each other, prevent fluid from leaking into intracellular space.
  - **Desmosomes** (found in animals): intercellular "rivets" that create tight bonds between cells, but allow fluids to pass through intracellular spaces.
  - **Gap junctions** (found in animals): formed by two connecting protein rings embedded in cell membrane of adjacent cells. Allows passage of water, small solutes, but not macromolecules (proteins, nucleic acids).
  - **Plasmodesmata** (found in plants): channels connecting cells; allow free passage of water and small solutes, but not macromolecules (proteins, nucleic acids).

STUDY QUESTIONS

1. Compare and contrast between prokaryotic and eukaryotic cells.
2. Describe briefly various components of a cell.
3. Write short note on the cell organells that are involved in energy transformation.
4. Describe briefly about origin and evolution of cells.

MCQs

Select one most appropriate answer given for each questions below:

1. **Endosymbiont hypothesis was proposed by:**
   A. Robert Brown
   B. Lynn Mangulis
   C. Theodore Schwann
   D. Virchow
2. **DNA content of a prokaryotic cell is approximatelyL:**
   A. Between 1×10^6 to 5×10^6
   B. Between 1×10^7 to 5×10^7
   C. Between 1×10^8 to 5×10^8
   D. Between 1×10^7 to 1×10^9
3. **Eukaryotic ribosomes are made up of:**
   A. 70 S ribosome
   B. 80 S ribosome
   C. 30 S and 60 S subunits
   D. 30 S and 50 S subunits
4. **Which of the following cellular proteins is motor protein:**
   A. Keratin
   B. β – actin
   C. Kinesin
   D. Cytokeratin.
5. **The spontaneous formation of organic molecules was first demonstrated by**
   A. Schleiden and Schwann
   B. Brown
   C. Virchow
   D. Miller
6. **The surface area of a spherical cell can be determined by the equation.**
   A. Area = 4 p r^2
   B. Area = 2 p r^2
   C. Area = p r^2
   D. Area = ½ p r^2

7. **Which of the following statements is false regarding a prokaryotic cell:**
   A. It has a nuclear envelope separating geneti materials from the cytoplasm
   B. It is comparatively simpler and smaller
   C. It does not have cytoplasmic material
   D. Its genome is less complex
8. **Prokaryotic cells:**
   A. Are generally haploid
   B. Contain introns.
   C. Proteins are generally histones
   D. Have linear chromosomes
9. **Size of ribosome is measured in:**
   A. Dalton unit (Da)
   B. Kilo Dalton unit (kDa)
   C. Svedberg unit (S)
   D. Microns (m)
10. **Eukaryotic ribosomes are made up of:**
    A. 30 S and 50 S subunits
    B. 40 S and 60 S subunits
    C. 30 S and 60 S subunits
    D. 40 S and 50 S subunits
11. **The endosymbiont hypothesis was proposal by:**
    A. Theodors Schwann
    B. Robert Brown
    C. Lynn Margulis
    D. Mathias J. Schleiden
12. **Golgi complex is mainly responsible for:**
    A. Protein folding
    B. Protein transport
    C. Protein synthesis
    D. Protein degradation

1 B **2** A **3** B **4** C **5** D **6** B **7** A **8** A
9 C **10** B **11** C **12** B

2

Basic Principles of Biochemical Reactions

I. INTRODUCTION

Cells are incredibly complex and diverse structure, capable not only of self-replication—the very essence of life - but also of performing a wide range of specialized tasks in multicellular organisms. Yet cells obey the same laws of chemistry and physics that determine the behaviour of non-living systems. Consequently, modern cell biology seeks to understand cellular processes in terms of chemical and physical reactions.

II. THE IONIC EQUILIBRIUM

K_{eq}, K_w and pH

As H_2O is the medium of biological systems one must consider the role of this molecule in the dissociation of ions from biological molecules. Water is essentially a neutral molecule but will ionize to a small degree. This can be described by a simple equilibrium equation:

$$H_2O \longleftrightarrow H^+ + OH^-$$

This equilibrium can be calculated as for any reaction:

$$K_{eq} = [H^+][OH^-]/[H_2O]$$

Since the concentration of H_2O is very high (55.5M) relative to that of the $[H^+]$ and $[OH^-]$, consideration of it is generally removed from the equation by multiplying both sides by 55.5 yielding a new term, K_w:

$$K_w = [H^+][OH^-]$$

This term is referred to as the ion product. In pure water, to which no acids or bases have been added:

$$K_w = 1 \times 10^{-14}\ M^2$$

As K_w is constant, if one considers the case of pure water to which no acids or bases have been added:

$$[H^+] = [OH^-] = 1 \times 10^{-7}\ M$$

This term can be reduced to reflect the hydrogen ion concentration of any solution. This is termed the pH, where:

$$pH = -\log[H^+]$$

pK_a

Acids and **bases** can be classified as proton donors and proton acceptors, respectively. This means that the conjugate base of a given acid will carry a net charge i.e. more negative than the corresponding acid. In biologically relevant compounds, various weak acids and bases are encountered, e.g. the acidic and basic amino acids, nucleotides, phospholipids, etc.

Weak acids and bases in solution do not fully dissociate and, therefore, there is an equilibrium between the acid and its conjugate base. This equilibrium can be calculated and is termed the **equilibrium constant** = K_a. This is also sometimes referred to as the dissociation constant as it pertains to the dissociation of protons from acids and bases.

In the reaction of a weak acid:

$$HA \longleftrightarrow A^- + H^+$$

the equlibrium constant can be calculated from the following equation:

$$K_a = [H^+][A^-]/[HA]$$

As in the case of the ion product:

$$pK_a = -\log K_a$$

Therefore, in obtaining the -log of both sides of the equation describing the dissociation of a weak acid we arrive at the following equation:

$$-\log K_a = -\log[H^+][A^-]/[HA]$$

Since as indicated above $-\log K_a = pK_a$ and taking into account the laws of logrithms:

$$pK_a = -\log[H^+] -\log[A^-]/[HA]$$

$$pK_a = pH -\log[A^-]/[HA]$$

From this equation it can be seen that the smaller the pK_a value the stronger is the acid. This is due to the fact that the stronger an acid the more readily it will give up H^+ and, therefore, the value of [HA] in the above equation will be relatively small.

The Henderson-Hasselbalch Equation:

By rearranging the above equation we arrive at the **Henderson-Hasselbalch equation**:

$$pH = pK_a + \log[A^-]/[HA]$$

It should be obvious now that the pH of a solution of any acid (for which the equilibrium constant is known, and there are numerous tables with this information) can be calculated knowing the concentration of the acid, HA, and its conjugate base $[A^-]$.

At the point of the dissociation where the concentration of the conjugate base $[A^-]$ = to that of the acid [HA]:

$$pH = pK_a + \log[1]$$

The log of 1 = 0. Thus, at the midpoint of a titration of a weak acid:

$$pK_a = pH$$

In other words, the term pK_a is that pH at which an equivalent distribution of acid and conjugate base (or base and conjugate acid) exists in solution.

Buffering

It should be noted that around the pK_a the pH of a solution does not change appreciably even when large amounts of acid or base are added. This phenomenon is known as **buffering**. In most biochemical studies, it is important to perform experiments, that will consume H^+ or OH^- equivalents, in a solution of a buffering agent that has a pK_a near the pH optimum for the experiment.

Ampholytes, Polyampholytes, pI and Zwitterion

Many substances in nature contain both acidic and basic groups as well as many different types of these groups in the same molecule. (e.g. proteins). These are called **ampholytes** (one acidic and one basic group) or **poly-ampholytes** (many acidic and basic groups). Proteins contain many different amino acids some of which contain ionizable side groups, both acidic and basic. Therefore, a useful term for dealing with the titration of ampholytes and polyampholytes (e.g. proteins) is the **isoelectric point, pI**. This is described as the pH at which the effective net charge on a molecule is zero.

For the case of a simple ampholyte like the amino acid glycine the pI, when calculated from the Henderson-Hasselbalch equation, is shown to be the average of the pK for the α-COOH group and the pK for the $\alpha\text{-}NH_2$ group:

$$pI = [pK_{\alpha\text{-}(COOH)} + pK_{\alpha\text{-}(NH3^+)}]/2$$

For more complex molecules such as polyampholytes the pI is the average of the pK_a values that represent the boundaries of the zwitterionic form of the molecule. The pI value, like that of pK, is very informative as to the nature of different molecules. A molecule with a low pI would contain a predominance of acidic groups, whereas a high pI indicates predominance of basic groups.

Solvation and Hydration Shells

Depending on the pH of a solution, macromolecules such as proteins which contain many charged groups, will carry substantial net charge, either positive or negative. Cells of the body and blood contain many **polyelectrolytes** (molecules that contain multiple same charges, e.g. DNA and RNA) and **polyampholytes** that are in close proximity. The close association allows these molecules to interact through opposing charged groups. The presence, in cells and blood, of numerous small charged ions (e.g. Na^+, Cl^-, Mg^{2+}, Mn^{2+}, K^+) leads to the interaction of many small ions with the larger macroions. This interaction can result in a **shielding** of the electrostatic charges of like-charged molecules. This electrostatic shielding allows macroions to become more closely associated than predicted based upon their expected charge repulsion from one another. The net effect of the presence of small ions is to maintain the solubility of macromolecules at pH ranges near their pI. This interaction between solute (e.g. proteins, DNA, RNA, etc.) and solvent (e.g. blood) is termed **solvation** or **hydration**. The opposite effect to solvation occurs when the salt (small ion) concentration increases to such a level as to interfere with the solvation of proteins by H_2O. This results from the H_2O forming **hydration shells** around the small ions.

III. LAWS OF THERMODYNAMICS

A. First Law of Thermodynamics

Stated simply; the total energy of the universe does not change. This does not mean that the form of the energy cannot change. Indeed, chemical energies of a molecule can be converted to thermal, electrical or mechanical energies.

The internal energy of a system can change only by work or heat exchanges. From this the change in the free energy of a system can be shown by the following equation:

$$\Delta E = q - w$$

When q is negative heat has been flowed from the system and when q is positive heat has been absorbed by the system. Conversely when w is negative work has been done on the system by the surrounding and when positive, work has been done by the system on the surroundings.

In a reaction carried out at a constant volume no work will be done on or by the system, only heat will be transferred from the system to the surroundings. The end result is that:

$$\Delta E = q$$

When the same reaction is performed at constant pressure the reaction vessel will do work on the surroundings. In this case:

$$\Delta E = q - w$$

$$\text{where, } w = P\Delta V$$

When the initial and final temperatures are essentially equal (e.g. in the case of biological systems):

$$\Delta V = \Delta n[RT/P]$$

therefore, $\mathbf{w = \Delta nRT}$

Thus, one can calculate the amount of heat released under constant pressure:

$$q = \Delta E + w = \Delta E + P\Delta V = \Delta E + \Delta nRT$$

where, Δn is the change in moles of gas per mole of substance oxidized (or reacted), R is the gas constant and T is absolute temperature.

Enthalpy

Since all biological reactions take place at constant pressure and temperature the state function of reactions defined to account for the heat evolved (or absorbed) by a system is **enthalpy** given the symbol, **H**.

The changes in enthalpy are related to changes in free energy by the following equation:

$$\Delta H = \Delta E + P\Delta V$$

Here, we are addressing the constant pressure situation. In the biological setting most all reaction occur in a large excess of fluid, therefore, essentially no gases are formed during the course of the reaction. This means that the value ΔV, is extremely small and thus the product $P\Delta V$ is very small as well. The values ΔE and ΔH are very nearly equivalent in biological reactions

Stated above was the fact that state functions, like ΔH and ΔE, do not depend on the path taken during a reaction. These functions pertain only to the differences between the initial and final states of a reaction. However, heat (q) and work (w) are not state functions and their values are affects by the pathway taken.

Reversible and Irreversible Reactions

In an idealized irreversible reaction such as one done by expanding an ideal gas against zero pressure, no work will be done by or on the system so the:

$$w = 0$$

In the case of an ideal gas (whose molecules do not interact) there will be no change in internal energy either so:

$$\Delta E = 0$$

since $\Delta E = q - w$, in this irreversible reaction $q = 0$ also.

In a reversible reaction involving an ideal gas, ΔE still will equal zero, however, the pressure will be changing continuously and work (w) is a function of P, work done must be determined over the entire course of the reaction. This results in the following mathematical reduction:

$$w = RT\ln[V_2/V_1]$$

Since in this situation $\Delta E = 0$, $q = w$. This demonstrates that some of the heat of the surroundings has to be absorbed by the system in order to perform the work of changing the system volume.

Reversible reactions differ from irreversible in that the former always proceeds infinitely slowly through a series of intermediate steps in which the system is always in the equilibrium state. Whereas, in the irreversible reaction no equilibrium states are encountered. Irreversible reactions are also spontaneous or favorable processes. Thermodynamic calculations do not give information as to the rates of reaction only whether they are favorable or not.

B. Second Law of Thermodynamics

Entropy

The second law of thermodynamics states that the universe (i.e. all systems) tend to the greatest degree of randomization. This concept is defined by the term **entropy, S.**

$$S = k\ln W$$

Where, k = Boltzmann constant (the gas constant, R, divided by Avagadros' number) and W = the number of substrates. For an isothermal reversible reaction the change in entropy can be reduced to the term:

$$\Delta S = \Delta H/T$$

Whereas, enthalpy is a term whose value is largely dependent upon electronic internal energies, entropy values are dependent upon translational, vibrational and rotational internal energies. Entropy also differs from enthalpy in that the values of enthalpy that indicate favored reactions are negative and the values of entropy are positive. Together the terms enthalpy and entropy demonstrate that a system tends toward the highest entropy and the lowest enthalpy.

In order to effectively evaluate the course (spontaneity or lack there of) of a reaction and taking into account both the first and second laws of thermodynamics, Josiah Gibbs defined the term, free energy. Free energy:

$$\Delta G = \Delta H - T\Delta S$$

Free energy is a valuable concept because it allows one to determine whether a reaction will proceed and allows one to calculate the equilibrium constant of the reaction, which defines the extent to which a reaction can proceed. The discussion above indicated that a decrease in energy, a negative ΔH, and an increase in entropy, a positive ΔS, are indicative of favorable reactions. These terms would, therefore, make ΔG a negative value. Reactions with negative ΔG values are termed exergonic and those with positive ΔG values endergonic. However, when a system is at equilibrium:

$$\Delta G = 0$$

Gibb's free energy calculations allows one to determine whether a given reaction will be thermodynamically favorable. The sign of ΔG states that a reaction as written or its reverse process is the favorable step. If ΔG is negative then the forward reaction is favored and visa versa for ΔG values that are calculated to be positive.

IV. STANDARD STATE CONDITIONS IN BIOLOGICAL REACTIONS

To effectively interpret the course of a reaction in the presence of a mixture of components, such as in the cell, one needs to account for the free energies of the contributing components. This is accomplished by calculating total free energy, which is comprised of the individual free energies. In order to carry out these calculations, one needs to have a reference state from which to calculate free energies. This reference state, termed the **Standard State**, is chosen to be the condition where each component in a reaction is at 1M. Standard state free energies are given the symbol:

$$G^o$$

The partial molar free energy of any component of the reaction is related to the standard free energy by the following:

$$G = G^o + RT\ln[X]$$

From this one can see that when the component X, or any other component, is at 1M the ln[1] term will become zero and:

$$G = G^o$$

The utility of free energy calculations can be demonstrated in a consideration of the diffusion of a substance across a membrane. The calculation needs to take into account the changes in the concentration of the substance on either side of the membrane. This means that there will be a ΔG term for both chambers and, therefore, the total free energy change is the sum of the ΔG values for each chamber:

$$\Delta G = \Delta G_1 + \Delta G_2 = RT\ln\{[A]_2/[A]_1\}$$

This tells one that if $[A]_2$ is less than $[A]_1$ the value of ΔG will be negative and transfer from region 1 to 2 is favored. Conversely if $[A]_2$ is greater than $[A]_1$ ΔG will be positive and transfer from region 1 to 2 is not favorable, the reverse direction is.

One can expand upon this theme when dealing with chemical reactions. It is apparent from the derivation of ΔG values for a given reaction that one can utilize this value to determine the equilibrium constant, $\mathbf{K_{eq}}$. As for the example above dealing with transport across a membrane, calculation of the total free energy of a reaction includes the free energies of the reactants and products:

$$\Delta G = G_{(products)} - G_{(reactants)}$$

Since this calculation involves partial molar free energies the ΔG^o terms of all the reactants and products are included. The end result of the reduction of all the terms in the equation is:

$$\Delta G = \Delta G^o + RT\ln\{[C][D]/[A][B]\}$$

When equation 20 is used for a reaction that is at equilibrium the concentration values of A, B, C and D will all be equilibrium concentrations and, therefore, will be equal to K_{eq}. Also, when at equilibrium $\Delta G = 0$. Therefore:

$$0 = \Delta G^o + RT\ln K_{eq}$$

$$K_{eq} = e^{-\{\Delta G^o/RT\}}$$

This demonstrates the relationship between the free energy values and the equilibrium constants for any reaction.

Coupled Reactions

Two or more reactions in a cell sometimes can be coupled so that thermodynamically unfavorable reactions and favorable reactions are combined to drive the overall process in the favorable direction. In this circumstance, the overall free energy is the sum of individual free energies of each reaction. This process of coupling reactions is carried out at all levels within cells. The predominant form of coupling is the use of compounds with high energy to drive unfavorable reactions.

The predominant form of high energy compounds in the cell are those which contain phosphate. Hydrolysis of the phosphate group can yield free energies in the range of -10 to -62 kJ/mol. These molecules contain energy in the phosphate bonds due to:

1. Resonance stabilization of the phosphate products
2. Increased hydration of the products
3. Electrostatic repulsion of the products
4. Resonance stabilization of products
5. Proton release in buffered solutions

The latter phenomenon indicates that the pH of the solution a reaction is performed in will influence the equilibrium of the reaction. To account for the fact that all cellular reactions take place in an aqueous environment and that the $[H_2O]$ and $[H^+]$ are essentially constant these terms in the free energy calculation have been incorporated into a free energy term identified as:

$$\Delta G^{o'} = \Delta G^o + RTln\{[H^+]/[H_2O]\}$$

Incorporation of this equation into a free energy calculation for any reaction in the cell yields:

$$\Delta G = \Delta G^{o'} + RTln\{[products]/[reactants]\}$$

STUDY QUESTIONS

1. Show that the transfer of heat from one object of higher temperature to one lower temperature, but not the reverse process, obeys the second law of thermodynamics.
2. In terms of thermodynamics concepts, why it is more difficult to park a car in a small space than it is to drive it out from such a space?

MCQs

Select one most appropriate answer given for each questions below:

1. **Chemical energy of a cell can be converted into:**
   A. Electrical energy
   B. Electrical and mechanical energy
   C. Electrical and thermal energy
   D. Electrical, mechanical and thermal energy.

2. **Energy in a cell is stored predominantly in the form of:**
   A. Carbonate
   B. Phosphate
   C. Sulphate
   D. Pyruvate
3. **DNA of a cell is typical example of:**
   A. Ampholytes
   B. Electrolytes
   C. Polyampholytes
   D. Polyelectrolytes
4. **Proteins of a cell are typical example of:**
   A. Electrolytes
   B. Polyelectrolytes
   C. Polyampholytes
   D. Ampholytes
5. **Isoelectric point of a protein can be described as pH at which net charge on the molecule is:**
   A. One
   B. Zero
   C. Positive
   D. Negative

---

1 D 2 B 3 D 4 C 5 B

3

Enzymes and Mechanisms of Enzyme Action

I. INTRODUCTION

Enzymes are **biological catalysts** responsible for supporting almost all of the chemical reactions that maintain animal homeostasis. Because of their role in maintaining life processes, the assay and pharmacological regulation of enzymes have become key elements in clinical diagnosis and therapeutics. The macromolecular components of almost all enzymes are composed of protein, except for a class of RNA modifying catalysts known as **ribozymes**. Ribozymes are molecules of ribonucleic acid that catalyze reactions on the phosphodiester bond of other RNAs.

Enzymes are found in all tissues and fluids of the body. Intracellular enzymes catalyze the reactions of metabolic pathways. Plasma membrane enzymes regulate catalysis within cells in response to extracellular signals, and enzymes of the circulatory system are responsible for regulating the clotting of blood. Almost every significant life process is dependent on enzyme activity.

II. CLASSIFICATION OF ENZYMES

Traditionally, enzymes were simply assigned names by the investigator who discovered them as knowledge expanded, systems of enzyme classification became more comprehensive and complex. Currently, enzymes are grouped into six functional classes by the International Union of Biochemists (IUB) as shown in Table 3.1.

These rules give each enzyme a unique number. The IUB system also specifies a textual name for each enzyme. The enzyme's name comprises the names of the substrate(s), the product(s) and the enzyme's functional class. Because many enzymes (e.g. alcohol dehydrogenase), are widely known in the scientific community by their common names, the change to IUB-approved nomenclature has been slow. In everyday usage, most enzymes are still called by their common names.

Table 3.1: Classification of enzymes and their properties

Number	*Classification*	*Biochemical properties*
1.	Oxidoreductases	Act on many chemical groupings to add or remove hydrogen atoms.
2.	Transferases	Transfer functional groups between donor and acceptor molecules. Kinases are specialized transferases that regulate metabolism by transferring phosphate from ATP to other molecules.
3.	Hydrolases	Add water across a bond, hydrolyzing it.
4.	Lyases	Add water, ammonia or carbon dioxide across double bonds, or remove these elements to produce double bonds.
5.	Isomerases	Carry out many kinds of isomerization: L to D isomerizations, mutase reactions (shifts of chemical groups) and others.
6.	Ligases	Catalyze reactions in which two chemical groups are joined (or **ligated**) with the use of energy from ATP

Enzymes are also classified on the basis of their composition. Enzymes composed wholly of protein are known as **simple enzymes** in contrast to **complex enzymes**, which are composed of protein plus a relatively small organic molecule. Complex enzymes are also known as **holoenzymes**. In this terminology, the protein component is known as the **apoenzyme**, while the non-protein component is known as the **coenzyme** or **prosthetic group** where ***prosthetic group*** describes a complex in which the small organic molecule is bound to the apoenzyme by covalent bonds; when the binding between the apoenzyme and non-protein components is non-covalent, the small organic molecule is called a **coenzyme**. Many prosthetic groups and coenzymes are water-soluble derivatives of vitamins. It should be noted that the main clinical symptoms of dietary vitamin insufficiency generally arise from the malfunction of enzymes, which lack sufficient cofactors derived from vitamins to maintain homeostasis.

The non-protein component of an enzyme may be as simple as a metal ion or as complex as a small non-protein organic molecule. Enzymes that require a metal in their composition are known as **metalloenzymes** if they bind and retain their metal atom(s) under all conditions, i.e. with very high affinity. Those enzymes which have a lower affinity for metal ion, but still require the metal ion for their activity, are known as metal-activated enzymes.

Role of Coenzymes

The functional role of coenzymes is to act as transporters of chemical groups from one reactant to another. The chemical groups thus carried, can be as simple as the hydride ion ($H^+ + 2e^-$) carried by NAD or the mole of hydrogen carried by FAD; or they can be even more complex than the amine ($-NH_2$) carried by pyridoxal phosphate.

Since coenzymes are chemically changed as a consequence of enzyme action, it is often useful to consider coenzymes to be a special class of substrates, or **second substrates**, which are common to many different holoenzymes. In all cases, the coenzymes donate the carried chemical grouping to an acceptor molecule and are thus regenerated to their original form. This regeneration of coenzyme and holoenzyme fulfills the definition of an enzyme as a chemical catalyst, since (unlike the usual substrates, which are used up during the course of a reaction) coenzymes are generally regenerated.

Enzyme Relative to Substrate Type

Although enzymes are highly specific for the kind of reaction they catalyze, the same is not always true of substrates they attack. For example, while ***succinic dehydrogenase*** **(SDH)** always catalyzes an oxidation-reduction reaction and its substrate is invariably succinic acid, ***alcohol dehydrogenase*** **(ADH)** always catalyzes oxidation-reduction reactions but attacks a number of different alcohols, ranging from methanol to butanol. Generally, enzymes having broad substrate specificity are most active against one particular substrate. In the case of ADH, ethanol is the preferred substrate.

Enzymes are generally specific for a particular steric configuration (optical isomer) of a substrate. Enzymes that attack D sugars will not attack its corresponding L isomer. Similarly enzymes that act on L amino acids will not employ the corresponding D optical isomer as a substrate. Some enzymes such as ***racemases*** provide a striking exception to these generalities; in fact, the role of racemases is to convert D isomers to L isomers and vice versa. Thus racemases attack both D and L forms of their substrate.

As enzymes have a more or less broad range of substrate specificity, it follows that a given substrate may be acted on by a number of different enzymes, each of which uses the same substrate(s) and produces the same product(s). The individual members of a set of enzymes sharing such characteristics are known as **isozymes**. These are the products of genes that vary only slightly; often, various isozymes of a group are expressed in different tissues of the body. The best studied set of isozymes is the ***lactate dehydrogenase*** **(LDH)** system. LDH is a tetrameric enzyme composed of all possible arrangements of two different protein subunits; the subunits are known as H (for heart) and M (for skeletal muscle). These subunits combine in various combinations leading to 5 distinct isozymes. The all H isozyme is characteristic of that from heart tissue, and all M isozyme is typically found in skeletal muscle and liver. All these isozymes catalyze the same chemical reaction, but they exhibit varying degree of efficiency. The detection of specific LDH isozymes in the blood is highly diagnostic of tissue damage such as myocardial infarction.

III. ENZYME-SUBSTRATE INTERACTIONS

The most favored model of enzyme substrate interaction is known as the **induced fit model**. This model proposes that the initial interaction between enzyme and substrate is relatively weak, but that these weak interactions rapidly induce conformational changes in the enzyme that strengthen binding and bring catalytic sites close to substrate bonds to be altered. After binding takes place, one or more mechanisms of catalysis generates transition—state complexes and reaction products. The four possible mechanisms of catalysis are as follows:

1. **Catalysis by bond strain:** In this form of catalysis, the induced structural rearrangements that take place with the binding of substrate and enzyme ultimately produce strained substrate bonds, which more easily attain the transition state. The new conformation often forces substrate atoms and bulky catalytic groups, such as aspartate and glutamate, into conformations that strain existing substrate bonds.
2. **Catalysis by proximity and orientation:** Enzyme-substrate interactions orient reactive groups and bring them into proximity with one another. In addition to inducing strain, groups such as aspartate are frequently chemically reactive as well, and their proximity and orientation toward the substrate thus favors their participation in catalysis.
3. **Catalysis involving proton donors (acids) and acceptors (bases):** Other mechanisms also contribute significantly to the completion of catalytic events initiated by a strain mechanism, e.g., the use of glutamate as a general acid catalyst (proton donor).
4. **Covalent catalysis:** In catalysis that takes place by covalent mechanisms, the substrate is oriented to active sites on the enzymes in such a way that a covalent intermediate forms between the enzyme or coenzyme and the substrate. One of the best-known examples of this mechanism is that involving proteolysis by **serine proteases**, which include both digestive enzymes (***trypsin***, ***chymotrypsin***, and ***elastase***) and several enzymes of the blood-clotting cascade. These proteases contain an active site serine whose R group hydroxyl forms a covalent bond with a carbonyl carbon of a peptide bond, thereby causing hydrolysis of the peptide bond.

IV. CHEMICAL REACTIONS AND RATES

According to the conventions of biochemistry, the rate of a chemical reaction is described by the number of molecules of reactant(s) that are converted into product(s) within a specified time period. Reaction rate is always dependent on the concentration of the chemicals involved in the process and on rate constants that are characteristic of the reaction. For example, the reaction in which A is converted to B is written as follows:

$$A \longrightarrow B$$

The rate of this reaction is expressed algebraically as either a decrease in the concentration of reactant A:

$$-[\mathbf{A}] = k[\mathbf{B}]$$

or an increase in the concentration of product B:

$$[\mathbf{B}] = k[\mathbf{A}]$$

In the second equation (of the 3 above), the negative sign signifies a decrease in concentration of A as the reaction progresses, brackets define concentration in molarity and the *k* is known as the rate constant. Rate constants are simply proportionality constants that provide a quantitative connection between chemical concentrations and reaction rates. Each chemical reaction has characteristic values for its rate constants; these in turn directly relate to the **equilibrium constant** for that reaction. Thus, reaction can be rewritten as an equilibrium expression in order to show the relationship between reaction rates, rate constants and the equilibrium constant for this simple case. The rate constant for the forward reaction is defined as k_{+1} and the reverse as k_{-1}.

At equilibrium the rate (v) of the forward reaction ($A \longrightarrow B$) is— by definition— equal to that of the reverse or back reaction ($B \longrightarrow A$), a relationship which is algebraically symbolized as:

$$v_{forward} = v_{reverse}$$

where, for the forward reaction:

$$v_{forward} = k_{+1}[A]$$

and for the reverse reaction:

$$v_{reverse} = k_{-1}[B]$$

In the above equations, k_{+1} and k_{-1} represent rate constants for the forward and reverse reactions, respectively. The negative subscript refers only to a reverse reaction, not to an actual negative value for the constant. To put the relationships of the two equations into words, we state that the rate of the forward reaction [$v_{forward}$] is equal to the product of the forward rate constant k_{+1} and the molar concentration of A. The rate of the reverse reaction is equal to the product of the reverse rate constant k_{-1} and the molar concentration of B.

At equilibrium, the rate of the forward reaction is equal to the rate of the reverse reaction leading to the **equilibrium constant** of the reaction and is expressed by:

$$[B]/[A] = k_{+1}/k_{-1} = K_{eq}$$

This equation demonstrates that the equilibrium constant for a chemical reaction is not only equal to the equilibrium ratio of product and reactant concentrations, but is also equal to the ratio of the characteristic rate constants of the reaction.

V. CHEMICAL REACTION ORDER

Reaction order refers to the number of molecules involved in forming a reaction complex that is competent to proceed to product(s). Empirically, order is easily determined by summing the exponents of each concentration term in the rate equation for a reaction. A reaction characterized by the conversion of one molecule of A to one molecule of B with no influence from any other reactant or solvent is a **first-order** reaction. The exponent on the substrate concentration in the rate equation for this type of reaction is 1. A reaction with two substrates forming two products would a **second-order** reaction. However, the reactants in second- and higher- order reactions need not be different chemical species.

An example of a second order reaction is the formation of ATP through the condensation of ADP with orthophosphate:

$$ADP + H_2PO_4 \longleftrightarrow ATP + H_2O$$

For this reaction the forward reaction rate would be written as:

$$v_{forward} = k_1[ADP][H_2PO_4]$$

VI. ENZYMES AS BIOLOGICAL CATALYSTS

In cells and organisms, most reactions are catalyzed by enzymes, which are regenerated during the course of a reaction. These **biological catalysts** are physiologically important because they speed up the rates of reactions that would otherwise be too slow to support life. Enzymes increase reaction rates—sometimes by as much as one millionfold, but more typically by about one thousand fold. Catalysts speed up the forward and reverse reactions proportionately so that, although the magnitude of the rate constants of the forward and reverse reactions is increased, the ratio of the rate constants remains the same in the presence or absence of enzyme. Since the equilibrium constant is equal to the ratio of rate constants, it is apparent that enzymes and other catalysts have no effect on the equilibrium constant of the reactions they catalyze.

Enzymes increase reaction rates by decreasing the amount of energy required to form a complex of reactants that is competent to produce reaction products. This complex is known as the activated state or **transition state complex** for the reaction. Enzymes and other catalysts accelerate reactions by lowering the energy of the transition state. The free energy required to form an activated complex is much lower in the catalyzed reaction. The amount of energy required to achieve the transition state is lowered; consequently, at any instance a greater proportion of the molecules in the population can achieve the transition state. The result is that the reaction rate is increased (Fig. 3.1).

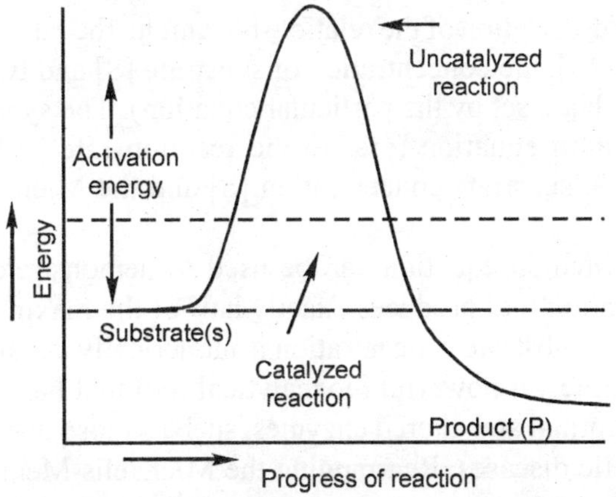

Fig. 3.1: Energy diagrams for catalyzed and uncatalyzed reactions

Michaelis-Menton Kinetics

In typical enzyme-catalyzed reactions, concentrations of reactant and product are usually hundreds or thousands of times greater than the enzyme concentration. Consequently, each enzyme molecule catalyzes the conversion to product of many reactant molecules. In biochemical reactions, reactants are commonly known as **substrates.** The catalytic event that converts substrate to product involves the formation of a transition state, that occurs most easily at a specific binding site on the enzyme. This site, called the **catalytic site** of the enzyme, has been evolutionarily structured to provide specific, high-affinity binding of substrate(s) and to provide an environment that favors the catalytic events. The complex formed during substrate(s) and enzyme combination is called the **enzyme substrate (ES) complex**. Reaction products arise when the ES complex breaks down releasing free enzyme.

Between the binding of substrate to enzyme, and the reappearance of free enzyme and product, a series of complex events must take place. At the minimum an ES complex must be formed; this complex must pass to the transition state (ES*); and the transition state complex must advance to an enzyme product complex (EP). The latter is finally competent to dissociate into product and free enzyme. The series of events can be shown as:

$$\mathbf{E + S \rightleftharpoons ES \rightleftharpoons ES^* \rightleftharpoons EP \rightleftharpoons E + P}$$

The kinetics of simple reactions like that above were first characterized by biochemists Michaelis and Menton. The concepts underlying their analysis of enzyme kinetics continue to provide the cornerstone for understanding metabolism today, and for the development and clinical use of drugs aimed at selectively altering rate constants and interfering with the progress of disease states. The **Michaelis-Menton equation:**

$$V_1 = \frac{V_{max}[S]}{\{K_m + [S]\}}$$

is a quantitative description of the relationship among the rate of an enzyme catalyzed reaction [v_1], the concentration of substrate [S] and two constants, V_{max} and K_m (which are set by the particular equation). The symbols used in the Michaelis-Menton equation refer to the reaction rate [v_1], maximum reaction rate (V_{max}), substrate concentration [S] and the Michaelis-Menton constant (K_m).

The Michaelis-Menton equation can be used to demonstrate that at the substrate concentration that produces exactly half of the maximum reaction rate, i.e. ½ V_{max}, the substrate concentration is numerically equal to K_m. This fact provides a simple yet powerful bioanalytical tool that has been used to characterize both normal and altered enzymes, such as those that produce the symptoms of genetic diseases. Rearranging the Michaelis-Menton equation leads to:

$$K_m = [S]\left\{\left[\frac{V_{max}}{V_1}\right] - 1\right\}$$

The latter is an algebraic statement of the fact that, for enzymes of the Michaelis-Menton type, when the observed reaction rate is half of the maximum possible reaction rate, the substrate concentration is numerically equal to the Michaelis-Menton constant. In this derivation, the units of K_m are those used to specify the concentration of S, usually molarity.

The Michaelis-Menton equation has the same form as the equation for a rectangular hyperbola; graphical analysis of reaction rate (v) versus substrate concentration [S] produces a hyperbolic rate plot (Fig. 3.2).

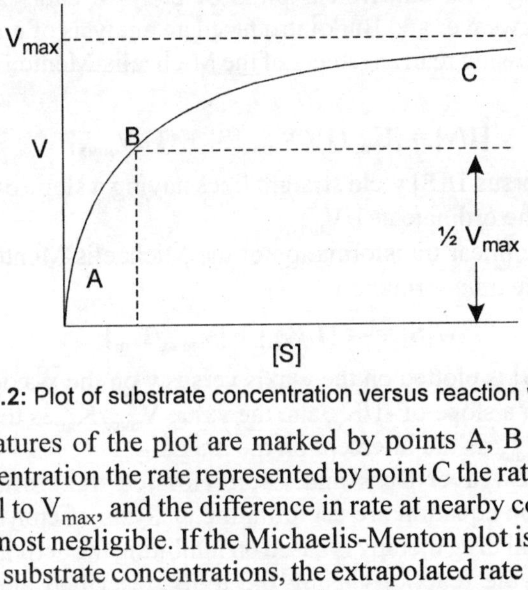

Fig. 3.2: Plot of substrate concentration versus reaction velocity

The key features of the plot are marked by points A, B and C. At high substrate concentration the rate represented by point C the rate of the reaction is almost equal to V_{max}, and the difference in rate at nearby concentrations of substrate is almost negligible. If the Michaelis-Menton plot is extrapolated to infinitely high substrate concentrations, the extrapolated rate is equal to V_{max}. When the reaction rate becomes independent of substrate concentration, or nearly so, the rate is said to be zero order. (Note that the reaction is zero order only with respect to this substrate. If the reaction has two substrates, it may or may not be zero order with respect to the second substrate). The very small differences in reaction velocity at substrate concentrations around point C (near V_{max}) reflect the fact that at these concentrations almost all of the enzyme molecules are bound to substrate and the rate is virtually independent of substrate, hence zero order. At lower substrate concentrations, such as at points A and B, the lower reaction velocities indicate that at any moment only a portion of the enzyme molecules are bound to the substrate. In fact, at the substrate concentration denoted by point B, exactly half the enzyme molecules are in an ES complex at any instant and the rate is exactly one half of V_{max}. At substrate concentrations near point A the rate appears to be directly proportional to substrate concentration, and the reaction rate is said to be first order.

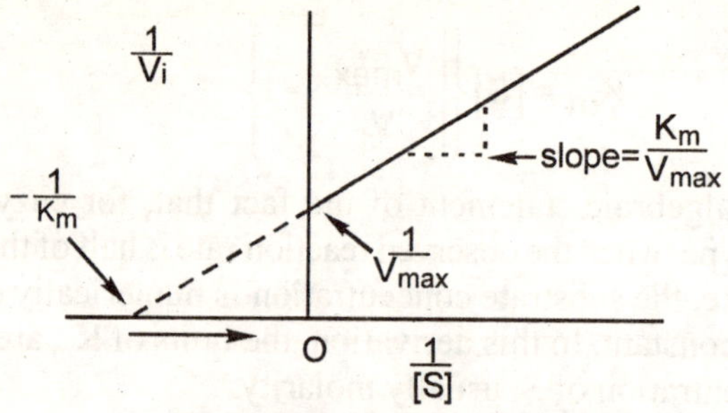

Fig. 3.3: A Lineweaver-Burk Plot

VII. INHIBITION OF ENZYME CATALYZED REACTIONS

To avoid dealing with curvilinear plots of enzyme catalyzed reactions, biochemists Lineweaver and Burk introduced an analysis of enzyme kinetics based on the following rearrangement of the Michaelis-Menton equation (Fig. 3.3):

$$[1/v] = [K_m (1)/ V_{max}[S] + (1)/V_{max}]$$

Plots of 1/v versus 1/[S] yield straight lines having a slope of K_m/V_{max} and an intercept on the ordinate at $1/V_{max}$.

An alternative linear transformation of the Michaelis-Menton equation is the Eadie-Hofstee transformation:

$$v/[S] = -v\ [1/K_m] + [V_{max}/K_m]$$

and when v/[S] is plotted on the y-axis versus v on the x-axis, the result is a linear plot with a slope of $-1/K_m$ and the value V_{max}/K_m as the intercept on the y-axis and V_{max} as the intercept on the x-axis.

Both the Lineweaver-Burk and Eadie-Hofstee transformation of the Michaelis-Menton equation are useful in the analysis of enzyme inhibition. Since most clinical drug therapy is based on inhibiting the activity of enzymes, analysis of enzyme reactions using the tools described above has been fundamental to the modern design of pharmaceuticals. Well-known examples of such therapy include the use of methotrexate in cancer chemotherapy to semi-selectively inhibit DNA synthesis of malignant cells, the use of aspirin to inhibit the synthesis of prostaglandins which are at least partly responsible for the aches and pains of arthritis, and the use of sulfa drugs to inhibit the folic acid synthesis that is essential for the metabolism and growth of disease-causing bacteria. In addition, many poisons—such as cyanide, carbon monoxide and polychlorinated biphenols (PCBs)—produce their life-threatening effects by means of enzyme inhibition.

Enzyme inhibitors fall into two broad classes: those causing irreversible inactivation of enzymes and those whose inhibitory effects can be reversed. Inhibitors of the first class usually cause an inactivating, covalent modification of enzyme structure. Cyanide is a classic example of an irreversible enzyme inhibitor: by covalently binding mitochondrial cytochrome oxidase, it inhibits all the reactions associated with electron transport. The kinetic effect of irreversible inhibitors is to decrease the concentration of active enzyme, thus decreasing the maximum possible concentration of ES complex. Since the limiting enzyme reaction rate is often k_2[ES], it is clear that under these circumstances the reduction of enzyme concentration will lead to decreased reaction rates. Note that when enzymes in cells are only partially inhibited by irreversible inhibitors, the remaining unmodified enzyme molecules are not distinguishable from those in untreated cells; in particular, they have the same turnover number and the same K_m. **Turnover number, related to V_{max}, is defined as the maximum number of moles of substrate that can be converted to product per mole of catalytic site per second.** Irreversible inhibitors are usually considered to be poisons and are generally unsuitable for therapeutic purposes.

Reversible inhibitors can be divided into two main categories— **competitive inhibitors** and **noncompetitive inhibitors**—with a third category, **uncompetitive inhibitors**, rarely encountered (Table 3.2).

Table 3.2: Reversible inhibitors of enzymes and their effects on enzyme kinetics

Inhibitor type	*Binding site on enzyme*	*Kinetic effect*
Competitive inhibitor	Specifically at the catalytic site, where it competes with substrate for binding in a dynamic equilibrium-like process. Inhibition is reversible by substrate.	V_{max} is unchanged; K_m, as defined by [S] required for ½ maximal activity, is increased.
Noncompetitive inhibitor	Binds E or ES complex other than to the catalytic site. Substrate binding unaltered, but ESI complex cannot form products. Inhibition cannot be reversed by substrate.	K_m appears unaltered; V_{max} is decreased proportion ately to inhibitor concentration.
Uncompetitive inhibitor	Binds only to ES complexes at locations other than the catalytic site. Substrate binding modifies enzyme structure, making inhibitor-binding site available. Inhibition cannot be reversed by substrate	Apparent V_{max} decreased; K_m, as defined by [S] required for ½ maximal activity, is decreased.

The hallmark of all the reversible inhibitors is that when the inhibitor concentration drops, enzyme activity is regenerated. Usually, these inhibitors bind to enzymes by non-covalent forces and the inhibitor maintains a reversible equilibrium with the enzyme. The equilibrium constant for the dissociation of enzyme inhibitor complexes is known as K_I:

$$\mathbf{K_I = [E][I]/[E—I—complex]}$$

The importance of K_I is that in all enzyme reactions where substrate, inhibitor and enzyme interact, the normal K_m and or V_{max} for substrate enzyme interaction appear to be altered. These changes are a consequence of the influence of K_I on the overall rate equation for the reaction. The effects of K_I are best observed in Lineweaver-Burk plots.

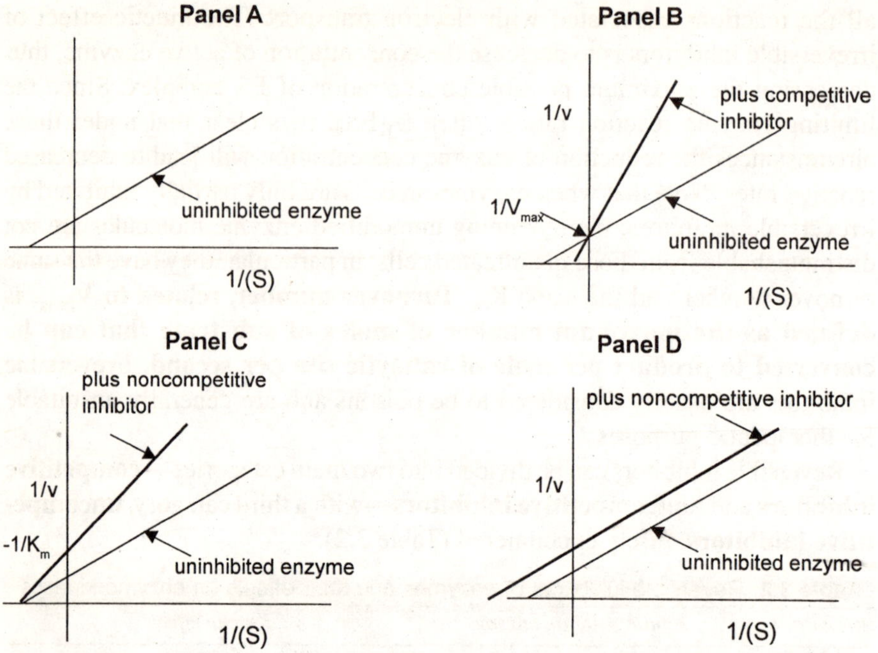

Fig. 3.4: Lineweaver-Burk plots of inhibited enzymes

Probably the best known reversible inhibitors are competitive inhibitors, which always bind at the catalytic or active site of the enzyme. Most drugs that alter enzyme activity are of this type. Competitive inhibitors are especially attractive as clinical modulators of enzyme activity because they offer two routes for the reversal of enzyme inhibition, while other reversible inhibitors offer only one. First, as with all kinds of reversible inhibitors, a decreasing concentration of the inhibitor reverses the equilibrium, regenerating active free enzyme. Second, since substrate and competitive inhibitors both bind at the same site, they compete with one another for binding

Raising the concentration of substrate (S), while holding the concentration of inhibitor constant, provides the second route for reversal of competitive inhibition. The greater the proportion of substrate, the greater the proportion of enzyme present in competent ES complexes.

As noted earlier, high concentrations of substrate can displace virtually all competitive inhibitor bound to active sites. Thus, it is apparent that V_{max} should be unchanged by competitive inhibitors. This characteristic of competitive inhibitors is reflected in the identical vertical-axis intercepts of Lineweaver-Burk plots, with and without inhibitor (Fig.3.4).

Since attaining V_{max} requires appreciably higher substrate concentrations in the presence of competitive inhibitor, K_m (the substrate concentration at half maximal velocity) is also higher, as demonstrated by the differing negative intercepts on the horizontal axis in panel B.

Analogously, panel C illustrates that noncompetitive inhibitors appear to have no effect on the intercept at the x-axis implying that noncompetitive inhibitors have no effect on the K_m of the enzymes they inhibit. Since noncompetitive inhibitors do not interfere in the equilibration of enzyme, substrate and ES complexes, the K_m's of Michaelis-Menton type enzymes are not expected to be affected by noncompetitive inhibitors, as demonstrated by x-axis intercepts in panel C. However, because complexes that contain inhibitor (ESI) are incapable of progressing to reaction products, the effect of a noncompetitive inhibitor is to reduce the concentration of ES complexes that can advance to product. Since $V_{max} = k_2[E_{total}]$, and the concentration of competent E_{total} is diminished by the amount of ESI formed, noncompetitive inhibitors are expected to decrease V_{max}, as illustrated by the y-axis intercepts in panel C.

A corresponding analysis of uncompetitive inhibition leads to the expectation that these inhibitors should change the apparent values of K_m as well as V_{max}. Changing both constants leads to double reciprocal plots, in which intercepts on the x and y axes are proportionately changed; this leads to the production of parallel lines in inhibited and uninhibited reactions.

VIII. REGULATION OF ENZYME ACTIVITY

While it is clear that enzymes are responsible for the catalysis of almost all biochemical reactions, it is important to also recognize that rarely, if ever, do enzymatic reactions proceed in isolation. The most common scenario is that enzymes catalyze individual steps of multistep metabolic pathways, as is the case with glycolysis, gluconeogenesis or the synthesis of fatty acids. As a consequence of these lock-step sequences of reactions, any given enzyme is dependent on the activity of preceding reaction steps for its substrate.

In humans, substrate concentration is dependent on food supply and is not usually a physiologically important mechanism for the routine regulation of enzyme activity. Enzyme concentration, by contrast, is continually modulated in response to physiological needs. Three principal mechanisms are known to regulate the concentration of active enzyme in tissues:

1. Regulation of gene expression controls the quantity and rate of enzyme synthesis.
2. Proteolytic enzyme activity determines the rate of enzyme degradation.
3. Covalent modification of pre-existing pool of inactive proenzymes produces active enzymes.

Enzyme synthesis and proteolytic degradation are comparatively slow mechanisms for regulating enzyme concentration, with response times of hours, days or even weeks. Proenzyme activation is a more rapid method of increasing enzyme activity but, as a regulatory mechanism, it has the disadvantage of not being a reversible process. Proenzymes are generally synthesized in abundance, stored in secretory granules and covalently activated upon release from their storage sites. Examples of important proenzymes include pepsinogen, trypsinogen and chymotrypsinogen, which give rise to the proteolytic digestive enzymes. Likewise, many of the proteins involved in the cascade of chemical reactions responsible for blood clotting are synthesized as proenzymes. Other important proteins, such as peptide hormones and collagen, are also derived by covalent modification of precursors.

Another mechanism of regulating enzyme activity is to sequester enzymes in compartments where access to their substrates is limited. For example, the proteolysis of cell proteins and glycolipids by enzymes responsible for their degradation is controlled by sequestering these enzymes within the lysosome.

In contrast to regulatory mechanisms that alter enzyme concentration, there is an important group of regulatory mechanisms that do not affect enzyme concentration, are reversible and rapid in action, and actually carry out most of the moment-to-moment physiological regulation of enzyme activity. These mechanisms include allosteric regulation, regulation by reversible covalent modification and regulation by control proteins such as calmodulin. Reversible covalent modification is a major mechanism for the rapid and transient regulation of enzyme activity. The best examples, again, come from studies on the regulation of glycogen metabolism where phosphorylation of ***glycogen synthase*** and ***glycogen phosphorylase kinase*** results in the stimulation of glycogen degradation while glycogen synthesis is coordinately inhibited. Numerous other enzymes of intermediary metabolism are affected by phosphorylation, either positively or negatively. These covalent phosphorylations can be reversed by a separate sub-subclass of enzymes known as ***phosphatases***. Recent research has indicated that the aberrant phosphorylation of growth factor and hormone receptors, as well as of proteins that regulate cell division, often leads to unregulated cell growth or cancer. The usual sites for phosphate addition to proteins are the serine, threonine and tyrosine R group hydroxyl residues.

IX. ALLOSTERIC ENZYMES

In addition to simple enzymes that interact only with substrates and inhibitors, there is a class of enzymes that bind small, physiologically important molecules and modulate activity in ways other than those described above. These are known as **allosteric enzymes**; the small regulatory molecules to which they bind are known as **effectors**. Allosteric effectors bring about catalytic modification by binding to the enzyme at distinct allosteric sites, well removed from the catalytic site, and causing conformational changes that are transmitted through the bulk of the protein to the catalytically active site(s).

The hallmark of effectors is that when they bind to enzymes, they alter the catalytic properties of an enzyme's active site. Those that increase catalytic activity are known as positive effectors. Effectors that reduce or inhibit catalytic activity are negative effectors.

Most allosteric enzymes are oligomeric (consisting of multiple subunits); generally they are located at or near branch points in metabolic pathways, where they are influential in directing substrates along one or another of the available metabolic paths. The effectors that modulate the activity of these allosteric enzymes are of two types. Those activating and inhibiting effectors that bind at allosteric sites are called **heterotropic effectors**. (Thus there exist both positive and negative heterotropic effectors). These effectors can assume a vast diversity of chemical forms, ranging from simple inorganic molecules to complex nucleotides such as cyclic adenosine monophosphate (cAMP). Their single defining feature is that they are not identical to the substrate.

In many cases, the substrate itself induces distant allosteric effects when it binds to the catalytic site. Substrates acting as effectors are said to be **homotropic effectors**. When the substrate is the effector, it can act as such, either by binding to the substrate-binding site, or to an allosteric effector site. When the substrate binds to the catalytic site it transmits an activity-modulating effect to other subunits of the molecule. Often used as the model of a homotropic effector is hemoglobin, although it is not a branch-point enzyme and thus does not fit the definition on all counts.

There are two ways that enzymatic activity can be altered by effectors: the V_{max} can be increased or decreased, or the K_m can be raised or lowered. Enzymes whose K_m is altered by effectors are said to be **K-type enzymes** and the effector a K-type effector. If V_{max} is altered, the enzyme and effector are said to be V-type. Many allosteric enzymes respond to multiple effectors with V-type and K-type behavior. Here again, hemoglobin is often used as a model to study allosteric interactions, although it is not strictly an enzyme.

In the preceding discussion, we assumed that allosteric sites and catalytic sites were homogeneously present on every subunit of an allosteric enzyme. While this is often the case, there is another class of allosteric enzymes that are comprised of separate catalytic and regulatory subunits. The archetype of this class of enzymes is ***cAMP-dependent protein kinase*** **(PKA)**, whose mechanism of activation is illustrated. The enzyme is tetrameric, containing two catalytic subunits and two regulatory subunits (Fig. 3.5). When intracellular

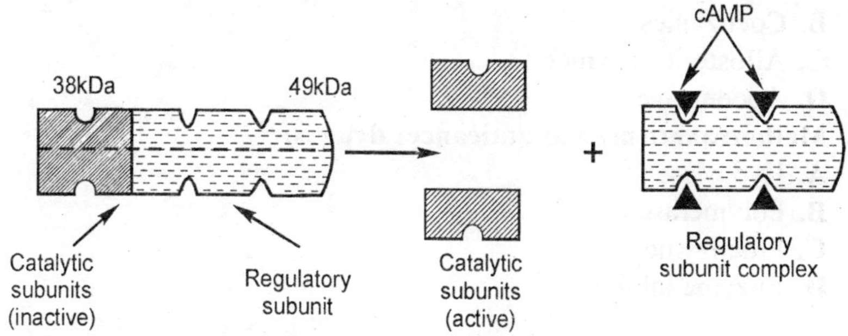

Fig. 3.5: Catalytic and regulatory subunits of cAMP-dependent protein kinase (PKA)

cAMP levels rise, one molecule of cAMP binds to each regulatory subunit, causing the tetramer to dissociate into one regulatory dimer and two catalytic monomers. In the dissociated form, the catalytic subunits are fully active; they catalyze the phosphorylation of a number of other enzymes, such as those involved in regulating glycogen metabolism. The regulatory subunits have no catalytic activity.

STUDY QUESTIONS

1. Describe different types of enzyme inhibitors with suitable examples
2. What are the different classes of enzymes? Describe how enzyme activity is regulateD.
3. Write short notes on the followings.
   A. Coenzymes
   B. Enzyme–substrate interaction
   C. Michaelis-Menton kinetics
   D. Allosteric enzymes.

MCQs

Select one most appropriate answer given for each questions below:

1. **High levels of LDH, isoenzyme in blood indicate:**
   A. Myocardial infarction
   B. Decalcification of bone
   C. Breast cancer
   D. Bone cancer
2. **Elastases are basically an example of:**
   A. Transferase enzyme
   B. Polymerase enzyme
   C. Ligase enzyme
   D. Protease enzyme
3. **Trypsinogen is a typical example of:**
   A. Proenyzmes
   B. Coenzymes
   C. Allosteric enzymes
   D. Apoenzymes
4. **Methotraxate used as anticancer drug is:**
   A. Hydrolase
   B. Polymerase
   C. Proenzyme
   D. Enzyme inhibitor

5. Succinic dehydrogenase catalyses:

A. Oxidation reaction
B. Oxidation-reduction reaction
C. Reduction reaction
D. Addition of hydrogen in the compund

6. The enzyme recemase:

A. Converts D isomers of amino acids to L isomers
B. Converts L isomers of amino acids to D isomers
C. Converts D isomers of amino acids to L isomers and vice versa
D. None of the above

7. LDH is a:

A. Tetramer of two different protein subunits
B. Dimer of two different protein subunits
C. Monomeric protein
D. Polymer of many protein subunits

8. Trypsinogen is a typical example of:

A. Coenzyme
B. Proenzyme
C. Apoenzyme
D. Holoenzyme

1 A 2 D 3 A 4 D 5 B 6 C 7 A 8 B

4

Molecular Composition of Cells

I. INTRODUCTION

Cells are composed of **water, inorganic ions, and carbon containing (organic) molecules.** Water is the most abundant molecule in cells, accounting for 70% or more of total cell mass. Consequently, the interactions between water and the other constituents of cells are of central importance in biological chemistry. The critical property of water in this respect is that it is a polar molecule, in which the hydrogen atoms have a slight positive charge and the oxygen has a slight negative charge. Because of their polar nature, water molecules can form hydrogen bonds with each other or with other **polar** molecules, as well as interacting with positively or negatively charged ions. As a result of these interactions, ions and polar molecules are readily soluble in water **(hydrophilic)**. In contrast, nonpolar molecules, which cannot interact with water, are poorly soluble in aqueous environment **(hydrophobic)**. Consequently, nonpolar molecules tend to minimize their contact with water by associating closely with each other instead. Such interactions play crucial role in formation of biological structures, e.g. cell membranes.

The inorganic ions of the cell, including sodium (Na^+), potassium (K^+), magnesium (Mg^{2+}), calcium (Ca^{2+}), phosphate (HPO_4^{2-}), chloride (Cl^-) and bicarbonate (HCO_3^-), constitute 1% or less of the cell mass. These ions are involved in a number of aspects of cell metabolism, and thus play critical roles in cell function.

It is, however, the organic molecules that are unique constituents of cells. Most of these organic compounds belong to one of four classes of molecules: **carbohydrates, lipids, proteins and nucleic acids**. Proteins, nucleic acids, and most carbohydrates (the polysaccharides) are macromolecules formed by joining (polymerization) of hundreds or thousands of low-molecular weight precursors: **amino acids, nucleotides**, and simple **sugars**, respectively. Such macromolecules constitute 80-90% of the dry weight of most cells. Lipids are the other major constituents of cells. The remainder of the cell mass is composed of a variety of small organic molecules, including macromolecular precursors. The basic chemistry of cells can thus be understood in terms of the structures and functions of these four major classes of organic molecules as described below.

II. CARBOHYDRATES

Carbohydrates are organic carbon compounds that contain large quantities of hydroxyl groups. The simplest carbohydrates also contain either an aldehyde moiety (these are termed **polyhydroxyaldehydes**) or a ketone moiety **(polyhydroxyketones).** All carbohydrates can be classified as either **monosaccharides, oligosaccharides or polysaccharides.** Anywhere from two to ten monosaccharide units, linked by **glycosidic bonds**, make up an oligosaccharide. Polysaccharides are much larger, containing hundreds of monosaccharide units. The presence of the hydroxyl groups allows carbohydrates to interact with the aqueous environment and to participate in hydrogen bonding, both within and between chains. Derivatives of the carbohydrates can contain nitrogens, phosphates and sulfur compounds. Carbohydrates also can combine with lipid to form glycolipids or with protein to form glycoproteins.

The major role of carbohydrates in the human body are as follows:

1. Simple sugars (glucose) act as major nutrients of cells. Their breakdown provides both a source of cellular energy and the starting material for the synthesis of other cell constituents.
2. Polysaccharides are storage forms of sugars are form structural components of the cell.
3. Polysaccharides and short polymers of sugar act as marker for a variety of cell recognition processes, including the adhesion of cells to their neighbors and the transport of proteins to appropriate intracellular destinations.

A. Carbohydrate Nomenclature

The predominant carbohydrates encountered in the body are structurally related to the aldotriose **glyceraldehyde** and to the ketotriose **dihydroxyacetone**. All carbohydrates contain at least one asymmetrical (chiral) carbon and are, therefore, optically active. In addition, carbohydrates can exist in either of two conformations, as determined by the orientation of the hydroxyl group about the asymmetric carbon farthest from the carbonyl. With a few exceptions, those carbohydrates that are of physiological significance exist in the **D-conformation.** The mirror-image conformations, called **enantiomers**, are in the **L-conformation.**

i. Monosaccharides

The monosaccharides commonly found in humans are classified according to the number of carbons they contain in their backbone structures. The major monosaccharides contain four to six carbon atoms.

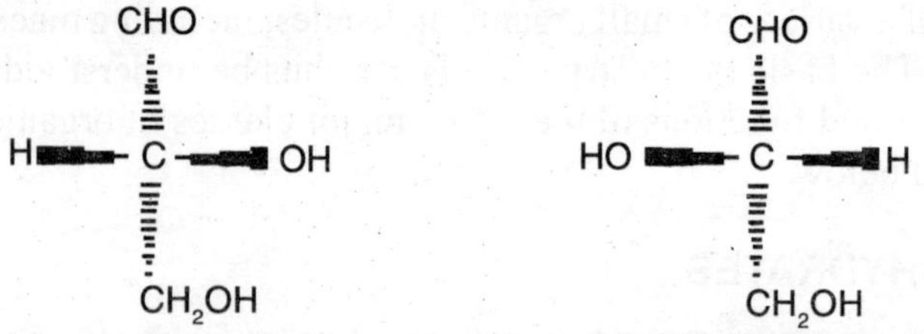

D Glyceraldehyde L Glyceraldehyde

Table 4.1: Carbohydrate classifications

# Carbons	*Category name*	*Relevant examples*
3	Triose	Glyceraldehyde, Dihydroxyacetone
4	Tetrose	Erythrose
5	Pentose	Ribose, Ribulose, Xylulose
6	Hexose	Glucose, Galactose, Mannose, Fructose
7	Heptose	Sedoheptulose
9	Nonose	Neuraminic acid also called sialic acid

The aldehyde and ketone moieties of the carbohydrates with five and six carbons will spontaneously react with alcohol groups present in neighboring carbons to produce intramolecular **hemiacetals** or **hemiketals**, respectively. This results in the formation of five- or six-membered rings. Because the five-membered ring structure resembles the organic molecule **furan**, derivatives with this structure are termed **furanoses**. Those with six-membered rings resemble the organic molecule **pyran** and are termed **pyranoses.**

Such structures can be depicted by either **Fischer** or **Haworth** style diagrams. The numbering of the carbons in carbohydrates proceeds from the carbonyl carbon, for aldoses, or the carbon nearest the carbonyl, for ketoses.

The rings can open and re-close, allowing rotation to occur about the carbon bearing the reactive carbonyl yielding two distinct configurations (α and β) of the hemiacetals and hemiketals. The carbon about which this rotation occurs

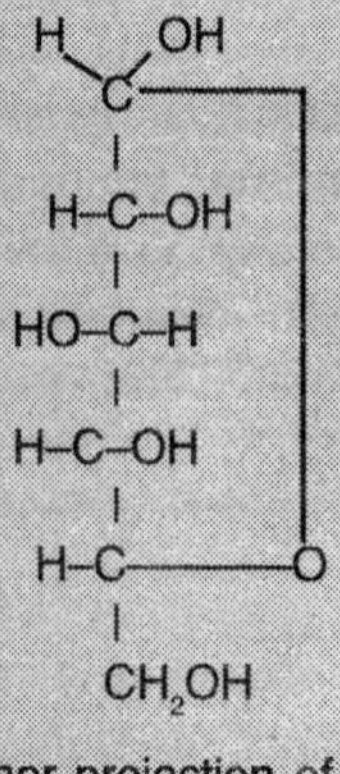

Cyclic Fischer projection of α-D-Glucose

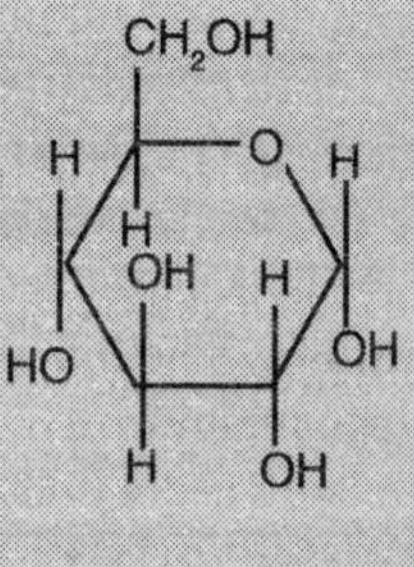

Haworth projection of α-D-Glucose

is the **anomeric carbon** and the two forms are termed **anomers**. Carbohydrates can change spontaneously between the α and β configurations—α process known as **mutarotation**. When drawn in the Fischer projection, the a configuration places the hydroxyl attached to the anomeric carbon to the right, towards the ring. When drawn in the Haworth projection, the α configuration places the hydroxyl downward.

The spatial relationships of the atoms of the furanose and pyranose ring structures are more correctly described by the two conformations identified as the **chair form** and the **boat form**. The chair form is the more stable of the two. Constituents of the ring that project above or below the plane of the ring are axial and those that project parallel to the plane are equatorial. In the chair conformation, the orientation of the hydroxyl group about the anomeric carbon of α-D-glucose is **axial** and **equatorial** in β-D-glucose.

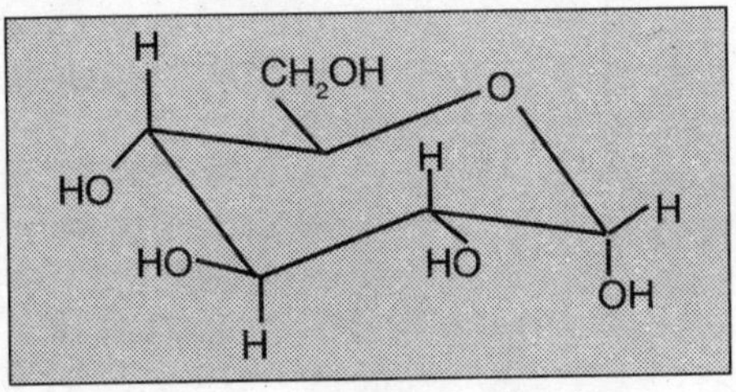

Chair form of α-D-Glucose

ii. Disaccharides

Covalent bonds between the anomeric hydroxyl of a cyclic sugar and the hydroxyl of a second sugar (or another alcohol containing compound) are termed **glycosidic bonds**, and the resultant molecules are **glycosides**. The linkage of two monosaccharides to form disaccharides involves a glycosidic bond. Several physiologically important disaccharides are sucrose, lactose and maltose.

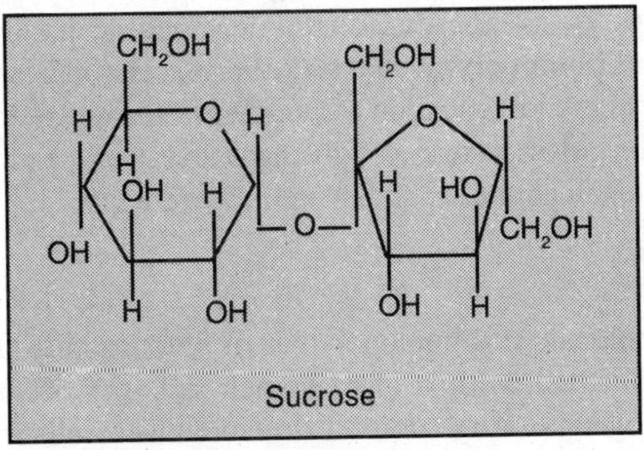

Sucrose

- **Sucrose:** prevalent in sugar cane and sugar beets, is composed of glucose and fructose through an α-(1,2)β-glycosidic bond.

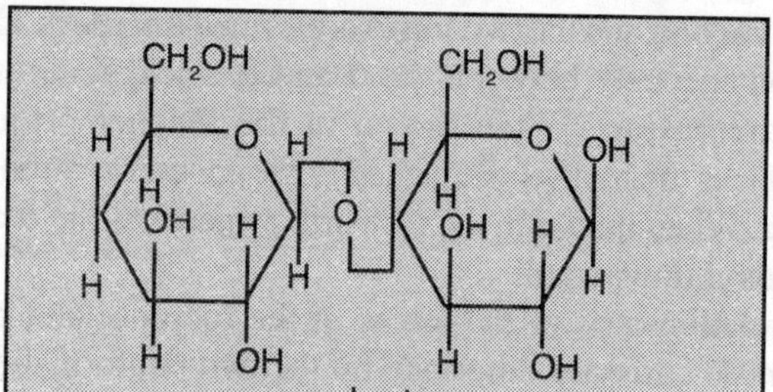

Lactose

- **Lactose:** It is found exclusively in the milk of mammals and consists of galactose and glucose in a β-(1,4) glycosidic bond.

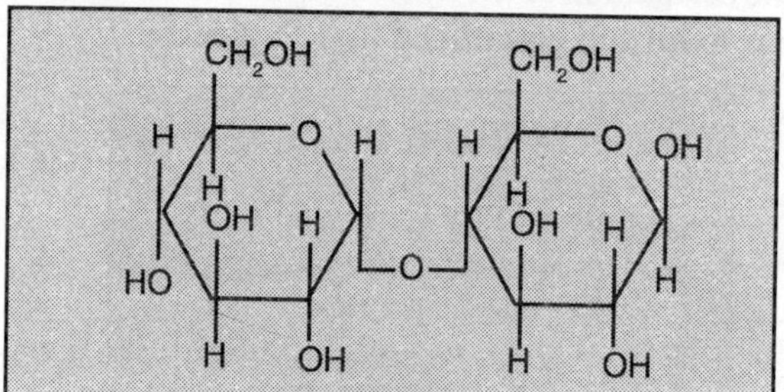

Maltose

- **Maltose:** The major degradation product of starch, is composed of 2 glucose monomers in an α-(1,4) glycosidic bond.

iii. Polysaccharides

Most of the carbohydrates found in nature occur in the form of high molecular weight polymers called polysaccharides. The monomeric building blocks used to generate polysaccharides can be varied; in all cases, however, the predominant monosaccharide found in polysaccharides is D-glucose. When polysaccharides are composed of a single monosaccharide building block, they are termed **homopolysaccharides.** Polysaccharides composed of more than one type of monosaccharide are termed **heteropolysaccharides.**

Two common polysaccharides—glycogen and starch—are the stroage forms of carbohydrates in animal and plant cell respectively.

- **Glycogen**

This crucial molecule is a homopolymer of glucose in α-(1,4) linkage; it is also highly branched, with α-(1,6) branch linkages occurring every 8-10 residues. Glycogen is a very compact structure that results from the coiling of the polymer chains. This compactness allows large amounts of carbon energy to be stored in a small volume, with little effect on cellular osmolarity.

• Starch

Its structure is identical to glycogen, except for a much lower degree of branching (about every 20-30 residues). Unbranched starch is called **amylose;** branched starch is called **amylopectin.**

• Cellulose

This polysacchardie is the principal structural component of the cell wall. The glucose residues in cellulse are in the β rather than α configuration, and it is an unbranded polysaccharide. The linkage of glucose residues by β-(1,4) rather than α-(1,4) bonds cause cellulose to form extend chains that pack side by side to form fibres of great mechanical strength.

III. LIPIDS

Biological molecules that are insoluble in aqueous solutions and soluble in organic solvents are classified as lipids. The lipids of physiological importance for humans have four major functions:

1. They serve as structural components of biological membranes.
2. They provide energy reserves, predominantly in the form of triacylglycerols.
3. Both lipids and lipid derivatives serve as vitamins and hormones.
4. Lipophilic bile acids aid in lipid solubilization.

a. Fatty Acids

Fatty acids are the simplest lipids, which fill two major roles in the body:

1. As the components of more complex membrane lipids.
2. As the major components of stored fat in the form of triacylglycerols.

Fatty acids are long-chain hydrocarbon molecules containing a carboxylic acid moiety at one end. The numbering of carbons in fatty acids begins with the carbon of the carboxylate group. At physiological pH, the carboxyl group is readily ionized, rendering a negative charge onto fatty acids in bodily fluids.

Fatty acids that contain no carbon-carbon double bonds are termed **saturated fatty acids**; those that contain double bonds are **unsaturated fatty acids**. The numeric designations used for fatty acids come from the number of carbon atoms, followed by the number of sites of unsaturation (e.g., palmitic acid is a 16-carbon fatty acid with no unsaturation and is designated by 16:0). The site of unsaturation in a fatty acid is indicated by the symbol Δ and the number of the first carbon of the double bond (e.g. palmitoleic acid is a 16-carbon fatty acid with one site of unsaturation between carbons 9 and 10, and is designated by $16{:}1^{\Delta 9}$).

Saturated fatty acids of less than eight carbon atoms are liquid at physiological temperature, whereas those containing more than ten are solid. The presence of double bonds in fatty acids significantly lowers the melting point relative to a saturated fatty acid.

Table 4.2 : Physiologically relevant fatty acids

Numerical symbol	*Common name*	*Structure*	*Comments*
14:0	Myristic acid	$CH_3(CH_2)_{12}COOH$	Often found attached to the N-term. of plasma membrane-associated cytoplasmic proteins
16:0	Palmitic acid	$CH_3(CH_2)_{14}COOH$	End product of mammalian fatty acid synthesis
$16{:}1^{D9}$	Palmitoleic acid	$CH_3(CH_2)_5C{=}C(CH_2)_7COOH$	
18:0	Stearic acid	$CH_3(CH_2)_{16}COOH$	
$18{:}1^{D9}$	Oleic acid	$CH_3(CH_2)_7C{=}C(CH_2)_7COOH$	
$18{:}2^{D9,12}$	Linoleic acid	$CH_3(CH_2)_4C{=}CCH_2C{=}C(CH_2)_7COOH$	Essential fatty acid
$18{:}3^{D9,12,15}$	Linolenic acid	$CH_3CH_2C{=}CCH_2C{=}CCH_2C{=}C(CH_2)_7COOH$	Essential fatty acid
$20{:}4^{D5,8,11,14}$	Arachidonic acid	$CH_3(CH_2)_3(CH_2C{=}C)_4(CH_2)_3COOH$	Precursor for eicosanoid synthesis

The majority of body fatty acids are acquired in the diet. However, the lipid biosynthetic capacity of the body (fatty acid synthase and other fatty acid modifying enzymes) can supply the body with all the various fatty acid structures needed. Two key exceptions to this are the highly unsaturated fatty acids known as **linoleic acid** and **linolenic acid**, containing unsaturation sites beyond carbons 9 and 10. These two fatty acids cannot be synthesized from precursors in the body, and are thus considered the **essential fatty acids**; essential in the sense that they must be provided in the diet. Since plants are capable of synthesizing linoleic and linolenic acid humans can aquire these fats by consuming a variety of plants or else by eating the meat of animals that have consumed these plant fats.

b. Triacylglycerides (Fig. 4.1)

Fatty acids are stored in the form of triacylglycerides, or fats, which are composed of a glycerol backbone to which 3 fatty acids are esterified.

c. Phospholipids

The basic structure of phospolipids, which are the principal components of cell membranes, is very similar to that of the triacylglycerides except that C3 (*sn*3)of the glycerol backbone is esterified to phosphoric acid. The building block of the phospholipids is **phosphatidic acid**, which results when the X substitution in the basic structure shown in the Figure 4.2 is a hydrogen atom. Substitutions include ethanolamine (**phosphatidylethanolamine**), choline (**phosphatidylcholine,** also called **lecithins**), serine (**phosphatidylserine**), glycerol (**phosphatidylglycerol**), *myo*-inositol (phosphatidylinositol, these

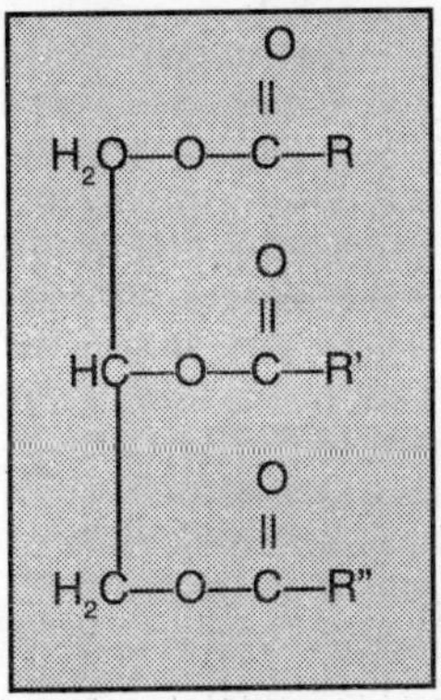

Fig. 4.1: Basic composition of a triacylglyceride

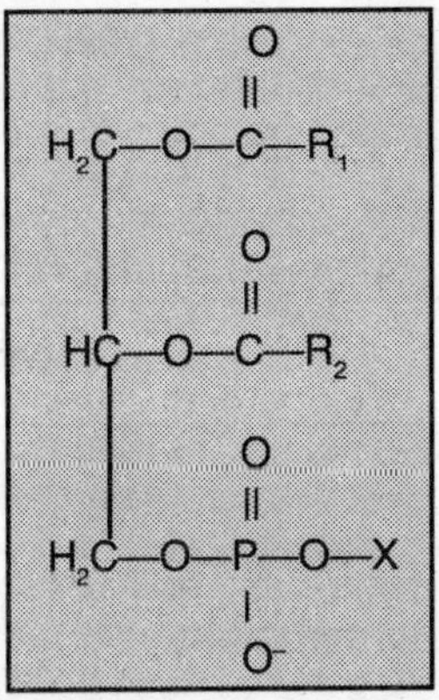

Fig. 4.2: Basic composition of a phospholipid. X can be a number of different substituents

compounds can have a variety in the numbers of inositol alcohols that are phosphorylated generating polyphosphatidylinositols), and phosphatidylglycerol (**diphosphatidylglycerol** more commonly known as **cardiolipins**).

d. Plasmalogens

Plasmalogens are complex membrane lipids that resemble phospholipids, principally phosphatidylcholine. The major difference is that the fatty acid at C1 (*sn*1) of glycerol contains either an *O*-alkyl or *O*-alkenyl ether species. A basic *O*-alkenyl ether species is shown in the Figure 4.3. One of the most potent biological molecules is **platelet activating factor** (PAF) which is a choline plasmalogen in which the C2 (*sn*2) position of glycerol is esterified with an acetyl group insted of a long chain fatty acid.

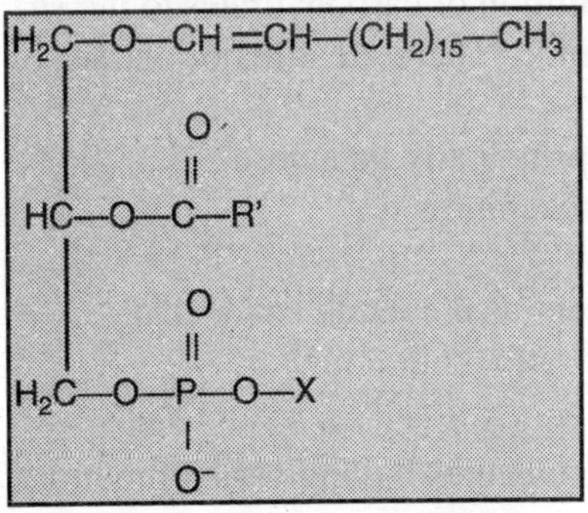

Fig. 4.3: Basic composition of *O*-alkenyl plasmalogens

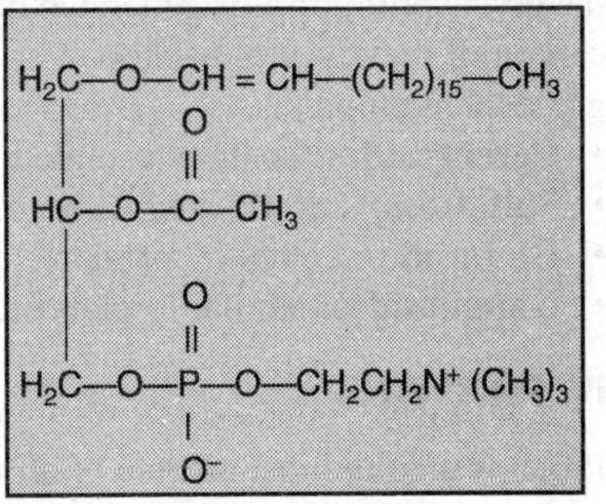

Structure of PAF

e. Sphingolipids

Sphingolipids are composed of a backbone of sphingosine which is derived itself from glycerol. Sphingosine is N-acetylated by a variety of fatty acids

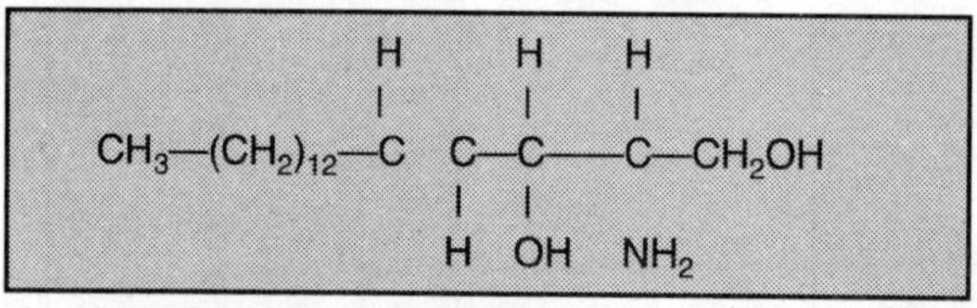

Sphingorine

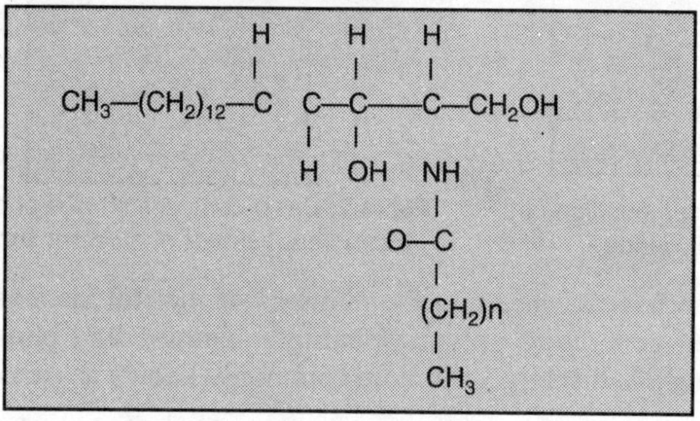

Basic composition of a ceramide
N indicates any fatty acid may be N-acetylated at this position

generating a family of molecules referred to as **ceramides**. Sphingolipids predominate in the myelin sheath of nerve fibers. **Sphingomyelin** is an abundant sphingolipid generated by transfer of the phosphocholine moiety of phosphatidylcholine to a ceramide, thus sphingomyelin is a unique form of a phospholipid.

The other major class of sphingolipids (besides the sphingomyelins) are the **glycosphingolipids** generated by substitution of carbohydrates to the *sn1* carbon of the glycerol backbone of a ceramide. There are 4 major classes of glycosphingolipids:

- **Cerebrosides:** contain a single moiety, principally galactose.
- **Sulfatides:** sulfuric acid esters of galactocerebrosides.
- **Globosides:** contain 2 or more sugars.
- **Gangliosides:** similar to globosides except also contain sialic acid.

IV. PROTEINS

Proteins are the most diverse of all macromolecules having varied functions in the cells as follows:

1. They serve as structural components of cells and tissues.
2. Act in the transport and storage of small molecules, e.g. the transport of oxygen by hemoglobin.
3. Transmit information between cells, e.g. protein hormones.
4. Provide a defence against infection, e.g. antibodies.
5. Act as enzymes.

Hence the Greek word *"proteios"* meaning "of the first rank". Proteins are polymers of **amino acids.**

A. AMINO ACIDS

All peptides and polypeptides are polymers of **alpha-amino acids**. There are 20 α-amino acids that are relevant to the make-up of mammalian proteins (see Table 4.3). Several other amino acids are found in the body free or in combined states (i.e. not associated with peptides or proteins). These non-protein associated amino acids perform specialized functions. Several of the amino acids found in proteins also serve functions distinct from the formation of peptides and proteins, e.g. tyrosine in the formation of thyroid hormones or glutamate acting as a neurotransmitter.

The α-amino acids in peptides and proteins (excluding proline) consist of a carboxylic acid (-**COOH**) and an amino (-$\mathbf{NH_2}$) functional group attached to the same tetrahedral carbon atom. This carbon is the α-**carbon**. Distinct **R-groups**, that distinguish one amino acid from another, also are attached to the alpha-carbon (except in the case of glycine where the R-group is hydrogen). The fourth substitution on the tetrahedral α-carbon of amino acids is **hydrogen**.

i. Amino Acid Classifications

Each of the 20 α-amino acids found in proteins can be distinguished by the R-group substitution on the α-carbon atom. There are two broad classes of amino acids based upon whether the R-group is **hydrophobic** or **hydrophilic**.

The hydrophobic amino acids tend to repel the aqueous environment and, therefore, reside predominantly in the interior of proteins. This class of amino acids does not ionize nor participate in the formation of H-bonds. The hydrophilic amino acids tend to interact with the aqueous environment, are often involved in the formation of H-bonds and are predominantly found on the exterior surfaces proteins or in the reactive centers of enzymes.

ii. Acid-Base Properties of the Amino Acids

The α-COOH and α-NH_2 groups in amino acids are capable of ionizing (as are the acidic and basic R-groups of the amino acids). As a result of their ionizability the following ionic equilibrium reactions may be written:

$$\mathbf{R\text{-}COOH \longleftrightarrow R\text{-}COO^- + H^+}$$

$$\mathbf{R\text{-}NH_3^+ \longleftrightarrow R\text{-}NH_2 + H^+}$$

The equilibrium reactions, as written, demonstrate that amino acids contain at least two weakly acidic groups. However, the carboxyl group is a far stronger acid than the amino group. At physiological pH (around 7.4) the carboxyl

Table 4.3: α-amino acids found in proteins

Amino acid	*Symbol*	*Structure**	*pK_1 (COOH)*	*pK_2 (NH_2)*	*pK_3 Group*
Amino acids with aliphatic R-groups					
Glycine	Gly - G	$H-CH-COOH$ | NH_2	2.4	9.8	
Alanine	Ala - A	$CH_3-CH-COOH$ | NH_2	2.4	9.9	
Valine	Val - V	H_3C \\ $CH-CH-COOH$ H_3C / | NH_2	2.2	9.7	
Leucine	Leu - L	H_3C \\ $CH-CH_2-CH-COOH$ H_3C / | NH_2	2.3	9.7	
Isoleucine	Ile - I	H_3C CH_2 \\ $CH-CH-COOH$ H_3O / | NH_2	2.3	9.8	
Non-aromatic amino acids with hydroxyl R-groups					
Serine	Ser - S	$HO-CH_2-CH-COOH$ | NH_2	2.2	9.2	~13
Threonine	Thr - T	H_3C \\ $CH-CH-COOH$ HO / | NH_2	2.1	9.1	~13
Amino acids with sulfur-containing R-groups					
Cysteine	Cys - C	$HS-CH_2-CH-COOH$ | NH_2	1.9	10.8	8.3
Methionine	Met - M	$H_3C-S-(CH_2)_2-CH-COOH$ | NH_2	2.1	9.3	
Acidic Amino Acids and their Amides					
Aspartic acid	Asp - D	$HOOC-CH_2-CH-COOH$ | NH_2	2.0	9.9	3.9
Asparagine	Asn - N	$H_2N-C-CH_2-CH-COOH$ ‖ | O NH_2	2.1	8.8	

Contd...

Contd...

Glutamic acid	Glu - E	$HOOC—CH_2—CH_2—CH(NH_2)—COOH$	2.1	9.5	4.1
Glutamine	Gln - Q	$H_2N—C(=O)—CH_2—CH_2—CH(NH_2)—COOH$	2.2	9.1	
Basic amino acids					
Arginine	Arg - R	$HN—CH_2—CH_2—CH_2—CH(NH_2)\ COOH$; HN bonded to $C=NH$, bonded to NH_2	1.8	9.0	12.5
Lysine	Lys - K	$H_2N—(CH_2)_4—CH(NH_2)—COOH$	2.2	9.2	10.8
Histidine	His - H	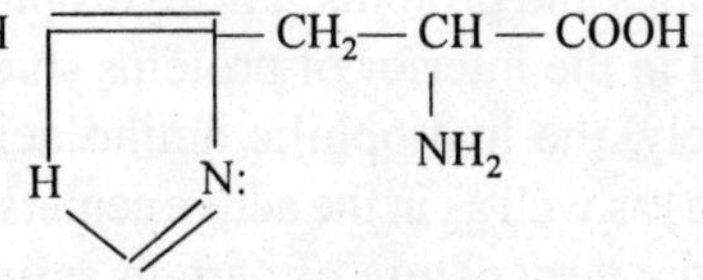	1.8	9.2	6.0
Amino acids with aromatic rings					
Phenylalanine	Phe - F	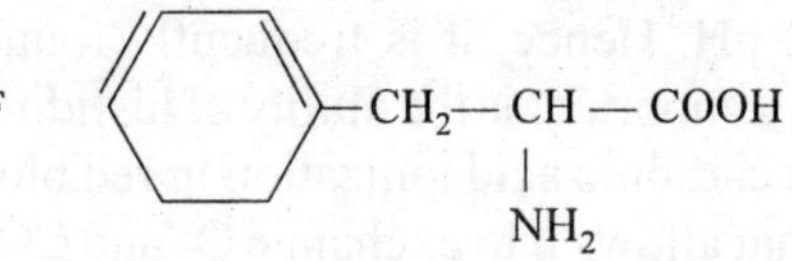	2.2	9.2	
Tyrosine	Tyr - Y	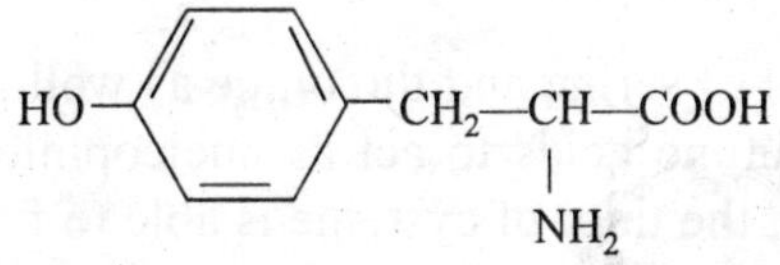	2.2	9.1	10.1
Tryptophan	Trp-W	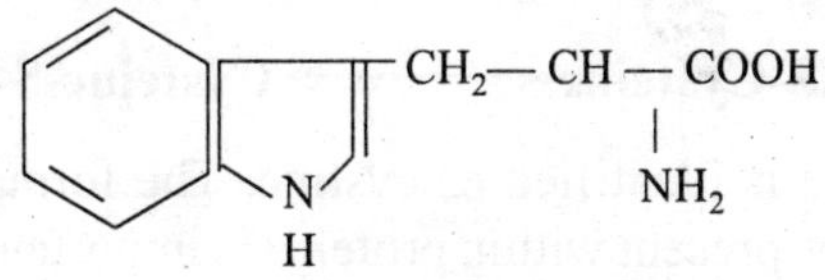	2.4	9.4	
Imino acids					
Proline	Pro - P	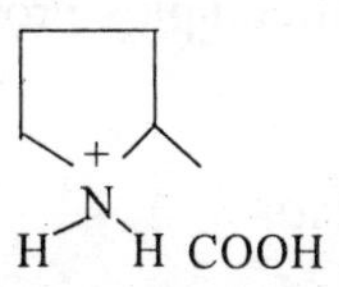	2.0	10.6	

group will be unprotonated and the amino group will be protonated. An amino acid with no ionizable R-group would be electrically neutral at this pH. This species is termed a **zwitterion**.

Like typical organic acids, the acidic strength of the carboxyl, amino and ionizable R-groups in amino acids can be defined by the association constant, K_a or more commonly the negative logrithm of K_a, the pK_a. The **net charge** (the algebraic sum of all the charged groups present) of any amino acid, peptide or protein, will depend upon the pH of the surrounding aqueous environment. As the pH of a solution of an amino acid or protein changes so too does the net charge. This phenomenon can be observed during the titration of any amino acid or protein. When the net charge of an amino acid or protein is zero the pH will be equivalent to the **isoelectric point, i.e. pI.**

iii. Functional Significance of Amino Acid R-Groups

In solution, it is the nature of the amino acid R-groups that dictate structure-function relationships of peptides and proteins. The hydrophobic amino acids will generally be encountered in the interior of proteins shielded from direct contact with water. Conversely, the hydrophilic amino acids are generally found on the exterior of proteins as well as in the active centers of enzymatically active proteins. Indeed, it is the very nature of certain amino acid R-groups that allow enzyme reactions to occur.

The imidazole ring of histidine allows it to act as either a proton donor or acceptor at physiological pH. Hence, it is frequently found in the reactive center of enzymes. Equally important is the ability of histidines in hemoglobin to buffer the H^+ ions from carbonic acid ionization in red blood cells. It is this property of hemoglobin that allows it to exchange O_2 and CO_2 at the tissues or lungs, respectively.

The primary alcohol of serine and threonine as well as the thiol (-SH) of cysteine allow these amino acids to act as nucleophiles during enzymatic catalysis. Additionally, the thiol of cysteine is able to form a disulfide bond with other cysteines:

Cysteine-SH + HS-Cysteine ⟷ Cysteine-S-S-Cysteine

This simple disulfide is identified as cystine. The formation of disulfide bonds between cysteines present within proteins is important to the formation of active structural domains in a large number of proteins. Disulfide bonding between cysteines in different polypeptide chains of oligomeric proteins plays a crucial role in ordering the structure of complex proteins, e.g. the insulin receptor.

iv. Optical Properties of the Amino Acids

A tetrahedral carbon atom with 4 distinct constituents is said to be chiral. The one amino acid not exhibiting chirality is glycine since its ""R-group" is a

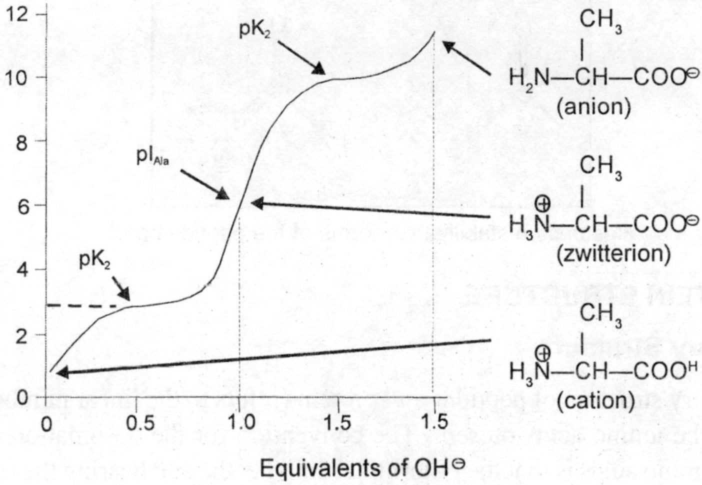

Fig. 4.4: Titration curve for alanine

hydrogen atom. Chirality describes the handedness of a molecule that is observable by the ability of a molecule to rotate the plane of polarized light either to the right (**dextrorotatory**) or to the left (**levorotatory**). All of the amino acids in proteins exhibit the same absolute steric configuration as **L-glyceraldehyde**. Therefore, they are all L-α-amino acids. D-amino acids are never found in proteins, although they exist in nature. D-amino acids are often found in polypetide antibiotics.

The aromatic R-groups in amino acids absorb ultraviolet light with an absorbance maximum in the range of 280 nm. The ability of proteins to absorb ultraviolet light is predominantly due to the presence of the tryptophan, which strongly absorbs ultraviolet light.

v. The Peptide Bond

Peptide bond formation is a condensation reaction leading to the polymerization of amino acids into peptides and proteins. Peptides are small consisting of few amino acids. A number of hormones and neurotransmitters are peptides. Additionally, several antibiotics and antitumor agents are peptides. Proteins are polypeptides of greatly divergent length. The simplest peptide, a **dipeptide**, contains a single peptide bond formed by the condensation of the carboxyl group of one amino acid with the amino group of the second with the concomitant elimination of water. The presence of the carbonyl group in a peptide bond allows electron resonance stabilization to occur such that the peptide bond exhibits rigidity not unlike the typical -C=C- double bond. The peptide bond is, therefore, said to have **partial double-bond character.**

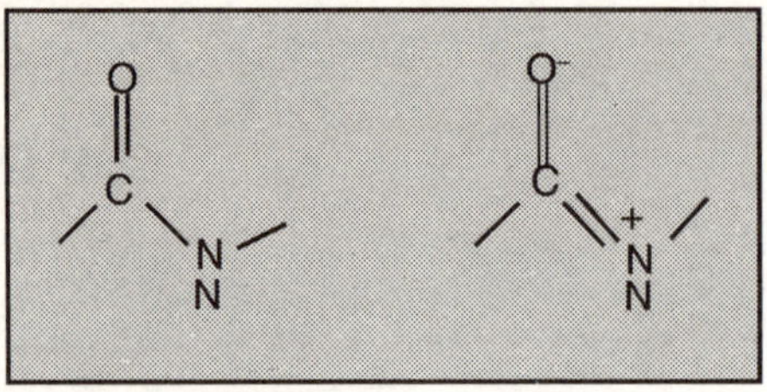

Resonance stabilization forms of the peptide bond

B. PROTEIN STRUCTURE

a. Primary Structure

The primary structure of peptides and proteins refers to the linear number and order of the amino acids present. The convention for the designation of the order of amino acids is that the N-terminal end (i.e. the end bearing the residue with the free α-amino group) is to the left (and the number 1 amino acid) and the C-terminal end (i.e. the end with the residue containing a free α-carboxyl group) is to the right.

b. Secondary Structure

The ordered array of amino acids in a protein confer regular conformational forms upon that protein. These conformations constitute the secondary structures of a protein. In general proteins fold into two broad classes of structure termed, **globular proteins** or **fibrous proteins**. Globular proteins are compactly folded and coiled, whereas, fibrous proteins are more filamentous or elongated. It is the partial double-bond character of the peptide bond that defines the conformations a polypeptide chain may assume. Within a single protein different regions of the polypeptide chain may assume different conformations determined by the primary sequence of the amino acids.

i. The Alpha-Helix

The α-helix is a common secondary structure encountered in proteins of the globular class. The formation of the α-helix is spontaneous and is stabilized by H-bonding between amide nitrogens and carbonyl carbons of peptide bonds spaced four residues apart.

This orientation of H-bonding produces a helical coiling of the peptide backbone such that the R-groups lie on the exterior of the helix and perpendicular to its axis.

Not all amino acids favor the formation of the α-helix due to steric constraints of the R-groups. Amino acids such as A, D, E, I, L and M favor the formation of α-helices, whereas, G and P favor disruption of the helix. This is particularly true for P since it is a pyrrolidine based imino acid (HN=) whose structure significantly restricts movement about the peptide bond in which it is present, thereby, interfering with extension of the helix. The disruption of the helix is important as it introduces additional folding of the polypeptide backbone to allow the formation of globular proteins.

ii. β -Sheets

Whereas an α-helix is composed of a single linear array of helically disposed amino acids, β-sheets are composed of 2 or more different regions of stretches of at least 5-10 amino acids. The folding and alignment of stretches of the polypeptide backbone aside one another to form β-sheets is stabilized by H-bonding between amide nitrogens and carbonyl carbons. However, the H-bonding residues are present in adjacently opposed stretches of the polypetide backbone as opposed to a linearly contiguous region of the backbone in the α-helix.

β-sheets are said to be pleated. This is due to positioning of the α-carbons of the peptide bond which alternates above and below the plane of the sheet.

β-sheets are either parallel or antiparallel. In parallel sheets, adjacent peptide chains proceed in the same direction (i.e. the direction of N-terminal to C-terminal ends is the same), whereas, in antiparallel sheets adjacent chains are aligned in opposite directions.

c. Super-secondary Structure

Some proteins contain an ordered organization of secondary structures that form distinct functional domains or structural motifs. Examples include the helix-turn-helix domain of bacterial proteins that regulate transcription and the leucine zipper, helix-loop-helix and zinc finger domains of eukaryotic transcriptional regulators. These domains are termed **super-secondary structures.**

d. Tertiary Structure

Tertiary structure refers to the complete three-dimensional structure of the polypeptide units of a given protein. Included in this description is the spatial relationship of different secondary structures to one another within a polypeptide chain and how these secondary structures themselves fold into the three-dimensional form of the protein. Secondary structures of proteins often constitute distinct **domains**. Therefore, tertiary structure also describes the relationship of different domains to one another within a protein. The interactions of different domains is governed by several forces: These include **hydrogen bonding, hydrophobic interactions, electrostatic interactions** and **van der Waals forces.**

e. Forces Controlling Protein Structure

i. Hydrogen Bonding

Polypeptides contain numerous proton donors and acceptors both in their backbone and in the R-groups of the amino acids. The environment in which proteins are found also contains the ample H-bond donors and acceptors of the water molecule. H-bonding, therefore, occurs not only within and between polypeptide chains but with the surrounding aqueous medium.

ii. Hydrophobic Forces

Proteins are composed of amino acids that contain either hydrophilic or hydrophobic R-groups. It is the nature of the interaction of the different R-groups with the aqueous environment that plays the major role in shaping protein structure. The spontaneous folded state of globular proteins is a reflection of a balance between the opposing energetics of H-bonding between hydrophilic R-groups and the aqueous environment and the repulsion from the aqueous environment by the hydrophobic R-groups. The hydrophobicity of certain amino acid R-groups tends to drive them away from the exterior of proteins and into the interior. This driving force restricts the available conformations into which a protein may fold.

iii. Electrostatic Forces

Electrostatic forces are mainly of three types; **charge-charge**, **charge-dipole** and **dipole-dipole.** Typical charge-charge interactions that favor protein folding are those between oppositely charged R-groups such as K or R and D or E. A substantial component of the energy involved in protein folding is charge-dipole interactions. This refers to the interaction of ionized R-groups of amino acids with the dipole of the water molecule. The slight dipole moment that exist in the polar R-groups of amino acid also influences their interaction with water. It is, therefore, understandable that the majority of the amino acids found on the exterior surfaces of globular proteins contain charged or polar R-groups.

iv. Van der Waals Forces

There are both attractive and repulsive van der Waals forces that control protein folding. Attractive van der Waals forces involve the interactions among induced dipoles that arise from fluctuations in the charge densities that occur between adjacent uncharged non-bonded atoms. Repulsive van der Waals forces involve the interactions that occur when uncharged non-bonded atoms come very close together but do not induce dipoles. The repulsion is the result of the electron-electron repulsion that occurs as two clouds of electrons begin to overlap.

Although van der Waals forces are extremely weak, relative to other forces governing conformation, it is the huge number of such interactions that occur in large protein molecules that make them significant to the folding of proteins.

f. Quaternary Structure

Many proteins contain two or more different polypeptide chains that are held in association by the same non-covalent forces that stabilize the tertiary structures of proteins. Proteins with multiple polypetide chains are termed oligomeric proteins. The structure formed by monomer-monomer interaction in an oligomeric protein is known as **quaternary structure.**

Oligomeric proteins can be composed of multiple identical polypeptide chains or multiple distinct polypeptide chains. Proteins with identical subunits are termed **homo-oligomers.** Proteins containing several distinct polypeptide chains are termed **hetero-oligomers.**

Hemoglobin, the oxygen carrying protein of the blood, contains two α and two β subunits arranged with a quaternary structure in the form, $\alpha_2\beta_2$. Hemoglobin is, therefore, a hetero-oligomeric protein.

g. Complex Protein Structures

Proteins are also found to be covalently conjugated with carbohydrates. These modifications occur following the synthesis (translation) of proteins and are, therefore, termed post-translational modifications. These forms of modification impart specialized functions upon the resultant proteins. Proteins covalently associated with carbohydrates are termed glycoproteins. Glycoproteins are of two classes, N-linked and O-linked, referring to the site of covalent attachment of the sugar moieties. N-linked sugars are attached to the amide nitrogen of the R-group of asparagine; O-linked sugars are attached to the hydroxyl groups of either serine or threonine and occasionally to the hydroxyl group of the modified amino acid, hydroxylysine.

There are extremely important glycoproteins found on the surface of erythrocytes. It is the variability in the composition of the carbohydrate portions of many glycoproteins and glycolipids of erythrocytes that determines blood group specificities. There are at least 100 blood group determinants, most of which are due to carbohydrate differences. The most common blood groups, A, B, and O, are specified by the activity of specific gene products whose activities are to incorporate distinct sugar groups onto RBC membrane glycoshpingolipids as well as secreted glycoproteins.

Structural complexes involving protein associated with lipid via noncovalent interactions are termed lipoproteins. Their major function in the body is to aid in the storage transport of lipid and cholesterol.

h. Clinical Significances

The substitution of a hydrophobic amino acid (V) for an acidic amino acid (E) in the β-chain of hemoglobin results in **sickle cell anemia** (HbS). This change of a single amino acid alters the structure of hemoglobin molecules in such a way that the deoxygenated proteins polymerize and precipitate within the erythrocyte, leading to their characteristic sickle shape.

Collagens are the most abundant proteins in the body. Alterations in collagen structure arising from abnormal genes or abnormal processing of collagen proteins results in numerous diseases, including Larsen syndrome, scurvy, osteogenesis imperfecta and Ehlers-Danlos syndrome.

Ehlers-Danlos syndrome is actually the name associated with at least ten distinct disorders that are biochemically and clinically distinct yet all manifest structural weakness in connective tissue as a result of defective collagen structure. Osteogenesis imperfecta also encompasses more than one disorder. At least four biochemically and clinically distinguishable maladies have been identified as **osteogenesis imperfecta**, all of which are characterized by multiple fractures and resultant bone deformities. Marfan's syndrome manifests itself as a disorder of the connective tissue and was originally believed to be the result of abnormal collagens. However, recent evidence has shown that **Marfan's syndrome** results from mutations in the extracellular protein, **fibrillin**, which is an integral constituent of the non-collagenous microfibrils of the extracellular matrix.

Several forms of **familial hypercholesterolemia** are the result of genetic defects in the gene encoding the receptor for low-density lipoprotein (LDL). These defects result in the synthesis of abnormal LDL receptors that are incapable of binding to LDLs, or that bind LDLs but the receptor/LDL complexes are not properly internalized and degraded. The outcome is an elevation in serum cholesterol levels and increased propensity toward the development of atherosclerosis.

A number of proteins can contribute to cellular transformation and carcinogenesis when their basic structure is disrupted by mutations in their genes. These genes are termed **proto-oncogenes**. For some of these proteins, all that is required to convert them to the **oncogenic** form is a single amino acid substitution. The cellular gene, c-**Ras**, is observed to sustain single amino acid substitutions at positions 12 or 61 with high frequency in colon carcinomas. Mutations in c-Ras are most frequently observed genetic alterations in colon cancer.

C. AMINO-TERMINAL SEQUENCE DETERMINATION

Prior to sequencing peptides it is necessary to eliminate disulfide bonds within peptides and between peptides. Several different chemical reactions can be used in order to permit separation of peptide strands and prevent protein conformations that are dependent upon disulfide bonds. The most common treatments are to use either **2-mercaptoethanol** or **dithiothreitol**. Both of these chemicals reduce disulfide bonds. To prevent reformation of the disulfide bonds the peptides are treated with **iodoacetic acid** in order to alkylate the free sulfhydryls.

There are three major chemical techniques for sequencing peptides and proteins from the N-terminus. These are the **Sanger**, **Dansyl chloride** and **Edman techniques.**

i. **Sanger's reagent:** This sequencing technique utilizes the compound, 2,4-dinitrofluorobenzene (DNF), which reacts with the N-terminal residue under alkaline conditions. The derivatized amino acid can be hydrolyzed and will be labeled with a dinitrobenzene group that imparts a yellow color to the amino acid. Separation of the modified amino acids (DNP-derivative) by electrophoresis and comparison with the migration of DNP-derivative standards allows for the identification of the N-terminal amino acid.

ii. **Dansyl chloride:** Like DNF, dansyl chloride reacts with the N-terminal residue under alkaline conditions. Analysis of the modified amino acids is carried out similarly to the Sanger method except that the dansylated amino acids are detected by fluorescence. This imparts a higher sensitivity into this technique over that of the Sanger method.

iii. **Edman degradation:** The utility of the Edman degradation technique is that it allows for additional amino acid sequence to be obtained from the N-terminus inward. Using this method it is possible to obtain the entire sequence of peptides. This method utilizes phenylisothiocyanate to react with the N-terminal residue under alkaline conditions. The resultant phenylthiocarbamyl derivatized amino acid is hydrolyzed in anhydrous acid. The hydrolysis reaction results in a rearrangement of the released N-terminal residue to a phenylthiohydantoin derivative. As in the Sanger and Dansyl chloride methods, the N-terminal residue is tagged with an identifiable marker, however, the added advantage of the Edman process is that the remainder of the peptide is intact. The entire sequence of reactions can be repeated over and over to obtain the sequences of the peptide. This process has subsequently been automated to allow rapid and efficient sequencing of even extremely small quantities of peptide.

Protease Digestion

Due to the limitations of the Edman degradation technique, peptides longer than around 50 residues cannot be sequenced completely. The ability to obtain peptides of this length, from proteins of greater length, is facilitated by the use of enzymes, *endopeptidases*, that cleave at specific sites within the primary sequence of proteins. The resultant smaller peptides can be chromatographically separated and subjected to Edman degradation sequencing reactions.

D. CARBOXY-TERMINAL SEQUENCE DETERMINATION

No reliable chemical techniques exist for sequencing the C-terminal amino acid of peptides. However, there are enzymes, *exopeptidases*, that have been identified that cleave peptides at the C-terminal residue which can then be

Table 4.4: Specificities of several endoproteases

Enzyme	*Source*	*Specificity*	*Additional points*
Trypsin	Bovine pancreas	Peptide bond C-terminal to R,	K, but not if next to P highly specific for positively charged residues
Chymotrypsin	Bovine pancreas	Peptide bond C-terminal to F, Y, W but not if next to P	Prefers bulky hydrophobic residues, cleaves slowly at N, H, M, L
Elastase	Bovine pancreas	Peptide bond C-terminal to A, G, S, V, but not if next to P	
Thermolysin	*Bacillus thermoproteolyticus*	Peptide bond N-terminal to I, M, F, W, Y, V, but not if next to P	Prefers small neutral residues, can cleave at A, D, H, T
Pepsin	Bovine gastric mucosa	Peptide bond N-terminal to L, F, W, Y, but when next to P	Exhibits little specificity, requires low pH
Endopeptidase V8	*Staphylococcus aureus*	Peptide bond C-terminal to E	

analyzed chromatographically and compared to standard amino acids. This class of exopeptidases are called, *carboxypeptidases*.

i. Chemical Digestion of Proteins

The most commonly utilized chemical reagent that cleaves peptide bonds by recognition of specific amino acid residues is **cyanogen bromide (CNBr).** This reagent causes specific cleavage at the C-terminal side of M residues. The number of peptide fragments that result from CNBr cleavage is equivalent to one more than the number of M residues in a protein.

The most reliable chemical technique for C-terminal residue identification is **hydrazinolysis**. A peptide is treated with **hydrazine, NH_2-NH_2**, at high temperature (90°C) for an extended length of time (20-100 hr). This treatment cleaves all of the peptide bonds yielding aminoacyl hydrazides of all the amino acids excluding the C-terminal residue which can be identified chromatographically compared to amino acid standards. Due to the high percentage of hydrazine induced side reactions this technique is only used on carboxypeptidase resistant peptides.

Table 4.5: Specificities of several exopeptidases

Enzyme	*Source*	*Specificity*
Carboxypeptidase A	Bovine pancreas	Will not cleave when C-terminal residue = R, K or P or if P resides next to terminal residue
Carboxypeptidase B	Bovine pancreas	Cleaves when C-terminal residue = R or K; not when P resides next to terminal residue
Carboxypeptidase C	Citrus leaves	All free C-terminal residues, pH optimum = 3.5
Carboxypeptidase Y	Yeast	All free C-terminal residues, slowly at G residues

ii. Size Exclusion Chromatography

This chromatographic technique is based upon the use of a porous gel in the form of insoluble beads placed into a column. As a solution of proteins is passed through the column, small proteins can penetrate into the pores of the beads and, therefore, are retarded in their rate of travel through the column. The larger proteins a protein is the less likely it will enter the pores. Different beads with different pore sizes can be used depending upon the desired protein size separation profile.

iii. Ion-exchange Chromatography

Each individual protein exhibits a distinct overall net charge at a given pH. Some proteins will be negatively charged and some will be positively charged at the same pH. This property of proteins is the basis for ion exchange chromatography. Fine cellulose resins are used that are either negatively (**cation exchanger**) or positively (**anion exchanger**) charged proteins of opposite charge to the resin are retained as a solution of proteins is passed through the column. The bound proteins are then eluted by passing a solution of ions bearing a charge opposite to that of the column. By utilizing a gradient of increasing ionic strength, proteins with increasing affinity for the resin are progressively eluted.

iv. Affinity Chromatography

Proteins have high affinities for their substrates or co-factors or prosthetic groups or receptors or antibodies raised against them. This affinity can be exploited in the purification of proteins. A column of beads bearing the high affinity compound can be prepared and a solution of protein passed through the column. The bound proteins are then eluted by passing a solution of unbound soluble high affinity compound through the column.

v. High Performance Liquid Chromatography (HPLC)

In column chromatography, the smaller and more tightly packed a resin is the greater separation capability of the column. In gravity flow columns, the limitation column packing is the time it takes to pass the solution of proteins through the column. HPLC utilizes tightly packed fine diameter resins to impart increased resolution and overcomes the flow limitations by pumping the solution of proteins through the column under high pressure. Like standard column chromatography, HPLC columns can be used for size exclusion or charge separation. An additional separation technique commonly used with HPLC is to utilize hydrophobic resins to retard the movement of non-polar proteins. The proteins are then eluted from the column with a gradient of increasing concentration of an organic solvent. This latter form of HPLC is termed **reversed-phase** HPLC.

E. ELECTROPHORESIS OF PROTEINS

Proteins also can be characterized according to size and charge by separation in an electric current (electrophoresis) within solid sieving gels made from polymerized and cross-linked acrylamide. The most commonly used technique is termed SDS polyacrylamide gel electrophoresis (SDS-PAGE). The gel is a thin slab of acrylamide polymerized between two glass plates. This technique utilizes a negatively charged detergent (sodium dodecyl sulfate) to denature and solubilize proteins. SDS denatured proteins have a uniform negative charge such that all proteins will migrate through the gel in the electric field based solely upon size. The larger the protein the more slowly it will move through the matrix of the polyacrylamide. Following electrophoresis the migration distance of unknown proteins relative to known standard proteins is assessed by various staining or radiographic detection techniques (Fig. 4.5).

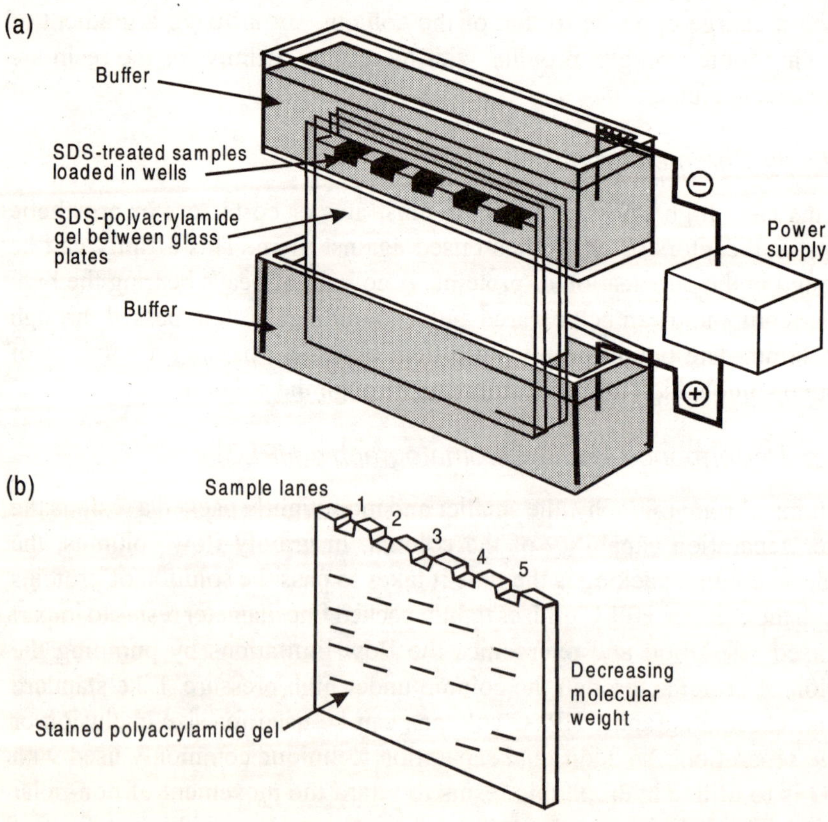

Fig. 4.5: (a) SDS-PAGE apparatus with power pack (b) SDS-PAGE Electrophotogram

The use of polyacrylamide gel electrophoresis also can be used to determine the isoelectric charge of proteins (pI). This technique is termed isoelectric focusing. Isoelectric focusing utilizes a thin tube of polyacrylamide made in the presence of a mixture of small positively and negatively charged molecules termed ampholytes. The ampholytes have a range of pIs that establish a pH gradient along the gel when current is applied. Proteins will, therefore, cease migration in the gel when they reach the point where the ampholytes have established a pH equal to the proteins pI (Fig. 4.6).

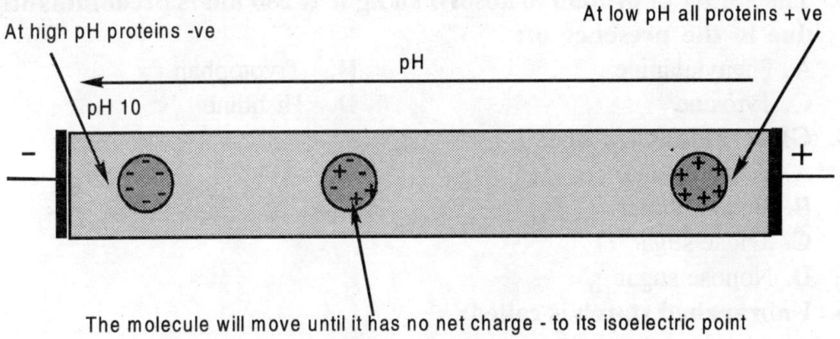

Fig. 4.6: Isoelectric focussing

F. CENTRIFUGATION OF PROTEINS

Proteins will sediment through a solution in a centrifugal field dependent upon their mass. Analytical centrifugation measures the rate that proteins sediment. The most common solution utilized is a linear gradient of sucrose (generally from 5-20%). Proteins are layered atop the gradient in an ultracentrifuge tube then subjected to centrifugal fields in excess of 100,000 × g. The sizes of unknown proteins can then be determined by comparing their migration distance in the gradient with those of known standard proteins.

STUDY QUESTIONS

1. Describe the forces determining protein structure.
2. Describe the general strategies for purification and characterization of protein from an organism.

MCQs

Select one most appropriate answer given for each questions below:

1. **Sialic acid is an example of:**
   A. Triose
   B. Pentose
   C. Hexose
   D. Mannose

2. **Which of the followings is an unbranched polysaccharide:**
   A. Cellulose B. Glycogen
   C. Starch D. Sucrose
3. **Sucrose is a disaccharide of:**
   A. Glucose and glucose
   B. Galactose and glucose
   C. Glucose and fructose
   D. Galactose and fructose
4. **The ability of protein to absorb un light at 280 nm is predominantly due to the presence of:**
   A. Phenylalanine B. Tryptophan
   C. Tyrosine D. Histidine
5. **Glyceraldehydes are:**
   A. Pentose sugar
   B. Hexose sugar
   C. Triose sugar
   D. Nonose sugar
6. **Unbranched starch is called:**
   A. Glycogen B. Amylose
   C. Cellulose D. Amylopectin
7. **Linoleic acid is an example of:**
   A. Saturated fatty acids
   B. Essential fatty acids
   C. Unsaturated fatty acids
   D. Nonessential fatty acid
8. **Which of the following fatty acids is a platelet activating factor:**
   A. Sphingolipid
   B. Plasmalogen
   C. Palmitic acid
   D. Myristic acid
9. **Which of the following lipid predominates in mylein sheath or nerve fibers:**
   A. Plasmalogen B. Myristic acid
   C. Sphingolipid D. Cleic acid
10. **Which of the following forces of chemical reaction is force:**
    A. Hydrogen bonding
    B. Hydrophobic force
    C. Electrostatic force
    D. Van der Waals force

---

1 D 2 A 3 C 4 B 5 C 6 B 7 C 8 B
9 C 10 D

5

Nucleic Acid

I. INTRODUCTION

As a class, the nucleotides may be considered as one of the most important metabolites of the cell. Nucleotides are found primarily as the monomeric units comprising the major nucleic acids of the cell, RNA and DNA. However, they also are required for numerous other important functions within the cell. These functions include:

1. Serving as energy stores for future use in phosphate transfer reactions. These reactions are predominantly carried out by ATP.
2. Forming a portion of several important coenzymes such as NAD^+, $NADP^+$, FAD and coenzyme A.
3. Serving as mediators of numerous important cellular processes such as **second messengers** in signal transduction events. The predominant second messenger is cyclic-AMP (cAMP), a cyclic derivative of AMP formed from ATP.
4. Controlling numerous enzymatic reactions through allosteric effects on enzyme activity.
5. Serving as activated intermediates in numerous biosynthetic reactions. These activated intermediates include S-adenosylmethionine (S-AdoMet) involved in methyl transfer reactions as well as the many sugar coupled nucleotides involved in glycogen synthesis.

II. NUCLEOSIDE AND NUCLEOTIDE STRUCTURE AND NOMENCLATURE

The nucleotides found in cells are derivatives of the heterocyclic highly basic, compounds, **purine** and **pyrimidine** (Fig.5.1a,b).

It is the chemical basicity of the nucleotides that has given them the common term "bases" as they are associated with nucleotides present in DNA and RNA. There are five major bases found in cells. The derivatives of purine are called **adenine** and **guanine**, and the derivatives of pyrimidine are called **thymine**, **cytosine** and **uracil**. The common abbreviations used for these five bases are, **A, G, T, C** and **U** (Table 5.1).

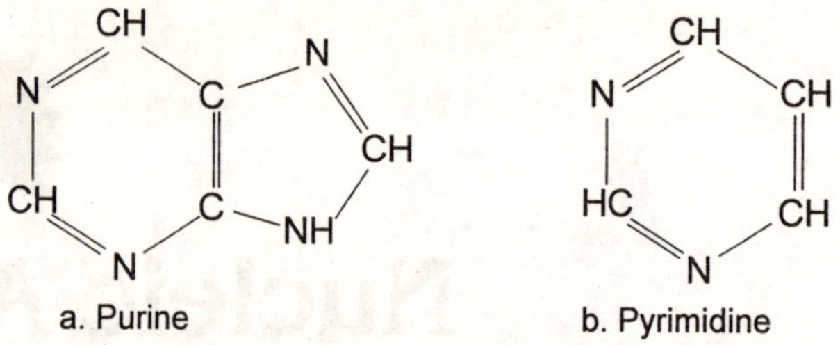

Fig. 5.1: Structure of purine and pyrimidine

The purine and pyrimidine bases in cells are linked to carbohydrate and in this form are termed, **nucleosides**. The nucleosides are coupled to D-ribose or 2'-deoxy-D-ribose through a β-N-glycosidic bond between the anomeric carbon of the ribose and the N^9 of a purine or N^1 of a pyrimidine.

The base can exist in 2 distinct orientations about the N-glycosidic bond. These conformations are identified as, ***syn*** and ***anti***. It is the anti-conformation that predominates in naturally occurring nucleotides (Fig. 5.2).

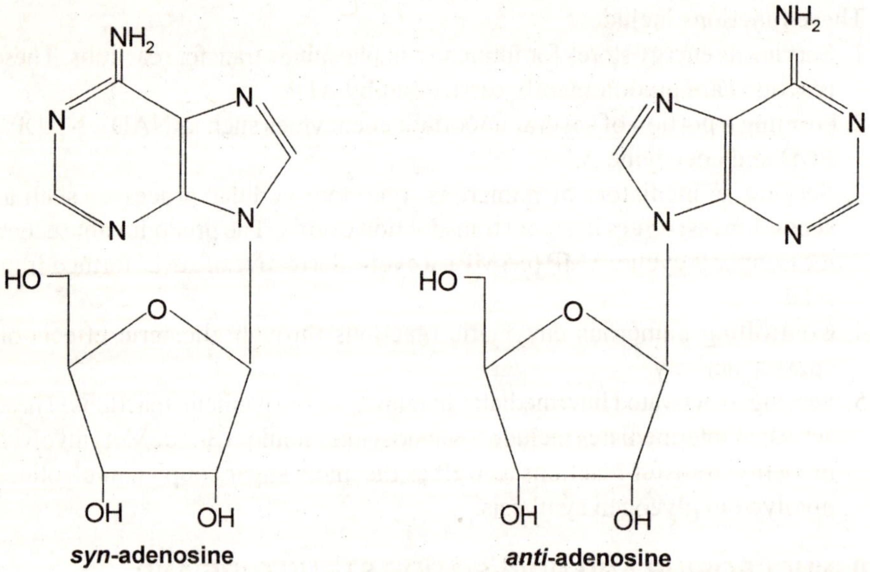

Fig. 5.2: Syn and anti-adenosine

Nucleosides are found in the cell primarily in their phosphorylated form. These are termed **nucleotides**. The most common site of phosphorylation of nucleotides found in cells is the hydroxyl group attached to the 5'-carbon of the ribose. The carbon atoms of the ribose present in nucleotides are designated with a prime (') mark to distinguish them from the backbone numbering in the bases. Nucleotides can exist in the mono-, di-, or tri-phosphorylated forms.

Nucleotides are given distinct abbreviations to allow easy identification of their structure and state of phosphorylation. The monophosphorylated form

Table 5.1: Bases of nucleic acid

Base formula	*Base (X=H)*	*Nucleoside X=ribose or deoxyribose*	*Nucleotide X=ribose phosphate*
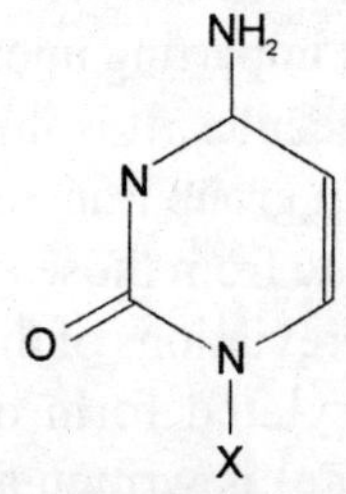	Cytosine, C	Cytidine, A	Cytidine monophosphate CMP
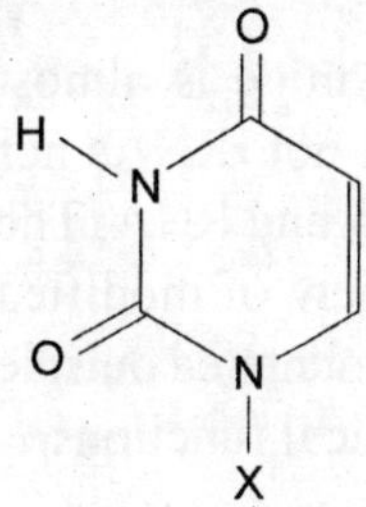	Uracil, U	Uridine, U	Uridine monophosphate UMP
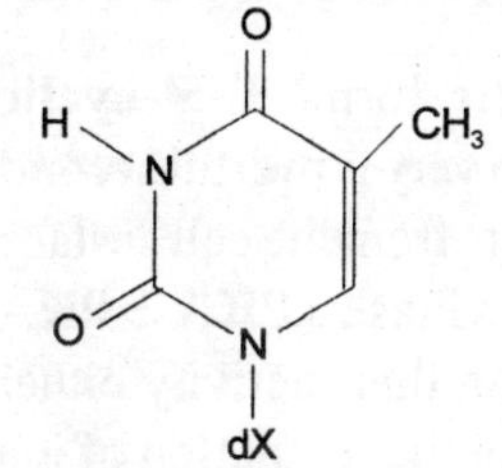	Thymine, T	Thymidine, T	Thymidine monophosphate TMP
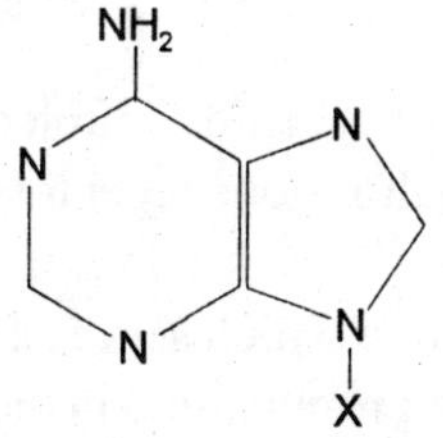	Adenine, A	Adenosine, A	Adenosine monophosphate AMP
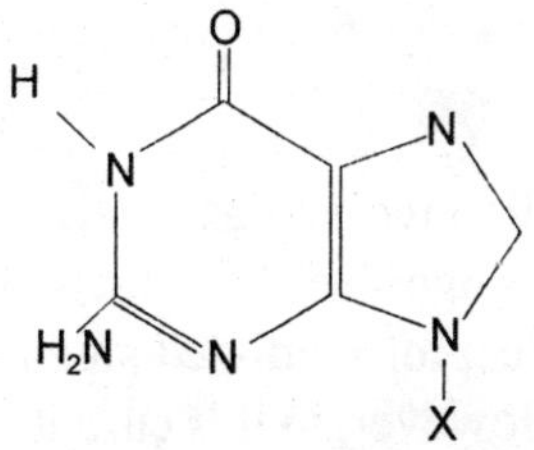	Guanine G	Guanosine, A	Guanosine monophosphate GMP

of adenosine (adenosine-5'-monophosphate) is written as, **AMP**. The di- and tri-phosphorylated forms are written as, **ADP** and **ATP**, respectively. The use of these abbreviations assumes that the nucleotide is in the 5'-phosphorylated form. The di- and tri-phosphates of nucleotides are linked by acid anhydride bonds. Acid anhydride bonds have a high $\Delta G^{0'}$ for hydrolysis imparting upon them a high potential to transfer the phosphates to other molecules. It is this property of the nucleotides that results in their involvement in group transfer reactions in the cell. The nucleotides found in DNA are unique from those of RNA in that the ribose exists in the 2'-deoxy form and the abbreviations of the nucleotides contain a **d** designation. The monophosphorylated form of adenosine found in DNA (deoxyadenosine-5'-monophosphate) is written as **cAMP**.

The nucleotide uridine is never found in DNA and thymine is almost exclusively found in DNA. Thymine is found in tRNAs but not rRNAs nor mRNAs. There are several less common bases found in DNA and RNA. The primary modified base in DNA is 5-methylcytosine. A variety of modified bases appear in the tRNAs. Many modified nucleotides are encountered outside of the context of DNA and RNA that serve important biological functions.

a. Adenosine Derivatives

The most common adenosine derivative is the cyclic form, **3'-5'-cyclic adenosine monophosphate, cAMP**. This compound is a very powerful **second messenger** involved in passing signal transduction events from the cell surface to internal proteins, e.g. cAMP-dependent protein kinase (**PKA**). PKA phosphorylates a number of proteins, thereby, affecting their activity either positively or negatively. Cyclic-AMP is also involved in the regulation of ion channels by direct interaction with the channel proteins, e.g. in the activation of odorant receptors by odorant molecules.

Formation of cAMP occurs in response to activation of receptor coupled ***adenylate cyclase***. These receptors can be of any type, e.g. hormone receptors or odorant receptors.

S-adenosylmethionine is a form of ***activated*** methionine which serves as a methyl donor in methylation reactions and as a source of propylamine in the synthesis of polyamines.

b. Guanosine Derivatives

A cyclic form of GMP (**cGMP**) also is found in cells involved as a second messenger molecule. In many cases its' role is to antagonize the effects of cAMP. Formation of cGMP occurs in response to receptor mediated signals similar to those for activation of adenylate cyclase. However, in this case it is ***guanylate cyclase*** that is coupled to the receptor.

The most important cGMP coupled signal transduction cascade is that photoreception. However, in this case activation of **rhodopsin** (in the rods) or other **opsins** (in the cones) by the absorption of a photon of light (through 11-*cis*-retinal covalently associated with rhodopsin and opsins) activates transducin which in turn activates a cGMP specific phosphodiesterase that hydrolyzes cGMP to GMP. This lowers the effective concentration of cGMP bound to gated ion channels resulting in their closure and a concomitant hyperpolarization of the cell.

c. Synthetic Nucleotide Analogues

Many nucleotide analogues are chemically synthesized and used for their therapeutic potential. The nucleotide analogues can be utilized to inhibit specific enzymatic activities. A large family of analogues are used as antitumor agents, for instance, because they interfere with the synthesis of DNA and thereby preferentially kill rapidly dividing cells such as tumor cells. Some of the nucleotide analogues commonly used in chemotherapy are 6-mercaptopurine, 5-fluorouracil, 5-iodo-2'-deoxyuridine and 6-thioguanine. Each of these compounds disrupts the normal replication process by interfering with the formation of correct Watson-Crick base-pairing.

Nucleotide analogs also have been targeted for use as antiviral agents. Several analogs are used to interfere with the replication of HIV, such as **AZT (azidothymidine)** and **ddI (dideoxyinosine)**.

Several purine analogs are used to treat gout. The most common is **allopurinol**, which resembles hypoxanthine. Allopurinol inhibits the activity of ***xanthine oxidase***, an enzyme involved in *de novo* purine biosynthesis. Additionally, several nucleotide analogues are used after organ transplantation in order to suppress the immune system and reduce the likelihood of transplant rejection by the host.

d. Polynucleotides

Polynucleotides are formed by the condensation of two or more nucleotides. The condensation most commonly occurs between the alcohol of a 5'-phosphate of one nucleotide and the 3'-hydroxyl of a second, with the elimination of H_2O, forming a **phosphodiester bond**. The formation of phosphodiester bonds in DNA and RNA exhibits directionality. The primary structure of DNA and RNA (the linear arrangement of the nucleotides) proceeds in the 5' ⟶ 3' direction. The common representation of the primary structure of DNA or RNA molecules is to write the nucleotide sequences from left to right synonymous with the 5' ⟶ 3' direction as shown:

5'-pGpApTpC-3'

III. STRUCTURE OF DNA

Utilizing X-ray diffraction data, obtained from crystals of DNA, James Watson and Francis Crick proposed a model for the structure of DNA. This model (subsequently verified by additional data) predicted that DNA would exist as a helix of two complementary antiparallel strands, wound around each other in a rightward direction and stabilized by H-bonding between bases in adjacent strands. In the **Watson-Crick model,** the bases are in the interior of the helix aligned at a nearly 90° angle relative to the axis of the helix. Purine bases form hydrogen bonds with pyrimidines, in the crucial phenomenon of **base pairing**. Experimental determination has shown that, in any given molecule of DNA, the concentration of adenine (A) is equal to thymine (T) and the concentration of cytidine (C) is equal to guanine (G). This means that A will only base-pair with T, and C with G. According to this pattern, known as **Watson-Crick base-pairing,** the base-pairs composed of G and C contain three H-bonds, whereas those of A and T contain two H-bonds. This makes G-C base-pairs more stable than A-T base-pairs (Fig. 5.4).

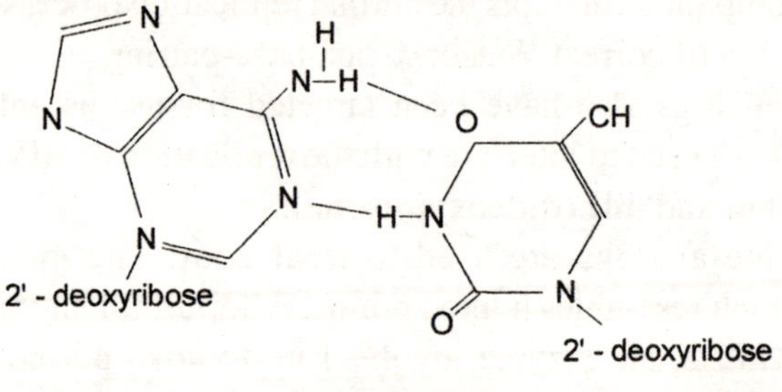

A-T Base pair

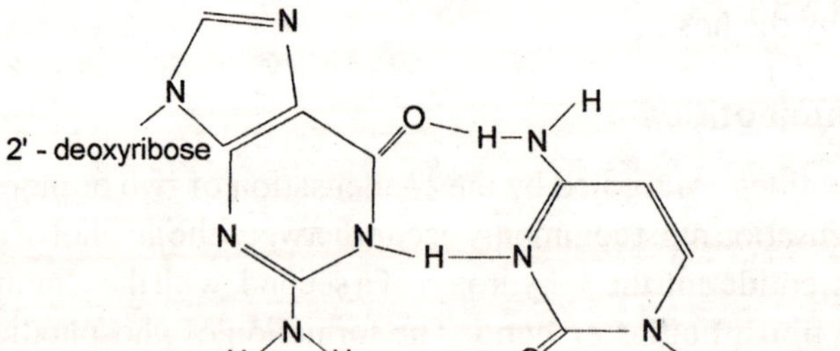

G-C Base pair

Fig. 5.4: A-T and G-C base pairs

The antiparallel nature of the helix stems from the orientation of the individual strands. From any fixed position in the helix, one strand is oriented

Table 5.2: Parameters of major DNA helices

Parameters	*A Form*	*B Form*	*Z-Form*
Direction of helical rotation	Right	Right	Left
Residues per turn of helix	11	10	12 base pairs
Rotation of helix per residue (in degrees)	33	36	-30
Base tilt relative to helix axis (in degrees)	20	6	7
Major groove	Narrow and deep	Wide and deep	Flat
Minor groove	Wide and shallow	Narrow and deep	Narrow and deep
Orientation of N-glycosidic bond	Anti	Anti	Anti for Py, Syn for Pu
Comments		Most prevalent within cells	Occurs in stretches of alternating purine pyrimidine base pairs

in the 5' ⟶ 3' direction and the other in the 3' ⟶ 5' direction. On its exterior surface, the double helix of DNA contains two deep grooves between the ribose-phosphate chains. These two grooves are of unequal size and termed the **major** and **minor grooves**. The difference in their size is due to the asymmetry of the deoxyribose rings and the structurally distinct nature of the upper surface of a base-pair relative to the bottom surface. Structure of B and Z-DNA have been shown in Figure 5.5. Parameters of major DNA helices has been shown in Table 5.2.

The double helix of DNA has been shown to exist in several different forms, depending upon sequence content and ionic conditions of crystal preparation. The **B-form** of DNA prevails under physiological conditions of low ionic strength and a high degree of hydration. Regions of the helix that are rich in pCpG dinucleotides can exist in a novel left-handed helical conformation termed **Z-DNA**. This conformation results from a 180° change in the orientation of the bases relative to that of the more common A- and B-DNA.

IV. THERMAL PROPERTIES OF DNA

As cells divide :• is a necessity that the DNA be copied (replicated), in such a way that each daughter cell acquires the same amount of genetic material. In order for this process to proceed the two strands of the helix must first be separated, in a process termed denaturation. This process can also be carried out *in vitro*. If a solution of DNA is subjected to high temperature, the H-bonds between bases become unstable and the strands of the helix separate in a process of **thermal denaturation**.

The base composition of DNA varies widely from molecule to molecule and even within different regions of the same molecule. Regions of the duplex

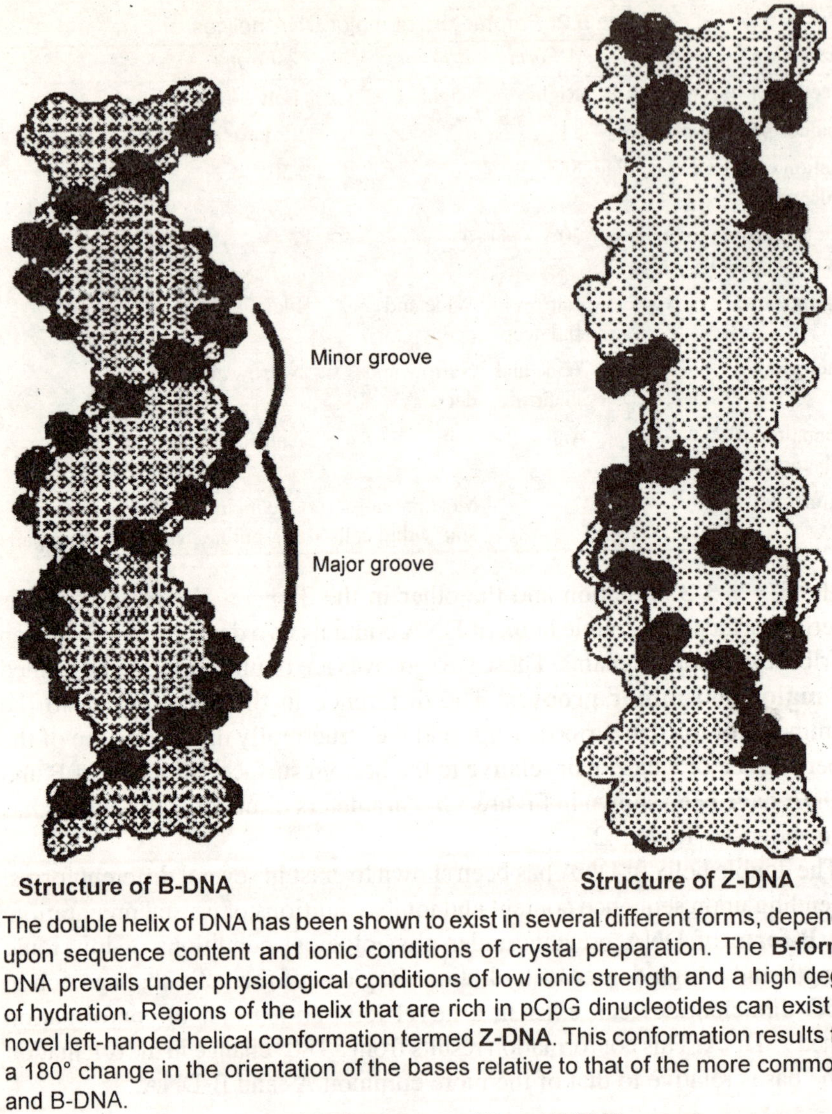

The double helix of DNA has been shown to exist in several different forms, depending upon sequence content and ionic conditions of crystal preparation. The **B-form** of DNA prevails under physiological conditions of low ionic strength and a high degree of hydration. Regions of the helix that are rich in pCpG dinucleotides can exist in a novel left-handed helical conformation termed **Z-DNA**. This conformation results from a 180° change in the orientation of the bases relative to that of the more common A- and B-DNA.

Fig. 5.5: Structures of B and Z-DNA

that have predominantly A-T base-pairs will be less thermally stable than those rich in G-C base-pairs. In the process of thermal denaturation, a point is reached at which 50% of the DNA molecule exists as single strands. This point is the **melting temperature (T_M)**, and is characteristic of the base composition of that DNA molecule. The T_M depends upon several factors in addition to the base composition. These include the chemical nature of the solvent and the identities and concentrations of ions in the solution.

When thermally melted DNA is cooled, the complementary strands will again re-form the correct base pairs, in a process is termed **annealing** or

hybridization. The rate of annealing is dependent upon the nucleotide sequence of the two strands of DNA.

V. ANALYSIS OF DNA STRUCTURE

Various techniques are used to characterize DNA. The important ones are as below.

A. Chromatography

Several of the chromatographic techniques available for the characterization of proteins can also be applied to the characterization of DNA. Paper chromatography and thin layer chromatography are useful in fractionating oligonucleotides. The most commonly used technique is HPLC (high performance liquid chromatography) and affinity chromatography. One common affinity matrix is hydroxyapatite (a form of calcium phosphate), which binds double-stranded DNA with a higher affinity than single-stranded DNA. Therefore, DNA can be rapidly isolated by passing a cell lysate through a hydroxyapatite column. Single-stranded DNA elutes from hydroxyapatite at a lower phosphate concentration than does the double-stranded DNA.

B. Electrophoresis

This procedure can serve the same function with regard to DNA molecules as it does for the analysis of proteins. However, since DNA molecules have much higher molecular weights than proteins, the molecular sieve used in electrophoresis of DNA must be different as well. The material of choice is agarose, a carbohydrate polymer purified from a salt water algae. It is a copolymer of mannose and galactose that when melted and re-cooled forms a gel with pores sizes dependent upon the concentration of agarose. The phosphate backbone of DNA is highly negatively charged, therefore DNA will migrate in an electric field. The size of DNA fragments can then be determined by comparing their migration in the gel to known size standards. Extremely large molecules of DNA (in excess of 106 base pairs) are effectively separated in agarose gels using **pulsed-field gel electrophoresis (PFGE)**. This technique employs two or more electrodes, placed orthogonally with respect to the gel, that receive short alternating pulses of current. PFGE allows whole chromosomes and large portions of chromosomes to be analyzed (Fig.5.6).

VI. CHEMICAL SYNTHESIS OF OLIGONUCLEOTIDES

Synthesis of oligonucleotides are widely used as probes in Southern transfer analysis for diagnosis of genetic diseases such as sickle-cell anemia. The basic

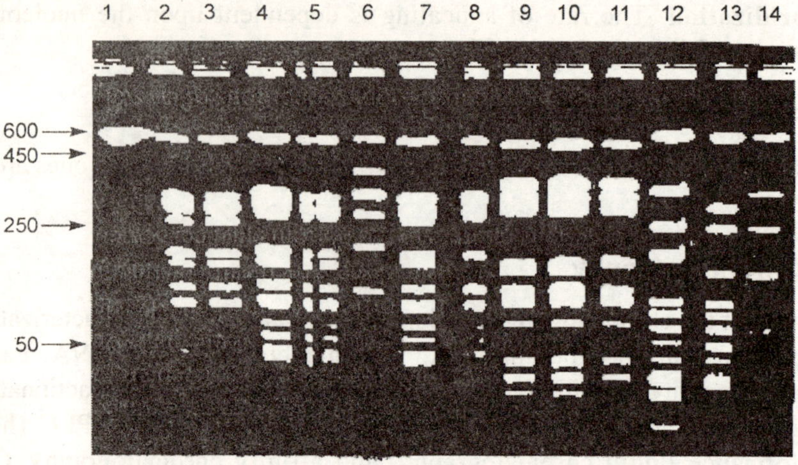

Fig. 5.6: DNA typing by pulsed -field gel electrophoresis

strategy of oligonucleotide synthesis is that a suitably protected nucleotide is coupled to growing end of the oligonucleotide chain, the protecting group is removed, and the process is repeated until the desired oligonucleotide has been synthesized. The first practical technique for DNA synthesis by **phosphodiester method** (a laborious aqueous method) was developed by H. Gobind Khorana in 1960s. The presently most widely used non-aqueous technique, known as **phosphoramidite method** was formulated by Robert Lestinger and further developed by Marvin Caruthers. This reaction sequence adds a single nucleotide to a growing oligonucleotide chain as follows (Fig.5.7):

1. The **dimethoxytrityl (DMTr)** protecting group at the 5' end of the growing oligonucleotide chain (which is anchored by a linking group at its 3'end to a solid support s) is removed by treatment with acid.
2. The newly liberated 5' end of the oligonucleotide is coupled to the 3'-phosphoramidite derivative of the next deoxynucleoside to be added to the chain. The coupling agent in this reaction is **tetrazole**.
3. Any untruncated 5' end is capped by acetylation so as to block its extension in subsequent coupling reactions. This prevents the extension of erroneous oligonucleotides.
4. The phosphite triester group resulting from the coupling step is oxidized to the phosphotriester, thereby yielding a chain that has been lengthened by one nucleotide.

This reaction has been automated in the commercially available **Oligonucleotide synthesizer**. This reaction sequence can be repeated upto > 150 times with a cycle time of <40 minutes on an automated synthesizer. Once an oligonucleotide of desired sequence has been synthesized, it is released from its support. The product can be purified by HPLC or gel electrophoresis.

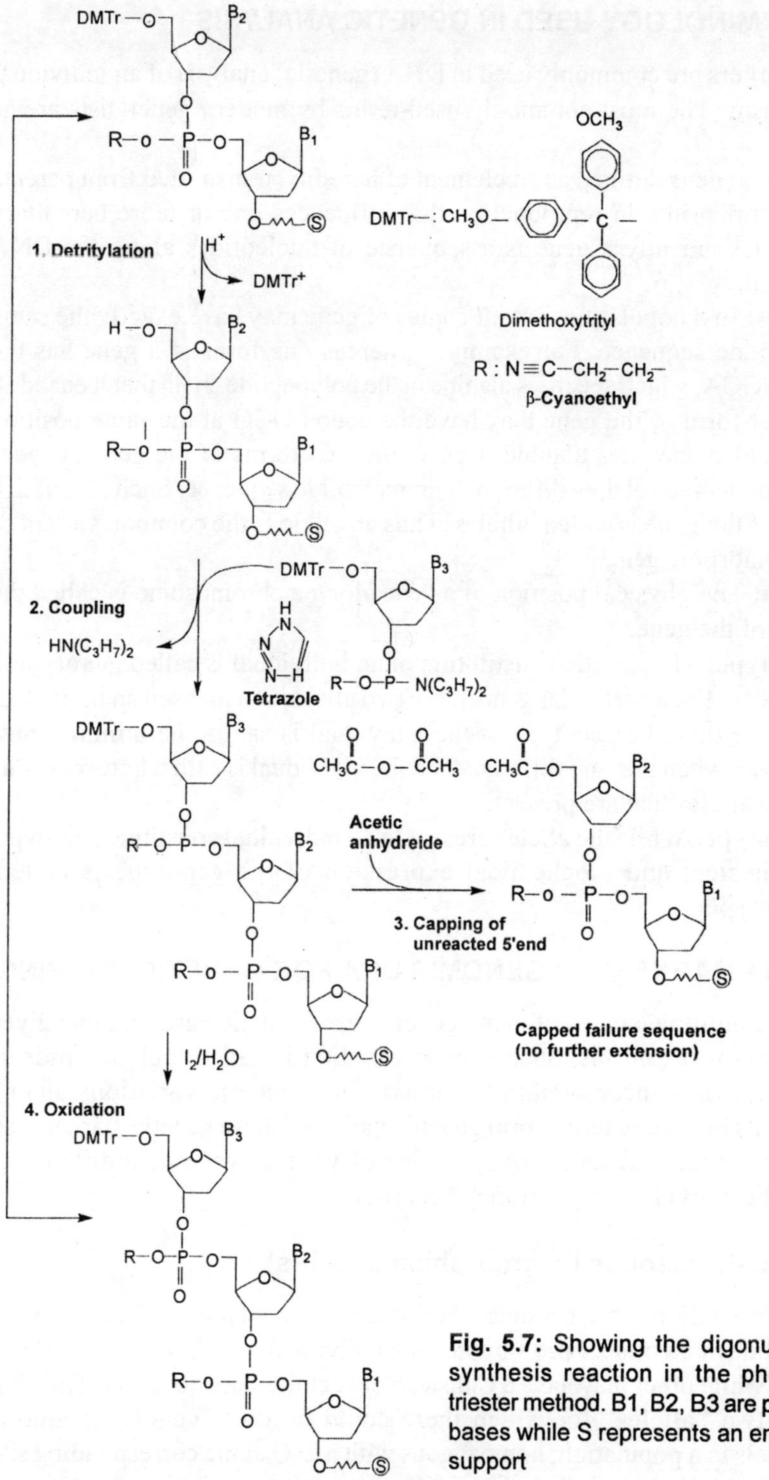

Fig. 5.7: Showing the digonucleotide synthesis reaction in the phosphite-triester method. B1, B2, B3 are protected bases while S represents an enert solid support

VII. TERMINOLOGY USED IN GENETIC ANALYSIS

DNA markers are commonly used in DNA (genetic) analysis of an individual or organism. The most commonly used terms by modern geneticists are as below.

a. **Gene:** Gene is defined as an element of heredity, transmitted from parents to the offspring, in reproduction that influences one or more hereditary traits. Chemically a gene is a sequence of nucleotides along the DNA molecule.
b. **Alleles:** In a population, not all copies of gene may have exactly the same nucleotide sequence. For example, whereas one form of a gene has the codon GCA, which specifies alanine in the polypeptide chain that it encodes, another form of the gene may have the codon GCG at the same position which also specifies alanine. Hence, the two forms of the gene encodes same molecule yet they differ in their nucleotide sequence. Such alternative form of the gene is called, **alleles**. Thus an allele is the common variant of a polymorphic gene.
c. **Locus:** The physical position of a gene along a chromosome is called the **locus** of the gene.
d. **Genotype:** The genetic constitution of an individual is called **genotype**.
e. **Zygosity:** For a particular gene, if the two alleles at a locus in an individual are same then the genotype of the individual is said to be **homozygous**, however, when they are different then the individual is called **heterozygous** for the alleles that are present.
f. **Phenotype:** While the alleles present in an individual constitute genotype, the physical and biochemical expression of the genotype is called, **phenotype.**

VIII. DNA MARKERS IN GENOMIC DNA FOR GENETIC MAPPING

In nature, multiple alleles of many genes causes genetic variation in a given population of organisms. Such genetic variation is called, **polymorphisms** (polymorphisms means multiple forms). Such genetic variations among individuals in genetic terms form genetic markers. These genetic markers can be studied by manipulating DNA in a variety of way to detect genetic differences among the individuals as mentioned below:

A. Single-Nucleotide Polymorphisms (SNPs)

These are position in the genome where one individual may differ from other by only one nucleotide. For example, one individual may have a G at a given location while other may have a C instead of G at the same location. The SNP defines two "alleles" for which there could be three genotypes among individuals in a population: homozygous with a C-G at the corresponding site in both the chromosomes, homozygous with T-A at the corresponding site in both homologous chromosomes: or a heterozygous with C-G in one and T-A in the other homologous chromosome. The SNP need not be present in the coding sequence or in a gene. In a human chromosome, there are approximately 1.42 million SNPs.

B. Restriction Fragment Length Polymorphism (RFLP)

Restriction enzymes cut a DNA molcecule at a specific site (recognition sequence). It means that a DNA molecule with a restriction enzyme should always produce same set of fragments. However, this is not the case with genomic DNA molecules because some restriction sites are polymorphic, existing two alleles, one allele displaying the correct sequence for the restriction site while other having a sequence alteration at that site so that the restriction site is not recognized by the enzyme. The result of sequence alteration is that the two adjacent fragments remain linked together after treatment with the enzyme , leading to length polymorphism (Fig.5.8). This is an RFLP. There are thought to be 10^5 RFLP in human genome.

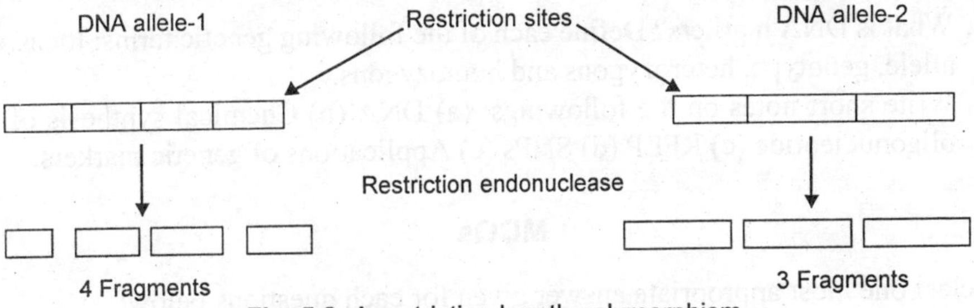

Fig. 5.8: A restriction length polymorphism

C. Single Sequence Length Polymorphisms (SSLPs)

SSLPs are arrays of repeat sequences that show length variations, different alleles containing different numbers of repeat units. Unlike RFLPs, SSLPs can be multi-allelic as each SSLP can have a different length variants. There are two types of SSLP:

a. **Minisatellites.** These are also called, **variable number of tandem repeats (VNTRs)**, in which repeat unit is upto 25 base pair in length.
b. **Microsatellites** or **simple tandem repeats (STRs)**, whose repeats are shorter, usually dinucleotide and tetranucleotide units.

Microsatellites are more popular than minisatellites as DNA markers for two reasons. First, minisatellites are not spread evenly around the genome but tend to be more frequently in the telomeric regions at the end of the chromosomes while microsatellites are more conveniently spaced troughout the genome. Second, PCR typing is much quicker and more accurate with microsatellites due to its shorter length. There are about 10×10^5 microsatellites in human genome.

IX. APPLICATIONS OF GENETIC MARKERS

Genetic markers are useful for following applications:

1. Genetic mapping and identification of genetic diseases.
2. Identification of individuals in paternity testing and forensic sciences.
3. Epidemiology and food sciences for tracking infectious organisms.
4. Human population history in anthropology.
5. Improvement of domesticated plants and animals through breeding of animals/ plants with favorable traits.
6. As an indicator of bio/genetic diversity .
7. It has applications in evolutionary genetics.
8. It may be highly useful in population genetics.

STUDY QUESTIONS

1. What are the bases commonly found in the nucleotides in DNA? Which form base pairs.
2. What is DNA markers? Define each of the following genetic terms: locus, allele, genotype, heterozygous and homozygous.
3. Write short notes on the followings: (a) DNA (b) Chemical synthesis of oligonucleotide (c) RFLP (d) SNPs (e) Applications of genetic markers.

MCQs

Select one most appropriate answer given for each questions below:

1. 5-fluorouracil is used in cancer chemotherapy because it:

A. Disrupts DNA replication
B. Causes DNA fragmentation
C. Inhibits transcription factors
D. Inhibits DNA polymerase binding

2. Annealing is the process of by which:

A. A new phosphodiester bond is formed
B. Complimentary DNA strands re-form the correct base pairs
C. Elongation of DNA strands take place
D. DNA strands are separated apart

3. Aqueous method of oligonucleotide synthesis was first used by:

A. Jacob and Monod
B. Robert Lesinger
C. H. Gobind Khorana
D. M. Caruthers

4. **Microsatellite genetic markers are more popular than minisatellites because:**
   A. Minisatellites are evenly distributed in the genome
   B. Microsatellites are present at the ends of a chromosomes
   C. PCR typing is much quicker and accurate in microsatellites
   D. Microsatellites are unevenly distributed in the genome
5. **Nucleosides are compounds which lack:**
   A. Nitrogenous base
   B. Pentose sugar
   C. Phosphate
   D. None of the above
6. **Which of the following nucleosides is purine?**
   A. Guanine
   B. Cytosine
   C. Thymine
   D. Uracil
7. **Which of the following is true for DNA molecule in a cell?**
   A. $A = T$
   B. $G = C$
   C. $A + G = T + C$
   D. All the above
8. **Which one of the following is true for DNA molecule:**
   A. $A\,^{\circ}\,T$
   B. $G = C$
   C. $A = T$
   D. $G - C$

1 A 2 B 3 C 4 C 5 C 6 A 7 D 8 C

PART II:

IMMUNOLOGY

•

Immune System: An Overview *85*

•

Innate Immunity *93*

•

Antigens Immunization Strategies *105*

•

MHC Structure, Genomic Organisation, Antigen Processing and Presentation *114*

•

Antibody Structure, Function, Anitbody Engineering and Uses *128*

•

Antigen-Antibody Interactions *151*

•

Biology of B-Lymphocytes and Generation of Humoral Immunity *172*

•

Biology of T-Lymphocytes and Generation of Cellular Immunity *180*

•

Cytokines and Cytokine Receptors *195*

•

Chemokines, Chemokine Receptors, Cell Migration and Homing *204*

•

Autoimmunity *214*

•

Transplantation Immunity *223*

•

Tumor Immunity *235*

•

Infection Immunity and Strategies for Vaccine Development *251*

6

Immune System: An Overview

Mother nature has provided all the living forms both animals and plants with a unique mechanism to discriminate between self and non-self (world). Self is accepted and non-self if useful is welcomed (symbiosis), useless is ignored and harmful is destroyed. This unique ability is imparted to the host by a complex system termed as **Immune System** that comprises a network of cells and their soluble mediators. Evidence of self and non-self discrimination in the animal kingdom is present right from protozoans to vertebrates. In vertebrates, it is more evolved and complex in order to prevent the host from a wide variety of infections and tumors.

HISTORY

The term immunity is derived from Latin word *immunitas*, meaning 'to exempt' that in classical terms means, 'to be exempted from infections'. The earliest written reference to immunity dates back to 430 BC when Thucydides described that during outbreak of plague in Athens only those persons who had recovered from the disease, were allowed to nurse the sick since they were unlikely to contract the disease again.

During 15th century Chinese and Turks had mastered the art of immunizing an individual with dried crust of smallpox pustules. In 1798, Edwards Jenner immunized an eight-year-old boy, with liquid from a cowpox pustule and inoculated him with smallpox, as expected the boy did not develop infection. The landmark experiment in immunology was demonstrated by a French physician Louis Pasteur in 1981 when he demonstrated that injection of attenuated *anthrax bacilli* protected a group of sheep from infection with the live anthrax bacilli while it killed the other group of sheep that were not immunized.

A Russian scientist, Elie Metchnikoff in 1883 observed in a transparent starfish larva that it was able to ingest foreign particle(a thorn). Subsequently he had shown that some white blood cells can ingest and destroy foreign particles as well as micro-organisms. He coined the term ***phagocytosis*** to describe such phenomena and the cells that carry such function was called ***phagocytes***. This observation of Metschinkoff laid the foundation of ***cellular basis of immunity***.

The clinical use of vaccine has shown by Louis Pasteur in 1885 when he administered the first vaccine to a young boy, George Meister who was repeatedly bitten by a rabied dog. The boy survived and later became custodian of the Pasteur Institute and sacrificed his life for the institute during the second World War. Later on Emil von Boehring and Shibasaburo Kitasato in 1890 demonstrated that serum (cell free component of the blood) from immunized animal with diphtheria could transfer protection to the naïve recipient from diphtheria infection. This observation laid the foundation of ***humoral basis of immunity*** and earned Boehring Nobel prize of medicine in 1901.

INNATE AND ADAPTIVE IMMUNITY

The cardinal feature of immune system is recognition of the invader and response to contain it. Recognition and response could be of two types. Firstly, recognition could be general for all foreign invaders as it comes in contact with the host hence the response is also general. Such type of immunity is called ***innate immunity***. The components of innate immunity are macrophage, natural killer (NK) cells, complement, lysozyme, defesins, etc. In the second situation, recognition could be specific for specific invader hence response is also specific and the system is educated to remember the invader so that the response is heightened in subsequent exposures to the same organism. This type of immunity is called ***adaptive immunity***. Components of the adaptive immunity are lymphocytes and their products.

Innate Immunity: Innate immunity is the first line of defense against environmental agents. Evidence of innate immune system is present through protozoans to vertebrates. It is a non-specific, immediate but brief response that comprises of four types of barriers: anatomic, physiologic, phagocytic and inflammatory. These barriers prevent entry, its containment and subsequently destruction of pathogens by specialized effector cells like. phagocytes and NK cells and their molecular products such like enzymes, oxygen radicals and lytic granules, etc.

Adaptive Immunity: Unlike the innate immunity, adaptive immunity is more evolved and is present only in vertebrates. Adaptive immune response is delayed, specific and long-lasting. The cardinal features of adaptive immune response are: antigen specificity, diversity of responding cells and molecules (antibody), memory and self/ non-self discrimination. Although each individual is born with an inherent capacity to mount an immune response to a foreign invader, acquired immunity develops after initial contact and is specific to that invader only. The initial contact with the foreign agent is called

sensitization or ***immunization*** that triggers a chain of events inside the lymphocytes leading to its activation. Unlike the innate immune system where the effector cells can ingest whole organism, cells of the innate immune system recognize a short stretch of peptide and glycoprotein termed as, ***antigen***. The specific region of an antigen that is recognized by lymphocytes (T cells/B cells) is called, ***epitope***. Thus a given protein derived from pathogens or parasites may have different epitopes that could be recognized by different T cells or B cells depending upon their antigen receptors.

Upon exposure to ***exogeneous antigens***, B lymphocytes with receptor specific to that antigen gets activated, differentiate into plasma cells that secretes antibody. The antibodies recognize specific antigen and neutralize it (toxins), it can precipitate the soluble antigen and agglutinate cells (blood cells, bacteria, parasite) and target them to phagocytosis. Since antibodies are present in the humoral components like serum or body secretions (saliva, mucosa, etc), immunity imparted by it is called ***humoral immune response*** or B cell-mediated immunity (due to involvement of B lymphocytes). Chemically, antibodies constitute globulin fraction of the serum protein hence these are also called ***immunoglobulins***. On the other hand, ***endogeneous antigens*** such as viral proteins or tumour antigens are recognized by T lymphocytes through T cell receptor (TCR) that differentiate into effector cells such as cytotoxic T lymphocytes (CTL/Tc) and helper T-lymphocytes (Th). CTLs destroy the target cells through cytotoxic mechanisms where as Th cells secrete soluble mediators (cytokines) that enhance immune response of the host. Since such of immune response is mediated by cells, it is called ***cell-mediated immune response*** or T cel- mediated immunity. The major components of the immune system and their characterisitics has been depicted in Figure. 6.1.

ANTIGENS, IMMUNOGENS AND IMMUNIZATION

Antigens that can induce immune response in the host is called immunogen as against some antigens that may bind to TCR or BCR but do not induce immunity. Therefore, all immunogens are antigens but all antigens may not be immunogens. Adaptive immune response can be induced in an individual by introducing antigens directly into the host (immunization). Thus ***vaccination*** is a process to immunize naïve host against pathogenic micro-organisms, which can be achieved in three different ways:

1. Immunization of the host by introducing antigens in the system, called ***active immunization***. It leads to generation of adaptive immunity against that organism from which the antigen is derived (e.g. BCG vaccination).
2. Immunity can be transferred by injecting serum containing antibodies from immunized host to the naïve recipient. This process is called ***passive immunization***. Natural immunization of foetus through maternal circulation is a suitable example of passive immunization.

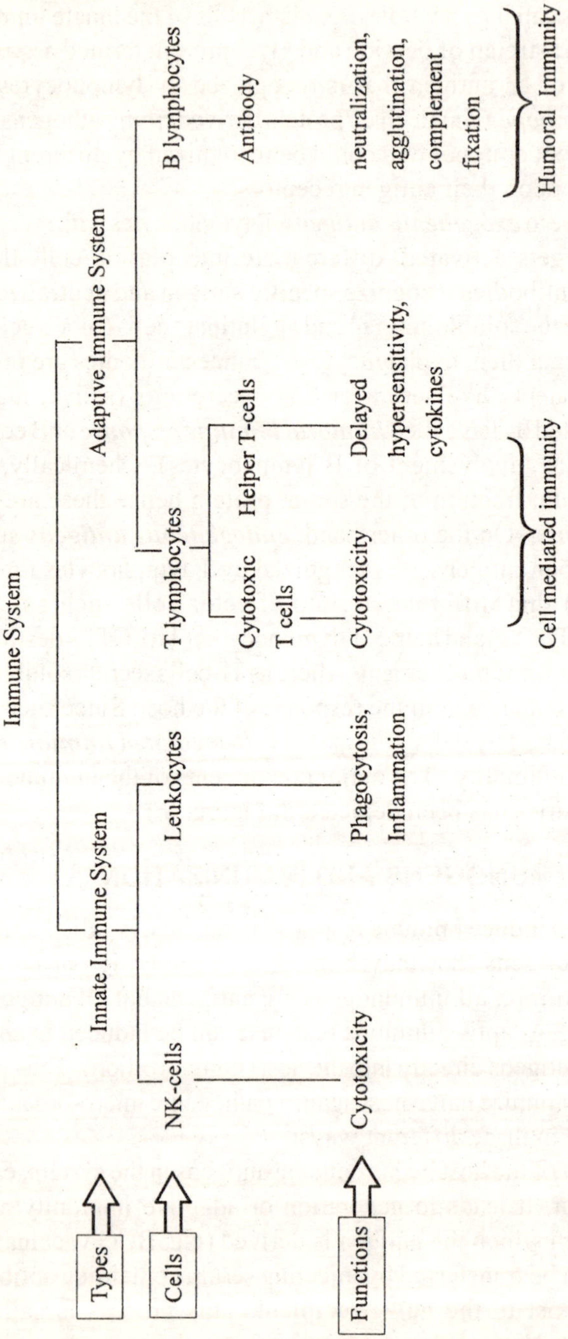

Fig. 6.1: Flow chart showing components of immune system

Table 6.1: Salient features of innate and adaptive immunity

Features	*Innante immunity*	*Acquired immunity*
Recognition (chemical pattern)	Non-specific	Specific (antigen)
Response	Immediate	Delayed
Diversity	Limited	Unlimited
Memory	No	Yes
Components:		
• Physical	Skin, mucosa	Tissue associated lymphocytes
• Chemical	Complements, enzymes	Antibodies
• Effector cells	NK cells, macrophages	T lymphocytes, B lymphocytes

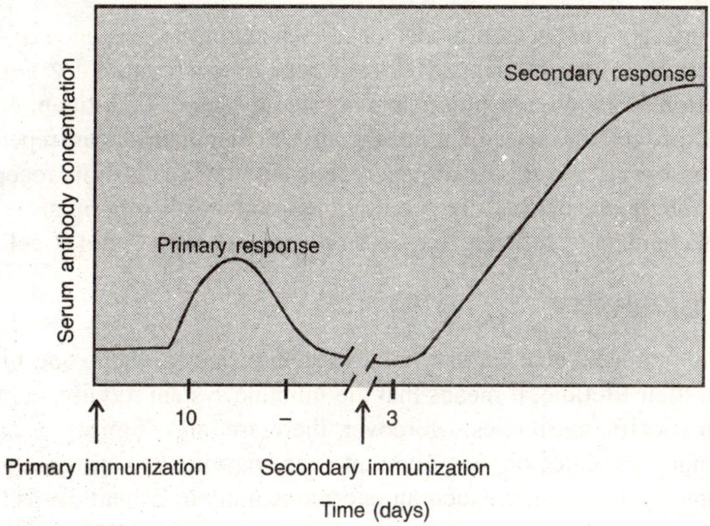

Fig 6.2 Antibody response following primary and secondary immunization

3. Immunity can be transferred by transferring lymphocytes from the immunized host, such immunization is called ***adoptive immunization***. Immunization leads to generation of two phases of immune response known as primary and secondary response.

Immune response generated to the first exposure of antigen is called ***primary immune response*** that takes place in about 2 weeks. When same antigen is introduced to the host second time the immune response is of higher magnitude than the first one and it is called ***secondary immune response*** (Fig. 6.2).

THEORY OF CLONAL SELECTION

Sir Mac Farlane Burnet in 1950 postulated the presence of pre-existing pool of a large number of different antibody producing cells, each having ability to produce antibody of a single specificity. Such antibody is displayed on the cell surface serving as receptor for antigen. On binding to antigen, the cell is activated to proliferate and produce many identical progeny, called clone, which produce antibody specific to that antigen. This theory was called, **theory of clonal selection.** It holds true for B and T lymphocytes both and suggests that:

1. Each host contains B and T lymphocytes of myriad specificities even before it contacts antigen.
2. Each lymphocytes carry receptors of only one specificity on its surface. T cells have T cell receptors (TCR) and B cells possess B cell receptor (BCR).
3. Antigen exposure triggers activation of only those lymphocytes (T or B) that bear receptors specific for that antigen, leading to clonal expansion of such cells.

The theory of clonal selection also describes about the existence of large pool of lymphocytes bearing different repertoire formed by genetic recombination during their development before the contact of antigen. At this stage only those cells are selected for further differentiation that bears repertoire with none or low affinity to self antigens. Cells bearing high affinity receptors for self-antigens are deleted by programmed call death (apoptosis). This deletion mechanism of self-reactive cells operates for both T and B-cells.

IMMUNE DIVERSITY

It has been estimated that human beings are exposed to about one billion antigens in their lifetime. It means that the immune system requires to make one billion specific antibodies. Moreover, there are large number of pathogens and many parasites or viruses mutate very rapidly. It is intriguing that how immune system combat such an enormous number of pathogens? Is it that the immune system have to educate itself to recognize all the antigens? The answer is, no The fact that immune system is capable of generating enormous diversity was elegantly shown by Tonegawa (Nobel laureate) and his colleague. Using molecular biology technique they could show that during immune response B cells undergo DNA synthesis and cell division resulting in production of cells with enormous diversity in their receptors (BCR) through unique mechanism of genetic recombination. This leads to production of large number of B cells capable of producing antibody of diverse specificity many of which may not encounter antigens bearing that specific epitope in its lifetime. Although the mechanism of genetic recombination was initially studied in B cells, the same mechanism holds true in case with T cell receptor (TCR) as well. The cells with newly acquired antigen specific (antigen-epitope specific) receptor generated during the immune response remains available to the immune system in the form of memory cells that could be expanded and transformed into effector cells at the time of exposure to the specific antigen.

EFFECTOR MECHANISMS OF THE IMMUNE SYSTEM

There are different ways by which immune system combat infections depending upon the nature of antigen. By producing specific antibody it can ***neutralize*** antigens (e.g. Toxins), ***agglutinate*** the parasites or coat the parasites with specific antibody (***opsonization***) for the subsequent elimination by complement-mediated lysis or by phagocytosis. In case with phagocytosis, the parasites are engulfed and contained in phagosomes in which reactive oxygen intermediates (ROIs) and nitric oxide (NO) is secreted. Finally, lysosome fuses with the phagosomes resulting in the digestion of the ingested materials by lysosomal enzymes. Antibody coated cells can also be recognized by NK cells that can bind to the target cells through Fc receptor and can be destroy it by the mechanism called, ***antibody dependent cell-mediated cytotoxicity*** (ADCC).

In case with the endogenous antigens, like virus infected or tumor cells, the target cells may be recognized by Tc cells through TCR. Tc cells get activated and produce large quantity of lytic granules which are injected into the target cells by punching holes using ***perforins*** (lytic granules)causing lysis of the target cells. Tc cells can also transmit signals to the target cells to undergo programmed cell death.

IMMUNOREGULATION

The ultimate aim of the immunoregulation is to shut down the ongoing immune response once the infection is contained. The mechanism by which active immune response is reverted back to normal level is called ***homeostasis***. There are a number of mechanisms to bring about homeostasis in the immune system. It includes negative feedback inhibition by soluble products, cell-cell communication and activation of suppressor T cells (Ts) or production of molecules (antibodies, cylokines, etc.) that can regulate the ongoing immune response. At the end of the immune response, effector (T or B) cells undergo programmed cell death while a population of antigen specific ***memory cells*** survive that is responsible for providing long lasting protection against that particular parasite upon natural exposure or in case with immunization.

IMMUNOPATHOLOGY

Although the purpose of immune response is to contain the invaders and protect the self, however, self and non-self discrimination is not absolute. During immune response a huge diversity in the lymphocytes are generated by genetic recombination. Some of the receptor repertoire may be directed against the self-components as well. In case with the failure of immunoregulation or inappropriate immune response, such autoreactive cells may lead to the development of ***autoimmune diseases*** such as rheumatoid arthritis, diabetes, pernicious anaemia, etc. On the other hand deficiency of the immune response may lead to ***infections*** and ***cancer***. Immunosuppression in case with AIDS has been found to result in a variety of infections and development of cancer such as Kaposi's sarcoma. It is important therefore, that the immune system keeps its delicate balance intact. When such balance is broken disease sets in. Hyperactivity of the immune response leads to ***hypersensitivity diseases*** such as common allergy, asthma and hay fever.

APPLICATIONS OF IMMUNOLOGY

Immunology has found diverse applications in the area of ***biomedicine***, particularly in the areas related to diagnosis, prevention, treatment and research related to pathogenesis of a number of diseases. The classical antigen-antibody reactions have been successfully used in the diagnosis of a large number of diseases including cancer and AIDS. Many of the deadly infections such as cholera, small-pox, measles, etc. have been eradicated by vaccination. Development of hybridoma technology have led to production of highly specific ***monoclonal antibody*** which have an enormous impact in the area of diagnosis and immunotherapy of wide variety of diseases as well as in the area of reproduction and transplantation. Monoclonal antibodies, ***cytokines***, ***immunotoxins*** and even genetically engineered ***immune cells*** are being extensively investigated for the treatment of cancer, neurological disorders, autoimmune disorders as well as in reproduction and transplantation.

7

Innate Immunity

I. INTRODUCTION

The fully developed human and most animals have two types of immune systems: the **innate immune system** and the **adaptive immune system**. Among these two, innate immune system has developed quite early in the evolution. It presents a first line of defense against pathogens and microorganisms. It is relatively non-specific and immediate response of the host to a variety of pathogens and is genetically determined.

II. DEFINITION

Innate immunity refers to the inherent antigen-nonspecific defense mechanisms that set in immediately within minutes or hours of the exposure of the foreign invaders. Its principal role is to provide immediate defense against any organism or substance that is harmful to the host.

III. COMPONENTS

Components of the innate immune system are: (i) anatomical barriers; (ii) physiological barriers (humoral components); and (iii) effector cells of the innate immune system (cellular components). Anatomical barriers prevents entry of the foreign invaders, while cellular and humoral components are designed to **recognize** a few highly **conserved pattern of molecules** present on them through receptors called, **pattern recognition** receptors. Effector cells of the innate immune system contain the infection by specialized effector mechanisms such as **cytotoxicity**, **phagocytosis** or by **inflammatory responses**.

1. ***Anatomical barrier*:** Anatomical barrier consists of **skin**, **mucosa**, and the lymphoid tissues associated with these structures. Skin provides a tough physical barrier to microbes. The external dead keratinized epithelium gets

sloughed off frequently, therefore, it prevents colonization of the pathogenic microbes. In addition, acidic pH of sweat and sebaceous secretions, lysozyme and fatty acids have antimicrobial activity. Skin associated lymphoid tissue **(SALT)** contains cells that become sensitized in presence of the microbes and trigger adaptive immune response as well as prevents colonization of the microorganisms.

Mucous membrane surrounds a vast area of the body such as alimentary tract, genitourinary tract and respiratory tract. It secretes mucus that entraps the foreign particles which are excluded out by the ciliary movement. Mucus also contains several chemicals such as **lysozyme** to degrade bacterial peptidoglycan, an antibody called **secretory IgA** that prevents microbes from attaching to mucosal cells and traps them in the mucous, **lactoferrin** to bind iron and keep it from being used by microbes, and **lactoperoxidase** to generate toxic superoxide radicals that kill microbes. Lymphoid cells associated with mucous membrane namely, **mucosa-associated lymphoid tissue (MALT)** contains cells for killing microbes and sampling antigens on the mucosa to start acquired immune responses against them. Intraepithelial **gamma delta T cells** and **B-1 cells** have limited diversity for antigen,therefore, they function more as effectors of innate immunity rather than adaptive immunity. B-1 cells produce IgM class of antibodies, often called as **natural antibodies**.

2. ***Physiological barrier.*** Physiological barriers include biologically active substances produced by various cells and tissues (Table 7.1). Therefore, certain tissues and organs have more resistance to microorganisms than others depending upon their capacity of producing such substances.

   Although anatomical and physiological barriers are efficient in preventing entry and colonization of the foreign invaders, infection may take place in case with physical injury and wounds. Some of the microorganisms have evolved mechanisms to evade anatomical and physical barriers of the host. For example, influenza virus has a surface molecule that helps the virus to attach to the mucous membrane firmly so that it is not swept away by the ciliary movement. Similarly, some bacterium have hair-like protrusions on its surface called **fimbriae** or **pilli** to facilitate its attachment to the mucous membrane (e.g. *Gonococcus*).

3. ***Effector cells.*** Once the pathogens evade anatomical and physiological barriers and enters the body, it is taken care of by a variety of cells that perform various specialized functions. The effector cells that participate in innate immune response are:

   A. Cytotoxic cells (NK cells and NK T cells);
   B. Phagocytic cells (neutrophils, monocytes and macrophages);
   C. Cells that participate in the inflammatory response (eosinophils, basophils, mast cells and platelets).

Table 7.1: Example of some of the biologically active molecules of innate immune system

Molecules	*Source*	*Functions*
Hydrochloric acid and enzymes	Gastric secretions	Microbicidal
Lysozyme	Tears, mucus, saliva, plasma, tissue fluid, etc.	Breakdown of bacterial peptidoglycan
Defensins*	Leukocyte granules	Perforation of bacterial membrane
Complements	Plasma	Cytotoxicity, opsonization and activation of leukocytes
Inflammatory proteins (C-reactive protein), coagulation factors	Plasma	Inflammation
Cytokines		
TNF, IL-1, Interferon-α and β, Chemokines	Leukocytes Macrophage, NK-cells and T cells;	Inflammation; activation antiviral

[**Defensins are a family of evolutionarily related vertebrate microbial peptides with a characteristic α-sheet rich fold and a framework of six disulphide-linked cysteins. These are very widely distributed in mammalian epithelial cells and phagocytes. Under low ionic concentrations most of the defensins show a very high microbicidal activity against bacteria and fungi. Recently, human neutrophil defensins have been implicated in the antiviral activity in patients with non-progressing HIV infection.*]

a. Cytotoxic Cells

Natural-Killer Cells. These are large granular lymphocytes that are able to lyse virus-infected cells and tumour cells. These cells can be recognized by CD16.56 monoclonal antibodies. CD16 (FcγRIII) is linked to the phosphatidylinositol glycan linkage that is involved in the activation of NK cells, while CD56 is a homophilic adhesion molecule of the immunoglobulin superfamily. Unlike cytotoxic T cells, NK cells lack antigen receptors on its surface. However, the cytotoxic reaction of NK cell is carried out through cell cell contact. NK cells possess non-TCR related molecules, called **killer cell inhibitory receptor (KIR)**. KIR serves as a ligand for MHC-I molecules present on the target cells. It protects normal cells from the cytotoxic activity of NK cells. Since almost all normal cells express MHC-I molecules, when ligated with KIR, cytotoxic function of NK cells is inhibited (Fig.7.1). However, there is down-regulation of MHC-I molecules in virus-infected cells and in tumour cells. Thereforc, when such cells encounter NK cells, they fail to engage KIR and become susceptible to NK cell mediated cytotoxicity. It seems therefore, that the NK cell mediated lysis of virus-infected and tumour cells are carried out by NK cells prior to the development of adaptive immune response of the host.

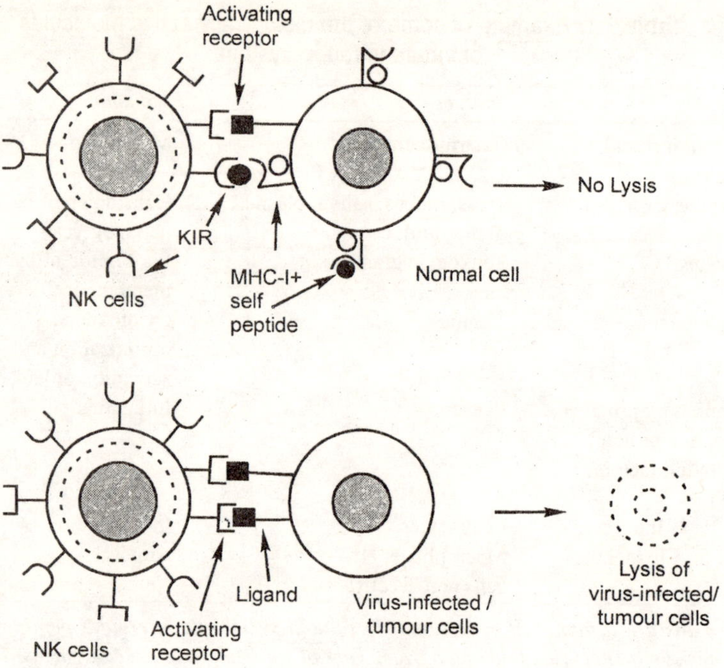

Fig. 7.1: Killer inhibitory and activating receptors and NK-cell mediated lysis of virus-infected tumour cells

Resting NK cells also express α-chain of IL-2 receptor as well γ-chain that is a common signaling molecule of many other cytokines. Therefore, cytotoxic activity of NK cells is highly increased in presence of IL-2, IL-12 and IFN-γ. Such cytokine activated NK cells are called **lymphokine activated killer (LAK)** cells that can lyse NK-resistant target cells. NK cells are able to kill IgG coated target cells through its IgG receptor (FcγRIII). Such killing mechanism is referred to as **antibody–dependent cell mediated cytotoxicity (ADCC)**. During ongoing immune response activated NK cells secrete some of the cytokines (IFN-γ, IL-1 and GM-CSF) that are important for haematopoiesis and triggering of adaptive immune response.

NK T cells. As the name indicates these are the thymus derived cells possessing T cell receptor (TCR) of limited variability and NK1.1 marker on its surface. These cells are capable of binding to the antigen presenting cells (APC) bearing MHC-associated CD1 ligand via NK1.1 receptor. Upon activation NK T cells secrete IL-4, IFN-γ and several other cytokines thus play an important role in immunoregulation. It also induces Fas-Fas L mediated apoptosis in the target cells.

b. Phagocytic Cells

Neutrophils. These are the largest subpopulation of leukocytes characterized by multilobed nucleus (Fig.7.2A). It contains two types of granules: (i) the

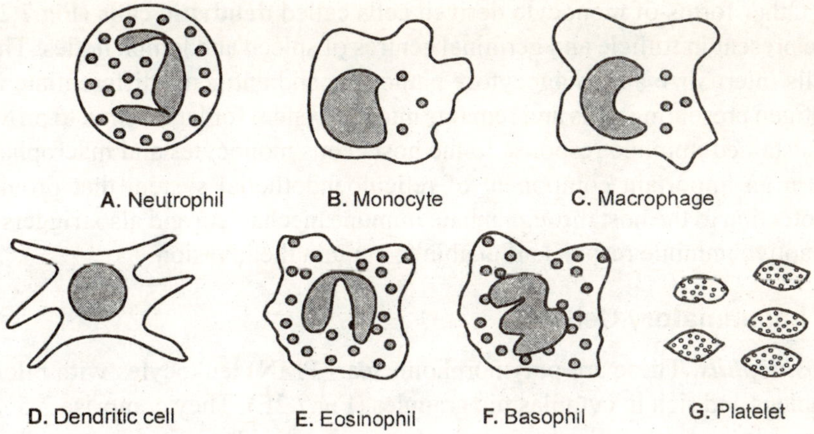

Fig. 7.2: Effector cells of innate immunity

primary azurophilic granules are lysosomes containing acid hydrolases, myeloperoxidase and muramidase; (ii) the secondary granule contains antibiotic proteins (defensin, seprocidins, cathelicidins and bacterial permeability inducing proteins (BPI).

Neutrophils are phagocytic cells. They ingest microorganisms and particulate matters by both non-immunological and immunological (antibody dependent) mechanisms. They are attracted by chemotactic factors that results in its margination and diapedesis.They release extracellular granules and cytotoxic substances when activated through FcγR.

Monocytes and macrophages. These cells are derived from myeloid progenitor cells in the bone marrow. Monocytes are characterized by its horseshoe-shaped nuclei (Fig.7.2B, C) and contain azurophilic granules. They remain in the circulation for about 6-8 hours where they enlarge in size and migrate into various tissues to become specific macrophages. In liver, they differentiate into **Kuffer cells** with large cytoplasmic projections. They are named as **histiocytes** in connective tissues, **mesangial cells** in kidney, **osteoclasts** in bone and **microglial cells** in the brain. In spleen, lung and peritoneum, these cells are called splenic, alveolar and peritoneal macrophages respectively. Thus macrophages have different names at different locations but all of them are equally efficient in phagocytosis.

Macrophages can be activated by a variety of mechanisms such as bacterial infection, inflammatory response and cytokines. Activated macrophages have enhanced phagocytic and microbicidal activity, enhanced expression of MHC-II molecule and it releases a variety of soluble mediators to activate other cells of the immune system. They play an important role in the induction of adaptive immune response by processing the larger antigen molecules into the smaller fragments and presenting it to T lymphocytes.

Other forms of monocyte derived cells called **dendritic cells** (Fig.7.2D), are present in follicle and germinal centres of spleen and lymph nodes. These cells internalize and phagocytose pathogens and antigens, differentiate into antigen presenting cells and remains into the system for larger time to provide a sustained immune response to the host. Thus monocytes and macrophages form an important component of reticuloendothelial system that provides protection to the host through innate immune mechanism and also triggers the adaptive immune response if unable to contain the invasion.

c. Inflammatory Cells

Eosinophils. These are polymorphonuclear (PMN) leukocytes with bilobed nucleus and rich in cytoplasmic granules (Fig.7.2E). They comprise 2-5% of blood leucocytes and appear red when stained with eosin. These cells are found in high numbers in allergic reactions and during parasitic infections. They use their granules against large parasitic worms. Eosinophils also release histaminase and aryl sulphatase that inactivate mast cell products.

***Basophils and mast cells*:** Basophils have granules that stain metachromatically due to their histamin and heparin contents. The tissue fixed mast cells, although display distinctive morphological features (Fig.7.2F), is indistinguishable from basophils. Both of these cells participate in allergic reactions.

Platelets. Platelets are derived from megakaryocytes in the bone marrow and contain granules (Fig. 7.2G). Although their primary role is in blood clotting, they are also involved in inflammatory response of innate immunity. Platelets contain FcγRII, receptor for IgE and factor VIII. During injury to the epithelial cells platelets aggregate at the surface of damaged vascular tissue. They release histamine and fibrinogen that increases vascular permeability, activate complement components and attract leukocytes.

IV. RECEPTORS

In order to recognize pathogenic microbes, various cells of innate immune system have developed receptors that can recognize conserved molecular patterns present on the microorganisms but not on human cells. The pathogen associated molecular patterns recognized by cells of the innate immune system are as below:

1. **Lipopolysaccharide (LPS)** on the gram-negative bacteria;
2. **Peptidoglycan** present on cell wall of gram-positive bacteria;
3. **Lipoteichoic acid** found in gram-positive bacteria;
4. **Mannose-rich glycans** in microbial glycoproteins;
5. **Flagellin** and **Pillin** found in bacterial flagella and pilli respectively;
6. **Nucleic acid** of bacteria;
7. **N-formyl methionine** present in bacterial protein;

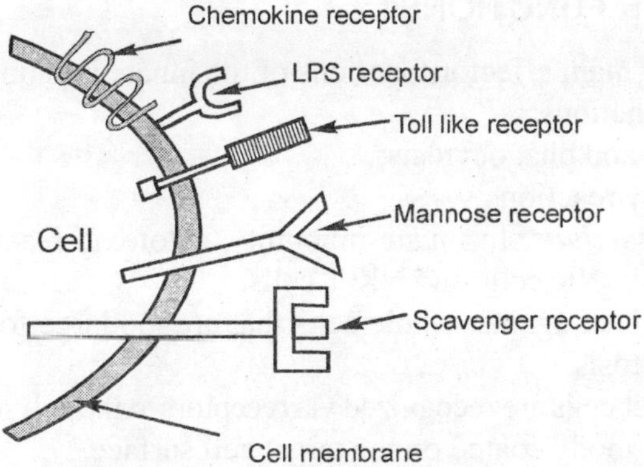

Fig. 7.3: Pattern recognition receptors

8. **Double-stranded RNA** present in viruses;
9. **Lipoteichoic acid, glycolipids** and **zymosan** on yeast cell wall.
10. **Phosphorylcholine and other lipids** present in bacterial membranes.

Such molecular patterns are recognized by **pattern recognition receptors** (Fig. 7.3). The pattern recognition receptors are chiefly of two types, i.**endocytic pattern recognition receptors (e.g. mannose receptors and scavenging receptors)** and ii.**signaling pattern-recognition receptors (e.g. toll-like receptors).**

i. ***Endocytic pattern-recognition receptors.*** These receptors are found on the surface of phagocytes and promote engulfment and internalization of microorganisms. They include **a. mannose receptor** that binds to the terminal mannose and fucose group of bacterial glycoproteins and glycolipids; **b. scavenger receptor** that binds to LPS, peptidoglycan and teichoic acids.

ii. ***Signalling pattern recognition receptors*:** These include: **a. toll-like receptors (TLRs)** and **b. CD14**. Binding of TLRs and CD14 to bacterial patterns leads to signal transduction mechanisms in the cell resulting in synthesis of cytokines. Thus these receptors are involved in skewing the host immune response from innate to adaptive immunity.

a. ***TLRs.*** A series of TLRs that directly or indirectly binds to different microbial components. For example, TLR-2 recognizes peptidoglycan and lipoproteins; TLR-4 recognize LPS and lipoteichoic acid; TLR 5 recognizes bacterial flagellin; and TLR 9 recognizes bacterial DNA.

b. ***CD14.*** Present on monocytes, macrophages and neutrophils, CD14 has synergistic effect when engaged along with TLR-4. Interaction of thesc two molecules leads to expression of proinflammatory cytokines such as IL-1, IL-6, IL-8 and TNF-α. These cytokines then bind to their receptor and induce inflammation, complement activation and coagulation.

IV. EFFECTOR FUNCTIONS

There are three main effector functions of the innate immune system:
1. Cytotoxic reactions;
2. Endocytosis and phagocytosis;
3. Inflammatory reactions.

1. ***Cytotoxic reactions.*** In innate immunity, cytotoxic reactions are mostly carried out by NK cells and NK T cells:
   i. It is directed against whole cells that are too large for endocytosis or phagocytosis.
   ii. The target cells are recognized via receptors to the cell surface molecules or by antibody coated on the target cell surface.
   iii. Effector cells direct its cytolytic granules toward target cells which perforate the cell membrane leading to leaking of the cellular contents and instant cell death.
   iv. Alternatively, the effector cells may also trigger a series of reactions in the target cells resulting to apoptosis.
2. ***Endocytosis and phagocytosis.*** These are the primary methods used by the body to remove microorganisms from the blood and tissue fluids.
   i. Inflammation due to infection or tissue injury allows movement of neutrophils to the site that are later replaced by phagocytes.
   ii. Lymph nodules located in the respiratory tract, the liver and the GI tract contain MALT that phagocytose organisms at these sites.
   iii. Beneath the skin surface SALT contains macrophages that ingest microbes.
   iv. Spleen, liver, kidneys, bones, brain and mucous membranes contain phagocytes where the organisms carried through blood are phagocytosed by tissue-fixed macrophages.

Endocytosis. Endocytosis is the term used for internalization of environmental materials by cells. In one form of endocytosis, extracellular materials binds to specific cell surface receptor, plasma membrane expands around the attached material that may include whole cells, and then it invaginate. Now fusion of the membrane takes place and the ingested material is internalized in the endosome/phagosome (Fig. 7.4A). The engulfed materials are digested by various mechanisms as described later. Such process is called, **receptor mediated endocytosis** as against non-receptor mediated phenomena such as **pinocytosis**.

Phagocytosis. Phagocytosis is a general term used for uptake of environmental materials by a variety of cell types. There are a number of distinct steps involved in phagocytosis (Fig. 7.4B).

a. Activation

Normally phagocytes are in the resting stage. Environmental assaults such as tissue injury and infections lead to release of humoral mediators like

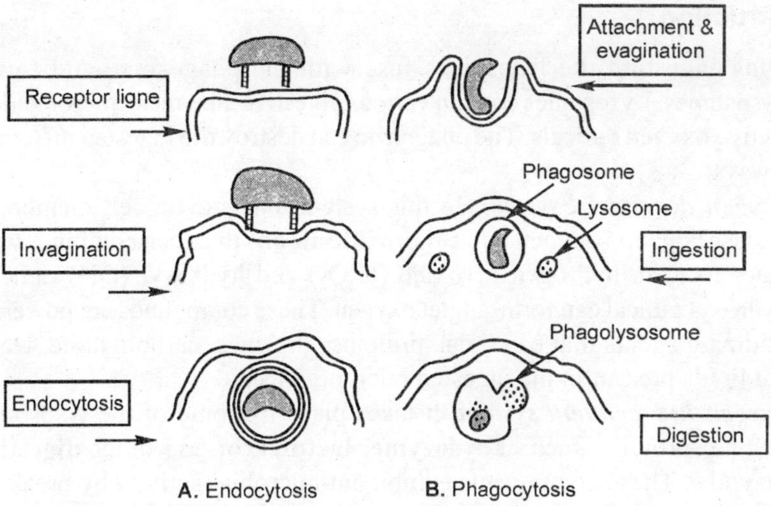

Fig. 7.4: Diagrammatic representation of endocytosis and phagocytosis

complement proteins, cytokines and prostaglandins that causes activation of phagocytes. Bacterial proteins may also activate phagocytes. The activated phagocytes produce endocytic pattern recognition receptors that enhances its attachment and binding to the pathogens. In addition, activated phagocytes show an increased metabolic and microbicidal activity.

b. Chemotaxis

The movement of phagocytes toward the increasing concentration of chemo-attractant is called chemotaxis. It is facilitated by some of the bacterial products such as bacterial cell wall proteins, endotoxins and by chemical mediators of inflammation.

c. Attachment

Attachment of phagocyte to the microorganism is a crucial step for its ingestion. In addition to pattern recognition receptors; opsonins such as antibodies (IgG) and complement components (C3b and C4b) enhance the specific attachment of phagocytes to the target cells.

d. Ingestion

Attachment of the target cells to phagocyte leads to activation of microfilaments. Polymerization and depolymerization of actin filaments result in the movement of pseudopods around the attached pathogen. The plasma membrane is sealed off in order to place ingested materials into the phagosome.

e. Destruction

Following ingestion the lysosomes fuse with the phagosome and forms phagolysosomes. Lysosomes contain various digestive enzymes, microbicides, and reactive oxygen radicals. The phagocyte can destroy the ingested materials in two ways:

i. ***Oxygen dependent system.*** In this system, phagocytic cell membrane oxidase converts oxygen into **superoxide anions** that then combines with water to form hydrogen peroxide (H_2O_2) and hydroxyl (OH) radical. Hydroxyl radical can form singlet oxygen. These compounds are powerful oxidizing agents that can break proteins, enzymes, carbohydrate, DNA and lipids present in the ingested microorganisms.

ii. ***Oxygen-independent system.*** It takes place by some of the lysosomal cationic proteins such as **lysozyme**, **lactoferrin** and other **digestive enzymes**. These components exhibit antimicrobial activity by breaking proteins, RNA, lipids and carbohydrates. Lactoferrin deprives the bacteria of iron that is most required by them.

iii. ***Inflammation.*** Injury or assault to the tissue leads to inflammatory response the purpose of which is to contain infection through migration of widely distributed immune cells to the affected site. Celsus, the Roman physician had described four cardinal signs of inflammation: *rubor* (redness), *tumor* (swelling), *calor* (heat), and *dolor* (pain). Inflammation involves:

1. Increased vascular permeability and dialation of blood vessel **(vaso-dialation)**.
2. Increased blood supply to the affected area.
3. Leukocyte migration (**extravasation**).
4. Accumulation of the immune cells at the site (**chemotaxis**).

The inflammatory event is mediated by a large number of chemical mediators such as **acute phase proteins, histamine, kinins** and **fibrin,** etc.

a. ***Acute phase proteins.*** It is a common name of proteins that are released during tissue injury and infection. C-reactive protein is a major acute phase protein produced by liver cells in response to injury. It binds to C-polysaccharide, a cell wall component of bacteria and fungi. The binding activates complement proteins leading to lysis and subsequent phagocytosis of microorganisms.

b. ***Histamine.*** It is released by a variety of cells such as PMNs, granulocytes and eosinophils. Histamine upon binding to its receptor leads to increase in vascular permeability and vasodilation.

c. ***Kinins.*** These are small peptide normally present in the plasma in an inactive form. It also increases vascular permeability and causes vasodilation. In addition, one form of kinin (**bradykinin**) stimulates pain receptor.

d. ***Fibrin.*** It is the important component of the blood clotting system. Inflammation results in release of a number of enzymes that through a cascade of reaction results in the deposition of fibrin strands that forms a wall around the injured area so that the infection do not spread in the body.

Finally cell debris are removed from the site by phagocytes and the recovery and repairing of the damaged area begins. If infection is not contained by the innate immune mechanism, adaptive immunity comes into play the role.

STUDY QUESTIONS

1. Name the cellular components of innate immunity. Describe the mechanism of action of any one of them.
2. Name the phagocytic cells of innate immune system. With suitable diagram describe the process of phagocytosis.
3. Describe briefly the role of soluble mediators of innate immune system and their mechanism of action.
4. What are pattern recognition receptors? Describe its role in host resistance to microorganisms.
5. Write short notes on followings:
   a. Toll like receptors
   b. Oxygen dependent and independent mechanism of bacteriolysis
   c. Receptor mediated endocytosis
   d. NK cells
   e. Pattern recognition receptors.

MCQs

Select one most appropriate answer given for each questions below:

1. **NK cells cannot be detected by one of the following monoclonal antibodies:**
   A. Anti CD3
   B. Anti CD16.56
   C. Anti CD16
   D. Anti CD56
2. **Which of the following cells are present in the brain?**
   A. Kuffer cells
   B. Histiocytes
   C. Microglial cells
   D. Mesangial cells

3. **Which of the following receptors is signatting receptors?**
   A. Mannose receptor
   B. Toll like receptor
   C. Scavanging receptor
   D. LPS receptor
4. **Which of the following cells when activated with cytokine become LAK cells (or, LAK cells are cytokine activated):**
   A. Macrophage
   B. B1 cells
   C. NK cells
   D. Dendritic cells
5. **Which of the following does not apply to innate immune mechanisms:**
   A. Absence of specificity
   B. Activation by a stimulus
   C. Involvement or multiple cell types
   D. A memory component
6. **Mature dendritic cells are capable of which of the following:**
   A. Activating naive Ag specific T cells
   B. Removing RBC
   C. Producing bradykinin
   D. Extracellular killing of target cells
7. **Which of the following is a correct statement about NK cells?**
   A. They proliferate in response to antigen
   B. They kill target cells by phagocytosis
   C. They are PMN cells
   D. They kill target cells in a extracellular fashin
8. **Removal or the bursa of fabricins results in:**
   A. A markedly decrease in circulating T cells
   B. A markedly decrease in B cells
   C. A delayed rejection of skin graft
   D. Anaemia

1 A 2 C 3 B 4 C 5 A 6 A 7 D 8 B

8

Antigens, Immunization Strategies

I. INTRODUCTION

Cells of immune system are capable of recognizing self and non-self (foreign) substances. The components of innate immune system generally recognize foreign organisms on the whole, while cells of adaptive immune system (T and B cells) recognize specific proteins, glycoproteins and carbohydrate component of the invader, commonly called as antigen. Exposure of foreign antigens may generate, adaptive immune response in the host, such antigens are called immunogens and this property of antigen is termed as immunogenicity. Thus antigens could be both immunogenic or non-immunogenic (if it fails to generate immune response upon binding with lymphocytes), while all immunogens are antigens (by virtue of its binding to antigen receptors). However, both antigens and immunogens are usually referred to as antigens for all practical purposes.

II. DEFINITION

Antigens are defined as substance that reacts with antibody molecules and is recognized by lymphocytes. Recognition of antigen initiates adaptive immune response in the host.

III. CHARACTERISTICS

An antigen has got following characteristics:
1. High molecular weight
2. Chemical complexity
3. Foreignness

High Molecular Weight

- Although many different substances can generate immune response, chemically antigens are larger molecular weight proteins or protein conjugates.

- High molecular weight proteins are more immunogenic compared to proteins of the low molecular weight.
- Generally, chemicals < 1,000 Da (dalton) are non-immunogenic, ≤ 10,000, Da, weakly immunogenic and > 10,000 Da highly immunogenic.

Chemical Complexity

- The immunogenicity of an antigen is directly related to molecular complexities.
- Polymerization, conjugation and presence of histidine, lysine and arginine repeats in an antigen contribute to strong antigenicity.

Foreignness

- Immune response cannot be generated to self-proteins.
- Foreignness to the host is prerequisite for immunogenicity of an antigen in that host.
- Viral, bacterial and parasite antigens are more immunogenic to human
- Serum/protein derived from one species is immunogenic to other species.

IV. CHEMICAL NATURE

Chemically, antigens are proteins, nucleic acids and polysaccharide.

A. ***Proteins:*** The proteins expressed by different cells such as enzymes, hormones, cytokines are usually immunogenic to other species.
   - Co-polymers of different amino acids and carbohydrates are more immunogenic than homopolymer
   - Structural complexity such as primary, secondary, tertiary and quarternary structures of proteins contributes structural complexity, hence immunogenicity.

B. ***Lipids:*** Lipid antigens can induce both T and B cells response if appropriately presented.
   - Lipids are used as haptens that can be immunogenic if complexed with carrier proteins (e.g. Kehole Limpet haemocyanin (KLH) complexed with Bovine serum albumin (BSA).

C. ***Polysaccharide:*** Polysaccharides either pure or glycoproteins, lipopolysaccharide and glycolipid-protein complex are immunogenic.
   - Many micro-organisms have polysaccharide-rich capsule or cell wall-antigens.
   - Human blood group antigen present on erythrocytes as a short chain polysaccharide is an example or high antigenic natural antigen.

D. ***Nucleic acids:*** Nucleic acids are generally non-immunogenic, however, it becomes highly immunogenic in the form of nucleoprotein.

- Some of the autoimmune diseases are associated with autoantibody directed against self-DNA and RNA (e.g. SLE [Systemic lupus erythematosus], Pamphigus etc).

V. ENDOGENOUS AND EXOGENOUS ANTIGENS

- Most antigens (such as bacterial, fungal, parasite, pollen and toxicants) to which a host is sensitized are exogenous antigens.
- Exogenous antigens mostly elicit humoral immune response as these are recognized by B cells.
- Antigens produced by self cells such as mutated (tumor) cells and viral infected cells are endogenous antigens.
- Endogenous antigens are recognized by T cells, hence they induce cell mediated immune response.
- Some endogenous antigens bear different antigenic determinants (epitopes) in different individuals of the same species they are called **alloantigens.**
- Alloantigens are mostly product of polymorphic genes (e.g. HLA, blood group Ag).
- Individuals can be sensitized by alloantigens in following ways.
  1. Through repeated pregnancy in woman.
  2. Through repeated blood transfusion.
  3. Through organ transplantation.

VI. ANTIGENIC DETERMINANTS (EPITOPES)

The actual portion or fragment of antigen that interacts with receptors on B or T lymphocytes are called **'antigenic determinants'** or epitopes (Fig. 8.1a)

1. B cells recognize epitopes expressed on native antigens directly (without MHC complex).
2. T cells recognize epitopes on processed/newly generated peptides bound to MHC molecule (Table 8.1).
3. The size of epitope is generally 5-15 amino acids or 3-4 carbohydrate residue.
4. Polysaccharide antigens generally have **many epitopes of same specificity** (Fig. 8.1b).
5. Protein antigens have many epitopes of different specificities (Fig. 8.1c).
6. Since a microbe has many different proteins and polysaccharides on its surface, it has many different epitopes of different specificity. Therefore, immune response against microbes is directed against different epitopes of same organisms.
7. Haptens have few epitopes thus they are antigenic but not immunogenic.
8. Therefore, epitope complexity increases antigenicity of the antigen.

Table 8.1: Antigen recognition by B and T lymphocytes

S.No.	*Features*	*B cells*	*T cells*
1.	Antigen interactions	B cell receptors	T cell receptor
2.	Nature of antigen	Protein, lipid, polysaccharide, nucleic acid	Peptides
3.	Binding to soluble antigens	Yes	No
4.	Epitopes recognized	Accessible	Internal

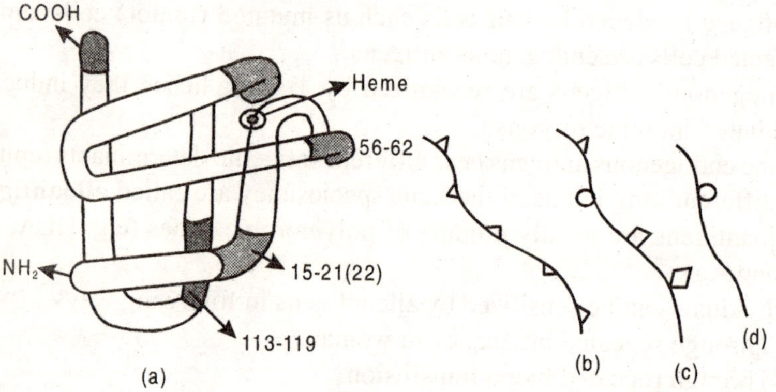

Fig. 8.1: (a) Antigen (sperm whale myoglobin) showing locations of five sequential B cell epitope; (b) Polysaccharide antigen with multiple epitopes of same specificity; (c) Protein antigen with multiple epitope of different specificity (d) Hapten with one epitope

Epitopes Recognized by B and T Cells

Antigen recognitions by B and T cells are different (Table 8.1). B-cells contain membrane bound antibody that binds to free antigen in solution where as T cells recognizes particulate antigens processed by antigen presenting cells, in combination with MHC molecules. Type of epitope and its recognition by B and T cells are as follows:

1. Terminal side chains of polysaccharides and hydrophilic portion of protein molecules constitute B cell epitopes.
2. These epitopes are typically on the outside of the molecules and easily accessible (e.g. myoglobin, Fig. 8.1a).
3. B cell epitopes are mostly linear but may also form as a result of folded conformation (called, conformational epitope in which folded conformation brings epitopes together (Fig. 8.1a).
4. T cell epitopes are mostly internal, continuous and linear (Fig. 8.1b and c) and they are composed of single polypeptide chain.
5. Therefore, polysaccharide antigens contain mostly B cell epitope while protein antigens contain both B and T cell epitopes.

6. Epitopes on an antigen may be single (hapten Fig. 8.1c), multiple, similar (Fig. 8.1b) or different (Fig. 8.1c).

VII. T-DEPENDENT AND T-INDEPENDENT ANTIGENS

In earlier experiments with mice subjected to sublethal dose of radiation in order to make them immunocompromised and reconstitution with T cells, B cells and mixture of T and B cells followed by immunization with various antigens showed that:

1. Most of the antigens including complex proteins elicit immune response in T and B cells mixture reconstituted mice.
2. Such mice mostly respond to complex proteins with large number of different epitopes, such antigens are called **T-dependent antigen.**
3. Polysaccharide antigens (e.g. Lipopolysaccharide, LPS) elicited antibody response in B cell 'only' reconstituted mice, such antigens do not require T cell help, these are called **T-independent antigens.**
4. T cell only reconstituted mice do not elicit antibody response suggesting that antibody is produced by B cells.
5. Some T-independents such as bacterial lipoploysaccharide antigens contain B cells epitope as well as lipid moieties that provides both antigenic as well as mitogenic (by virtue of lipid component) signal, i.e. dual signals to stimulate large number B cells and their differentiation into antibody producing plasma cells.
6. The antibody produced in response to T-independent antigens are mostly IgM.

VIII. HAPTEN AND CARRIER

In 1830, Karl Landsteiner demonstrated that it is possible to raise antibody against almost any substances including synthetic chemicals never found in the natural environment. He showed that many simple substances such as arsonates and nitrophenols are antigenic but not immunogenic but if these **haptens** (greek word means to fasten) are bound to carrier molecules (large molecular weight antigens) they induce antibody synthesis.

- Hapten-carrier system suggests that an individual have unlimited antibody repertoire.
- Hapten-carrier conjugate can induce antibody production with three distinct specificities:
  a. Specificty for epitope on hapten.
  b. Specificity for epitope on carrier
  c. Specificity for new epitope generated due to hapten-carrier conjugation.
- Hapten-carrier conjugate has high molecular weight, increased complexity and enhanced immunogenicity.

- Many biologically active substances such as drugs, peptide hormones, steroid hormones dinitrophenols (DNP), etc. are common haptens.
- The most commonly used carrier molecules are bovine serum albumin (BSA) and ovalbumin (OVA).

IX. HAPTEN-CARRIER EFFECT

Hapten-carrier system have shown some of the unique property of immune system known as hapten-carrier effect (Fig. 8.2). It is interesting to note that:

1. While immune response to hapten takes place only when it is conjugated with carrier molecules, both hapten and carrier is recognized independently by the immune cells (Fig. 8.2a).
2. Immunological memory depends upon the presence of carrier molecule only.
3. Once memory response sets in by carrier, exposure of hapten even first time will lead to secondary antibody response to both hapten and carrier (Fig. 8.2b).
4. Hapten can elicit antibody response only in immunogenic form with same carrier molecule (Fig. 8.2c).

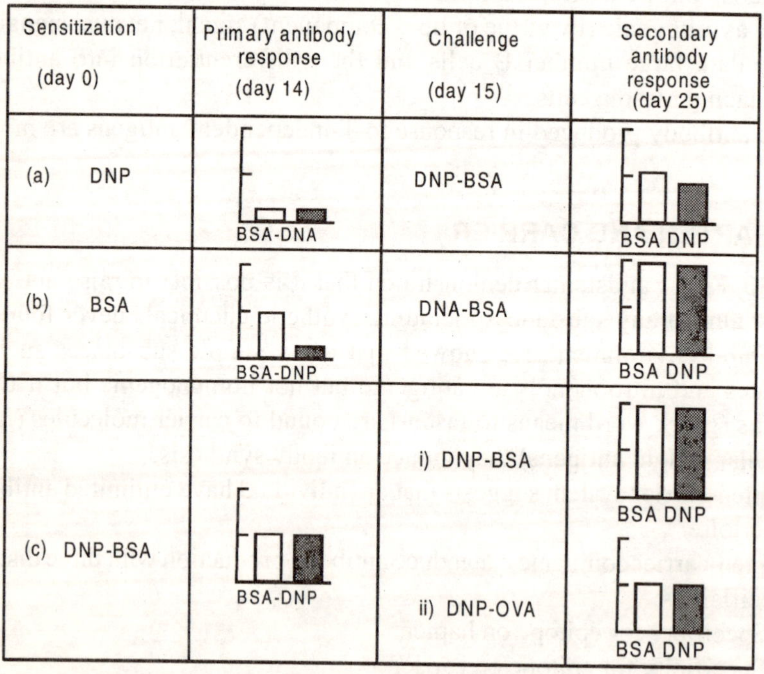

Fig. 8.2: The hapten-carrier effect: (a) Animal immunized with hapten (DNP) and challenged with hapten-carrier conjugate (DNP-BSA). (b) Animal immunized with carrier (BSA) and challenged with DNP-BSA. (c) Animal immunzied with DNP-BSA and challenged with (i) DNP-BSA and (ii) Hapten carrier 2 (DNP-OVA)

X. BIOLOGICAL FACTORS CONTRIBUTE TO IMMUNOGENICITY

Certain biological factors contributes to immunogenicity of an antigen irrespective of its prerequisite immunogenic potential. These are: genetic background of recipient animals, dose of antigen, route of administration and adjuvant or vehicle which carries antigen to the system.

A. **Genotype of recipient:** Hugh Mc Devitt showed that different strains of mice respond differently to the same antigen, some produce high level of antibody while others produce low level. It was shown that:
   1. MHC gene product play a central role in antigen recognition and response.
   2. B and T cells receptor genes are also involved in recognition and response to different antigens dose or:

B. **Immunogen:** Immunogen dose plays a critical role in stimulation and response of the animals.
   1. Low or insufficient dose fails to stimulate enough lymphocytes.
   2. Low immunogen dose generates tolerance.
   3. Single dose of antigen induces low antibody response.
   4. Repeated (booster) dose increases clonal expansion of antigen specific cells.
   5. Very high dose of immunogen fails to elicit immune response

C. **Route of immunogen delivery:** The routes of immunogen administration influence organs and cell population involved in immune response due to differences in local lymphocyte population. The commonly used routes of immunization are as follows:
   - Intravenous
   - Intradermal
   - Subcutaneous
   - Intramuscular
   - Intraperitoneal
   - Oral and nasal routes are commonly used for delivery of vaccines

D. **Adjuvants:** Adjuvants (Latin word *adjuvare* means to help) are the vehicles often mixed with antigens to enhance its immunogenicity. Some of the commonly used adjuvants have been shown in Table 8.2. It has following properties:
   1. Increases half-life of antigens when introduced into an animal
   2. Increases local production of inflammatory cytokines.
   3. Provides chronic stimulation due to slow release of antigens into the system.
   4. Bacterial component of the adjuvant (Freund's complete adjuvant containing killed *Mycobacterium tuberculosis*) activates dendritic cells that are efficient antigen presenting cells.

Table 8.2: Some of the common adjuvants and their mechanisms of action

	Adjuvants	*Mechanism of action*
1.	Freund's incomplete adjuvant (Mineral oil)	Ag persistence, ↑ co-stimulatory signal, granuloma formation
2.	Freund's complete adjuvant (Mineral oil + *M. tuberculosis*)	Ag persistence, ↑↑ co-stimulatory signal, granuloma formation
3.	Alum (Aluminium potassium sulphate)	Ag persistence, ↑↑ co-stimulatory signal, granuloma formation
4.	*M. tuberculosis* (heat killed)	Granuloma formation
5.	LPS	↑↑ co-stimulatory signal, non-specific stimulation of lymphocytes
6.	Synthetic polynucleotides (poly IC/Poly Au)	↑↑ co-stimulatory signal, lymphocyte activation

5. Microbial components are recognized by toll-like receptors and provide accessory signal to cells to express cytokines as well as co-stimulatory molecules.

STUDY QUESTIONS

1. What are the characteristics of an antigen? Compare and contrast antigen recognition by B and T-lymphocytes
2. Write short note on the followings:
   a. Epitope
   b. T-dependent and T-independent antigens
   c. Hapten carrier effect
3. What are the factors responsible for immunogenicity of an antigen? What strategy would you adopt to enhance antigenicity of an antigen to obtain high titre antibody?

MCQs

Select one most appropriate answer given for each questions below:

1. Haptens:
- **A.** Can stimulate secondary antibody response
- **B.** Are immunogenic
- **C.** Have multiple epitope of single specificity
- **D.** Interact with specific antibody

2. Adjuvants:
- **A.** Enhance immunogenicity of haptens
- **B.** Increase size of the immunogen
- **C.** Enhance immunologic response to an antigen
- **D.** Enhance immunologic cross-reactivity

3. **Which of the following statements is *true*?**
   A. All antigens are immunogens
   B. All immunogens are antigens
   C. Adjuvant increases chemical complexity of an antigen
   D. Haptens cannot bind to antibody in the absence of carrier molecule
4. **Lipopolysaccharide (LPS):**
   A. Can be recognized by T cells through TLR
   B. Can be recognized by B cells through BCR
   C. Can be recognized both by T and B cells
   D. Is a weak antigen
5. **Converting a toxin to a toxoid:**
   A. Makes the toxin more immunogenic
   B. Reduces the pharmacologic activity of the toxin
   C. Enhances binding with antitoxin
   D. Induces only innate immunity
6. **B cells recognize epitope of an antigen:**
   A. In association with MHC I
   B. In association of MHC II
   C. In association with complement
   D. Directly via antigen receptor
7. **T cell epitope in an antigen is generally:**
   A. Internal
   B. External
   C. Discontinuous
   D. Folded
8. **B cell epitope on an antigen is generally:**
   A. Internal
   B. Continuous
   C. Folded
   D. Single

1 D 2 C 3 B 4 B 5 B 6 D 7 A 8 C

9

MHC Structure, Genomic Organization, Antigen Processing and Presentation

I. INTRODUCTION

It has been observed that tissue or organ graft is always rejected when grafted in gentically unrelated individuals whereas genetically identical individuals such as twins and inbred strain of mice never reject the graft. Such acceptance and rejection of tissues (histocompatibility) depend upon the product of a set of genes called, major histocompatibility complex (MHC) in mouse; HLA (human leukocyte antigen) gene in human and DLA (dog leukocyte antigen) gene in dogs. Now it is known that every vertebrate species have MHC genes and its product expressed on the cell surface, however, most detailed study has been done in case with mouse and human. MHC molecules play a central role in the adaptive immune response as these molecules are involved in intercellular recognition, processing and presentation of antigens to T cells and also in self-non-self discrimination. MHC molecules also play an important role in the development of both cellular and humoral immune response.

II. BRIEF HISTORY

1. Karl Landsteiner in his Nobel Lecture in 1931 emphasized that a second (to blood group antigen) antigen on tissues, similar to blood group antigens, could be responsible for survival of transplanted tissues in the recipient.
2. This idea led Gorer to discover in 1940 a '*group of antigens*' in mice which when matched between donor and recipients, facilitated survival of the skin graft.

3. Almost at the same time, George Snell by classical genetic technique using inbred strain of mice, discovered that the above group of antigens are the product of one particular region of genome, he named it '**major histo-compatibility complex** ' or H-2 complex in mice.
4. George Snell was awarded Nobel Prize in 1980 for this work.
5. In 1960, Baruj Benacerraf and Hugh Mc Devitt reported that different strains of mice did or did not produce antibody in response to simple polypeptide antigens, and this immune responsiveness was **'autosomal dominant trait'** that mapped to the MHC region. The gene controlling such immune response was called **'Ir (Immune response) genes.**
6. In 1970, the central role of MHC molecules in immune response was explained with the demonstration that T cells do not recognize soluble protein antigens directly but recognize a portion of protein (peptides) only when it is covalently bound to MHC gene products.

III. SALIENT FEATURES

1. CD_4^+ (helper/inducer) T cells recognize antigens in association with MHC II molecules while CD_8^+ (suppresser/cytotoxic) T cells recognise antigen in association with MHC I molecules. In other sense, pattern of antigen association with MHC I or II molecules, decides recruitment of subsets of T cells and consequently the type of effector immune response.
2. MHC II molecules are directly involved in interaction between antigen presenting cells, regulatory T-lymphocytes (CD_4^+ T cells) and effector B lymphocytes, thus play a major role in immuneregulation.
3. MHC genes are highly polymorphic resulting the production of different MHC antigens in different members of the same species.
4. Different MHC molecules have selective affinity for various foreign antigens that may be responsible for variable antibody and cytotoxic response against a given antigen in different individuals. In other sense, particular set of MHC molecules expressed by an individual influences the repertoire of antigens to which CD_4^+ and CD_8^+ T cells can respond.
5. Consequently, MHC antigens are partly responsible for susceptibility or resistance to a particular infection and in the development of autoimmunity.

IV. CLASSIFICATION AND GENOMIC ORGANIZATION

1. MHC is a collection of genes arrayed within 4000 Kb stretch of DNA (*E. coli* genome is 3600 Kb in length) on chromosome 6 in human and 17 in mice.
2. It is organized in three regions (Fig. 9.1) called Class I, Class II and Class III.
3. MHC I gene encodes a single polypeptide chain. In mice, it localizes both at centromeric region and telomeric region. MHC II and III gene complex remains intercalated between Class I.

Genomic organization of murine and human MHCs

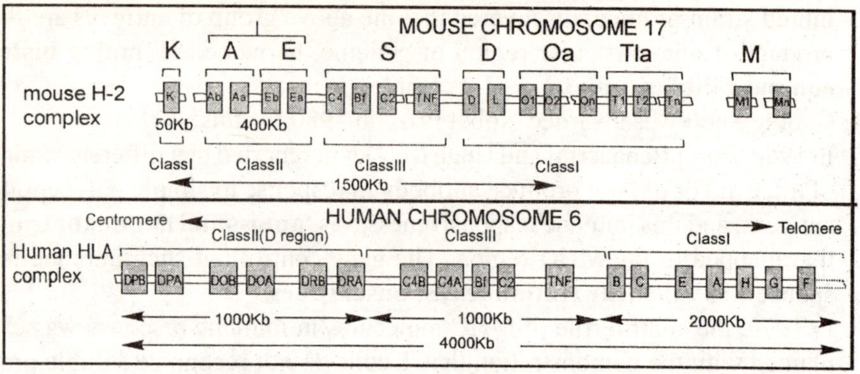

Fig. 9.1: Genomic map of the mouse and human MHCs: the positions of the major genes within these subregions

4. In human, MHC II is present in centromeric region and MHC I in telomeric region while MHC III is intercalated between these two.
5. MHC I is expressed on almost all nucleated cells; the major function being presentation of endogenomal antigen to CD^{+}_{8} T cells.
6. MHC II encodes αβ heterodimer peptides that is expressed primarily on the antigen presenting cells (macrophage, dendritic cells and B lymphocytes) where they present processed exogenous antigens to CD^{+}_{4} T cells.
7. MHC III genes encode various secreted proteins including complement, heat shock protein, lymphotoxin and molecules for antigen processing.
8. Mouse MHC I gene contains loci K and D and L that encodes glycoprotein while in human the equivalent loci are A, B and C that encode functional protein whereas DEF and G are pseudogenes (Fig. 9.2).

Genomic organization of the murine class I region

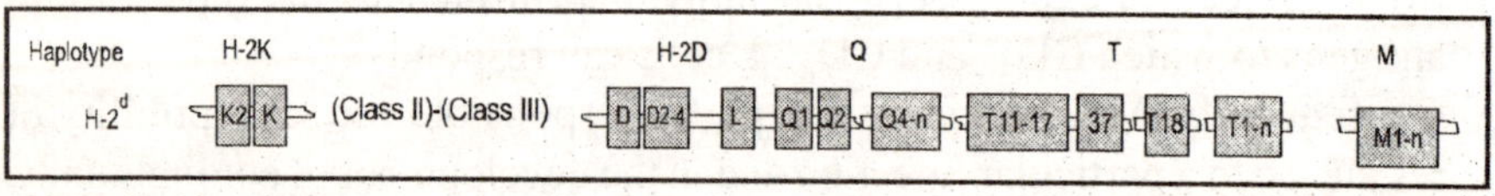

Fig. 9.2: The organization of the mouse MHC class I regin of H-2^d haplotype. Five class I genes map to the D/L region of BALB/c (H-2^d) mice. Two of these genes encode the serologically detectable H-2D^d and H-2L^d antigens. Three additional calss I genes are found in the region between the proximal H-2D^d and distal H-2L^d genes. These genes, called D2^d, D3^d and D4^d and distal H-2L^d genes. These genes, called D2^d, D3^d and D4^d, are of unknown function

9. In mice, MHC II region contains A and E loci while human MHC II region contains DP,DQ and DR as well as few less defined loci (Fig 9.3).
10. These genes have different exons which encode intracytoplasmic, transmembrane and extracellular regions of MHC molecules.

Genes within the human and murine class II regions

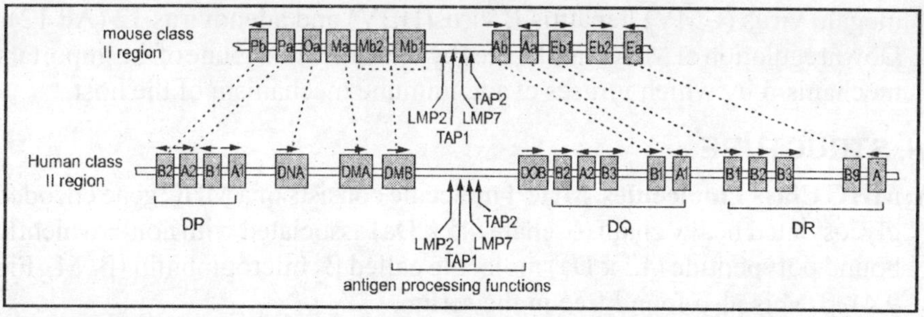

Fig. 9.3: The arrangement of the genes within the human and mouse MHC II region. Homologous genes between the two species are indicated. Pseudogenes are shown in open boxes

11. MHC I genes are less polymorphic as compared to MHC II genes.
12. MHC II region contains multiple α and β chain genes both in human and in mice. All the α and β chain genes can be expressed in a cell giving rise to multiple α β heterodimer molecules.
13. Genes encoding non-classical MHC II molecules have also been identified in human and mice.
14. MHC genes show higher degree of **'linkage disequilibrium'.**

V. REGULATION OF MHC GENE EXPRESSION

Although little is known about the regulation of differential expression of MHC molecules in various cells, followings are the important factors that regulate MHC gene expression.

1. Both MHC I and II genes are flanked by 5'promoter sequence which binds to specific transcription factors, that play an important role in MHC gene expression.
2. Defects in these transcription factors causes one form of **'bare lymphocyte syndrome,** characterized by deficient expression of HLA-A, B and C and absence of β_2M.
3. The expression of MHC molecules is also regulated by **cytokines.** TNF and all classes of interferons enhance expression of MHC I and MHC II molecules.
4. Corticosteroids and prostaglandins downregulate expression of MHC II molecules.
5. β_2 **microglobulin** and **variant chain protein** codominantly regulate expression of MHC I and MHC II molecules respectively.
6. TAP 1 and TAP2 gene (transporter genes) products play an imporant role in transport of peptide from cytoplasm to rough endoplasmic reticulum. Thus blocking of TAP gene products inhibit expression of MHC molecules on the cell surface.

7. MHC expression is downregulated by certain viral infections such as cytomegalo virus (CMV) hepatitis B virus(HBV) and adenovirus 12 (Ad 12).
8. Downregulation of MHC I molecules by viral infection is one of the important mechanism by which viruses evade immune mechanism of the host.

VI. STRUCTURE

A. MHC Class I molecules. MHC I molecule consists of a MHC gene encoded glycosylated heavy chain (α-chain, 45 κ Da) associated with non-covalently bound polypeptide (12 κ Da) molecule called β_2 microglobulin (β_2 M, Fig 9.4). β_2M is also found free in the serum.

1. The class I heavy chain consists of a long polypeptide chain comprising of three folded domains designated as α_1, α_2, and α_3, one transmembrane region and a cytoplasmic tail (Fig. 9.5).

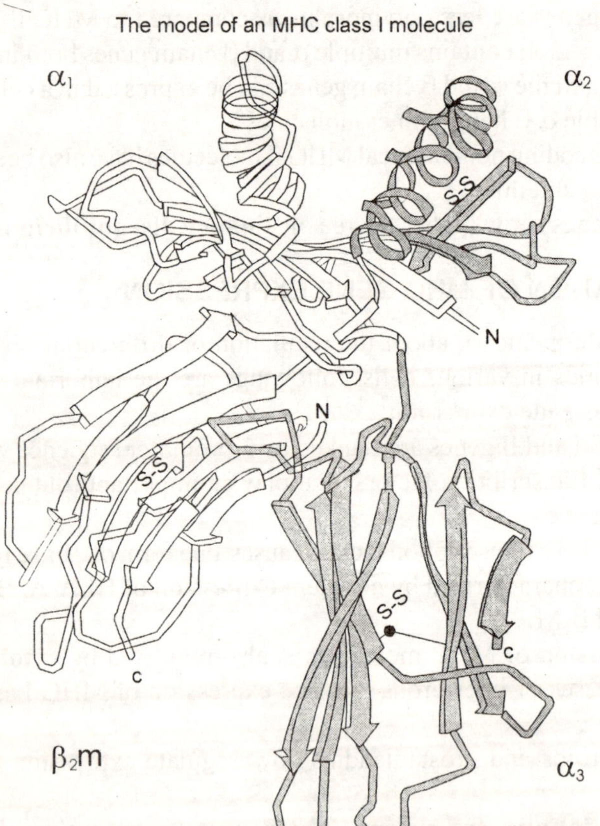

Fig. 9.4: The peptide backbone of the extracellular portion of MHC I is shown. The three globular domains (α_1, α_2 and α_3) of the heavy chain and closely associated with the non-MHC encoded peptide, β_2-microglobulin (grey). β_2-microglobulin is stablized by an intrachain disulphide (-ss-) bond and has a similar tertiary structure to an immunoglobulin domain

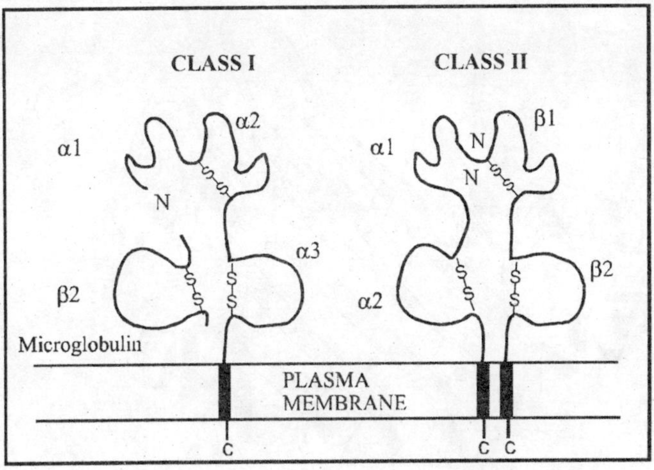

Fig. 9.5: Diagrammatic representation of MHC I and MHC II molecules

2. The three extracellular domains and each comprises about 90 (ninety) amino acids that can be cleaved from the surface by proteolytic enzymes.
3. α_2 and α_3 domains have interchain disulphide bonds enclosing loops of 63 and 86 amino acids respectively.
4. α_3 domain is homologous to immunoglobulin C domain and contains a site which interacts with CD_8 molecules on Tc cells.
5. Extracellular portion of MHC I heavy chain is glycosylated. The degree of glycosylation depends on MHC haplotype and species of the animal.
6. Hydrophobic transmembrane region consists of 25 amino acids.
7. The hydrophilic cytoplasmic domain is 30-40 amino acid long.
8. MHC I molecule remains non-covalently associated with non-polymorphic **β_2 microglobulin** encoded by a gene located on chromosome 15.
9. β_2 M provides stability to the newly formed MHC I glycoprotein and is essential for its expression on the cell surface.
10. Animal deficient in β_2M gene does not express MHC I molecules on the cell surface.
11. X-ray crystallographic studies have shown that α_1 and α_2 domain constitute a **plateform of eight antiparallel β strands supporting two antiparallel α helices** (Fig 9.6).
12. The disulphide bond in α_2 domain connects the N-terminal β strand to α helix of α_2 domain.
13. α_1 and α_2 domains are separated by a long groove that is binding site for endogenous antigens.
14. Majority of polymorphic residues and T cell epitopes on MHC I molecules are located near the groove.

B. MHC Class II molecules: MHC II molecule consists of a pair of MHC II gene encoded heavy (α-chain) and light (β-chain) chain heterodimer glycoprotein (Fig. 9.5).

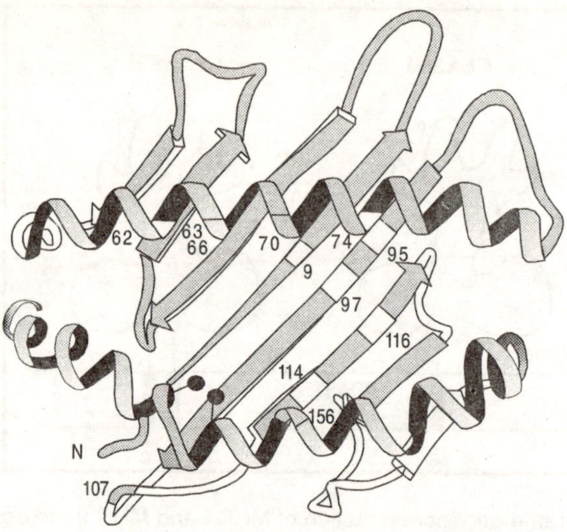

Fig. 9.6: The view of the peptide antigne-binding groove in MHCI as 'seen' by the TCR. The α_1 and α_2 domains each consist of four antiparallel β strands followed by a long helical region. The domains pair to form a single eight-stranded β sheet topped by α helices. The locations of the most polymorphic residues are hightlighted. Residues around the binding site are highly polymorphic

1. The α-chain has molecular weights 30-34 kDa and β-chain ranges from 26-29 kDa, depending upon the loci involved encoding this region.
2. Both α and β-chain polypeptides are similar in structure and consist of two extracellular domains (α_1 and α_2; β_1 and β_2) connected by a short 30 amino acid residue transmembrane and a cytoplasmic domain of about 10-15 amino acid residue.
3. The α_1 β_1 domains contain hypervariable region, α_1, α_2 and β_1 domains are N-glycosylated.
4. The β_2 domain is non-glycosylated , however, it contains binding site for CD 4 molecule on Th cells.
5. Overall three-dimensional structure of MHC I and MHC II molecules is similar despite of differences in their length and organization.
6. Despite the similarity, the MHC II groove is more open as compared to that of MHC I molecule therefore, it can accommodate larger peptides. Overall characteristics of MHC I and MHC II molecules have been shown in Table 9.1.

VII. FUNCTION

1. ***The main function of MHC molecules is antigen processing and presentation.*** During infection, the invader microorganisms such as bacteria, viruses, etc. are engulfed by antigen presenting cells; their protein (antigen) is cleaved into smaller peptides and some of these peptides get associated with MHC I or MHC II molecules. These are carried to the surface of

Table 9.1: Geneal characteristics of MHC I and MCH II genes and their products.

	MHC I	*MCH II*
Major loci (mouse)	K, D, L	I (A-E)
Major loci (human)	A, B, C	DP, DQ, DR
Allele numbers	> 100	> 20
Structure	α chain + β_2 M	α and β chains
Domains	$\alpha_1, \alpha_2, \alpha_3 + \beta_2$ M	α_1, α_2, and β_1, β_2
Distribution	All nucleated cells	Antigen presenting cells
Peptide presentation	Endogenous antigen cross-priming Ag.	Exogenous antigen
Peptide presented to	CD_8^+ T cells	CD_4^+ T cells

antigen presenting cells where these are presented to T cells. Exogenous peptides derived form bacteria, parasites, etc. are expressed along with MHC II molecule and presented to CD_4^+ T cells while endogenous peptide derived from virus-infected and tumor cells are expressed along with MHC I molecules and presented to CD_8^+ T cells. Therefore, CD_4^+ T. cells are called MHC II restricted while CD_8^+ T cells are called MHC I restricted. This phenomenon is known as **MHC restriction.**

Chiefly, there are two pathways of antigen processing and presentation. Endogenous antigens (generated within the cell) are processed in cytosolic pathway while exogenous antigens are processed in the endocytic pathway of antigen processing.

A. Endogenous Antigens: Cytosolic Pathway

Endogenous antigens are synthesized within the cell although they may be derived from pathogens like viruses or intracellular bacteria. The processing of viral proteins has been shown in Figure 9.7.

1. Intracellular proteins are cleaved into smaller peptides by proteolytic enzymes.
2. The target proteins often have a smaller protein attached to it called 'ubiquitin'.
3. Ubiquitin-protein complex can be degraded by a giant protein complex called a *proteasome*.

   Each proteasome is a lrge (265) cylindrical complex consisting four rings of protein subunits and a central channel of 10-50° in diameter. Degradation of ubiquitin-protein complex by proteasome is an energy dependent process that takes place in the central hollow of proteasome).
4. The proteasome involved in antigen processing constitutes two protein subunits (LMP2 and LMP 7) encoded within MHC complex and a third (LMP 10) non-protein subunit.
5. Proteins are cleaved into smaller peptides of 8-15 amino acids in length.
6. Peptides generated in the proteasome are translocated to the endoplasmic reticulum by transporter proteins, (TAP1 and TAP2), the product of TAP (transporter associated with antigen processing) gene.

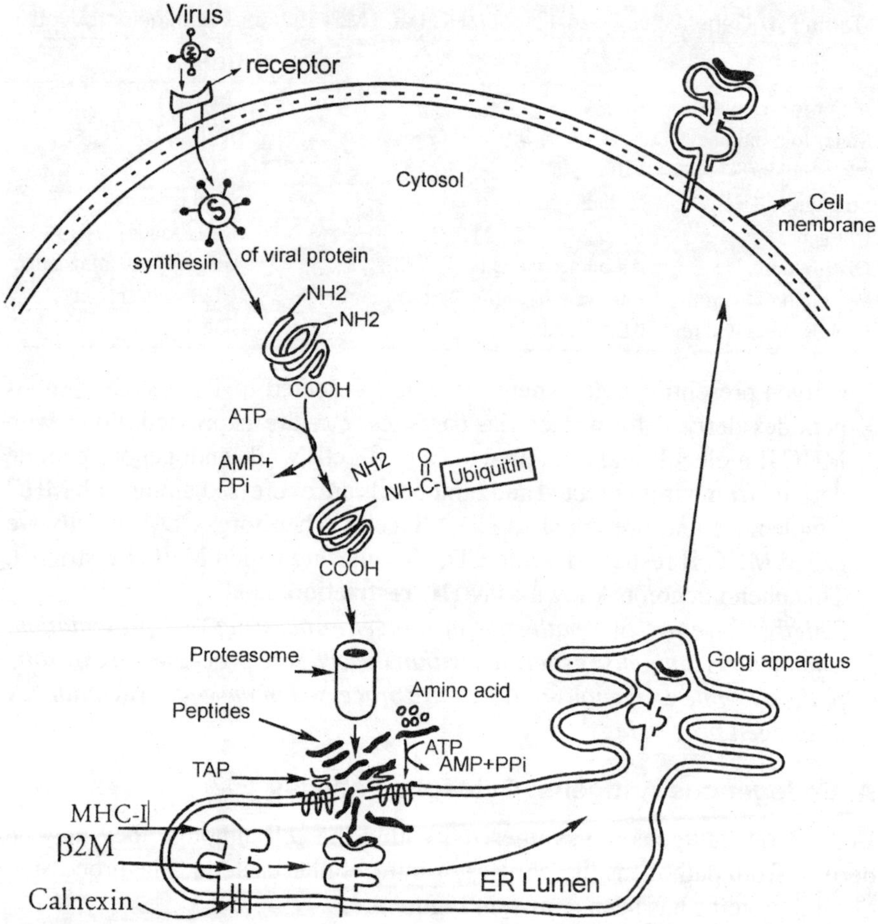

Fig. 9.7: Generation of antigenic peptide-MHC I complex by cytosolic pathway

[TAP1 and TAP 2 belong to the family of ATP binding casette proteins found in the cell membrane of many cells. The TAP 1 and TAP 2 genes map within MHC II region adjacent to LMP2 and LMP1 genes. TAP deficiency can lead to immunodeficiency as well as autoimmunity].

7. In the ER, peptides transported from the cytosol bind with newly synthesized MHC I in the peptide binding groove.

   (MHC I a chain and b_2M are synthesized in ER and their assembly involve several steps including the molecular chaperones called calnexin, which facilitate folding of MHC molecules. Newly formed MHC I requires assembly with b_2 M and peptide binding for its stability).

8. A peptide that binds to MHC I molecule in the ER moves via Golgi complex to the cell surface, where it is presented to CD_8^+ T cells. The unbound peptides are rapidly degraded.

B. Exogenous Antigens: Endocytic Pathway

Exogenous antigens derived from pathogens and foreign proteins are nromally internalized by antigen presenting cells (APCs) through endocytosis/phagocytosis or both. Once internalized, it is degraded into the endosome by proteolytic enzymes. Endocytic pathway appears to involve following steps (Fig. 9.8).

1. Internalized antigens move into early to late endosome and finally to lysosome which contains more than 40 hydrolytic enzymes and a lower pH in each compartment (also see Chapter 7).
2. Within this compartment antigen is degraded into oligopeptides of about 13-18 amino acid residues.
3. Protease known as **cathepsins**, that functions at low pH is involved in degrdation of antigenic proteins in these vesicles.
4. Because the hydrolytic enzymes are optimally active in acidic pH, antigen processing can be inhibited by chemical agents that increase pH of the environment or by protease inhibitors.

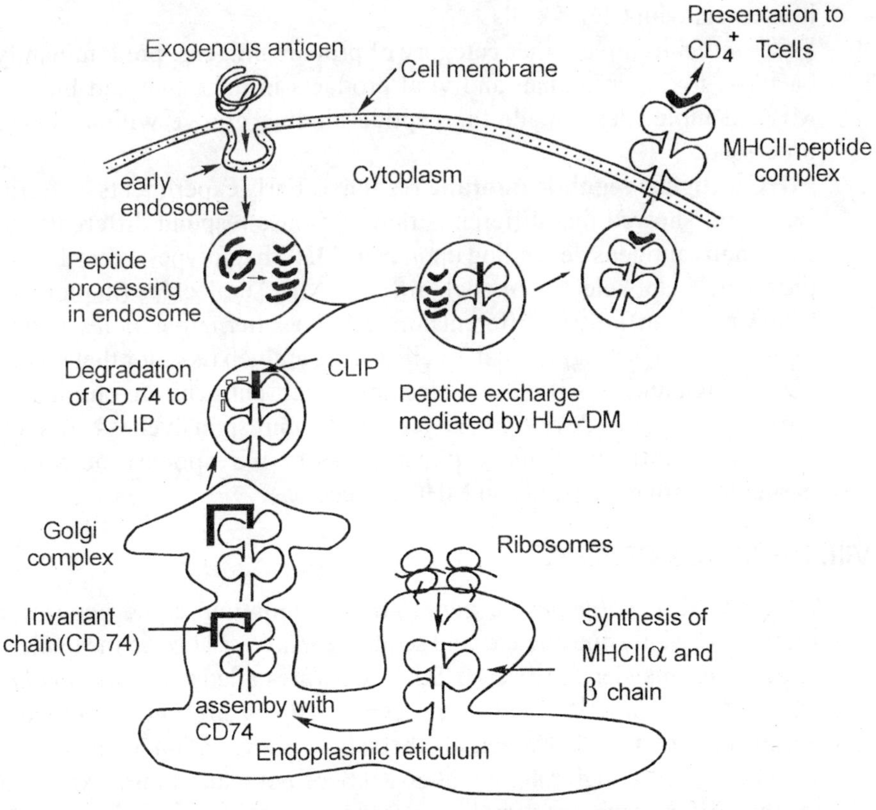

Fig. 9.8: Processing of exogenous antigens by endocytic pathway

5. MHC II molecules are synthesized in the rough ER. The αβ chain of MHC II associate in the ER with preassembled trimeric protein called, **invariant chain** (CD 74). Invariant chain has reverse orientation.
6. The trimeric protein CD 74 interacts with the peptide binding groove of newly formed MHC II molecules preventing binding of pre-existing endogenously synthesized peptides present in ER.
7. The invariant chain also appears to be involved the folding of the α and β chains thereby it acts as molecular **chaperone** for newly synthesized MHC II molecules.
8. Interaction of invariant chain directs MHC II α and β chains to leave ER and enter the Golgi and finally to the endocytic compartment.
9. In endocytic vesicle, invariant chain is removed and digested by proteolytic enzymes while a short fragment **CLIP (class II-associated invariant chain peptide)** physically occupies the peptide-binding groove of MHC II preventing premature binding of antigenic peptide.
10. In this compartment, a non-classical MHC molecule called **HLA-DM** catalyzes CLIP-antigenic peptide exchange in MHC II molecule, generating MHC II peptide-complex which moves to cell surface where it interacts with CD_4^+ T cells.
11. *Super antigens* are another category of protein antigens, predominantly bacterial toxins in human and viral products in mice that can bind to MHC II molecules outside the peptide binding groove without being processed.
12. **MHC antigens regulate immune response.** Early experiments by Baruj Bencerraf showed that different strain of mouse respond differently to exogenous antigens depending upon their MHC haplotype. According to the determinant-selection model, different MHC molecules (haplotype) have variable affinity to different antigens. The alternative '*holes-in the-repertoire model*' suggest that T cell bearing antigen receptor that recognize foreign antigens if closely resembling self-antigens are eliminated during embryonic development leading to unresponsiveness to such exogenous antigen. Thus responsiveness or unresponsiveness to a particular antigen depends on MHC molecules.

VIII. GENETICS OF MHC

A. *MHC genes are polymorphic genes:* In a population, some genes are represented by only one nucleic acid sequence. Any variation in the nucleic acid sequence may lead to disease. Such genes are called *non-polymorphic genes* and the normal or wild type genes will be present in both the chromosomes of a set. However, in others there is a variation in nucleotide sequence at a very high rate, i.e. almost 1% of population may express a gene that differs from the *homologous gene* (gene with similar sequence), such genes are called '*polymorphic genes*'.

B. *MHC genes have different alleles:* Each common variant of a polymorphic genes in an individual is called allele. An individual may have *same alleles* at a genetic locus on both the chromosomes (homozygous) or two different alleles on both the chromosomes (heterozygous).

C. *Inbred strain of mice has single MHC Haplotype:* The total set of alleles present on each chromosome is called MHC haplotype. Thus a heterozygous individual has two haplotypes whereas inbred strain of mice being homozygous have single haplotype.

D. *MHC gene remain in linkage disequilibrium:* Under ideal condition, the frequency of two gene combination on the same chromosome in a population is represented by the product of individual gene frequency. For example, human HLA-A1 is present in 10% individuals while B8 in 16%. Therefore, chances of A1 being linked to B8 is 10% × 16% = 1.6%. However, prevalence of A1. B8 linkage in a population is 8.8% which is much higher than the expected frequency. Such paired specificity is said to be in linkage disequilibrium.

IX. MHC GENE POLYMORPHISM AND DISEASE SUSCEPTIBILITY

Since MHC genes are polymorphic, certain MHC haplotypes have been associated with disease susceptibility and resistance.

1. Many autoimmune diseases such as insulin dependent diabetes, rheumatoid arthritis, multiple sclerosis, myasthenia gravis, SLE (systemic lupus erythmatosus), etc. have been linked to a particular MHC haplotype (Table 9.2).

Table 9.2: Disease association with MHC allele in human

Diseases	*Associated allele*	*Description/Symptoms*
1. Ankylosing spondylitis	B 27	Inflammation of spine, stiffness of neck (vertebral joints)
2. Insulin dependent diabetes	DR 4/DR 3	Deficiency of insulin due to destruction of pancreatic islet cells
3. Multiple sclerosis	DR 2	Progressive inflammation of brain, destruction of myelin sheath
4. Myasthenia gravis	DR 3	Destruction of muscle acetyl choline receptor
5. Rheumatoid arthritis	DR 4	Inflammation of joints
6. Coeliac disease	B 8, DW 3	Chronic inflammation of small intestine
7. Addison's disease	DR 3	Deficiency of ACTH by adrenal gland
8. Systemic lupus erythematosus (SLE)	DR 3	Redness (erythema) of skin

2. Allergic disorders, neurological disease and disorders of the complement system has been linked to MHC gene polymorphism.
3. Certain viral infections are more common in particular MHC haplotype. In case with Cheetah due to their secluded habitat and fewer breeding partners,

MHC gene is less polymorphic, that makes them susceptible to viral infections.

4. Various hypotheses have been put foreward to explain association between disease susceptibility and MHC gene polymorphism. The reasons are possibly-variable affinity of MHC molecules to antigens, defects in antigen presentation, defect in central and peripheral tolerance to self-antigens and molecular mimicry, etc.
5. Diversity of MHC molecules on the other hand appears to protect a species from a wide range of infections and a variety of infectious diseases.

STUDY QUESTIONS

1. Diagrammatically explain the genomic organization of MHC gene complex. What are the factors associated with regulation of MHC gene expression.
2. With suitable diagram explain the processing and presentation of antigens to CD_4^+ T cells by MHC complex.
3. Write short notes on:
   a. Generation of endogenous peptide MHC complex
   b. β_2 microglobulin
   c. Invariant chain
   d. Linkage disequilibrium.
4. With suitable diagrams describe the structure of MHC I and MHC II molecules.

MCQs

Select one most appropriate answer given for each questions below:

1. Which of the following statements is true?

A. Generation of peptide in endocytic vesicle can be inhibited by increasing pH of the vesicle.

B. A monoclonal antibody to β_2M can detect MHC haplotype of an individual.

C. MHC II is constitutively expressed on epithelial cells.

D. MHC I binds to relatively longer peptide as compared to MHC II.

2. Which of the following cellular antigens is not coded by MHC gene?

A. Tumour necrosis factor-α

B. Complement protein

C. TAP 1 and TAP 2 proteins

D. β2 microglobulin

3. Which of the following is required for expression of MHC I molecules on the cell surface?

A. Invariant chain

B. Proteasome

C. β_2 microglobulin
D. TAP 1 protein

4. In cytosolic pathway of antigen processing, antigenic proteins are cleaved in:
A. Endoplasmic reticulum
B. Golgi complex
C. Proteasome
D. Cytoplasm

5. Cheetah are more susceptible to viral infections due to:
A. Less polymorphism in MHC I gene
B. Low degree of polymorphism in MHC II gene
C. It is a carnivore
D. It lacks complement proteins

6. Products or TAP 1 and TAP 2 genes:
A. Bind β_2 microglobulin
B. Prevent peptide binding to MHC molecule
C. Are part of the proteasome
D. Transport peptides into the endoplasmic reticulum

7. The synthesis of which of the following molecules does not require VDJ recombination:
A. TCR α
B. TCR β
C. Ig heavy chain
D. MHC II α–chain

8. Which of the following statements about MHC is incorrect:
A. It codes for complement component
B. It codes for both chains of MHC I molecules
C. It codes for both chians or MHC II molecule

9. The human MHC genes can be described as all of the following *except*:
A. Autosomal
B. X-linked
C. Co-dominant
D. Polymorphic

10. Endosomal antigen processing:
A. Results in production of antigenic peptides that could associate with same MHC molecule
B. Result in activation of CD_4^+ T cells
C. May result in a peptide MHC II complex formation
D. All the above

1 A 2 D 3 C 4 C 5 A 6 D 7 D 8 B
9 B 10 D

10

Antibody Structure, Function, Genetics and Antibody Engineering

I. INTRODUCTION

Antibodies are glycoproteins made as a part of humoral arm of the adaptive immune system to eliminate immunogenic substances and infectious agent. The recognition of antigen leads to proliferation of antigen specific B cell clones which in turn differentiate into plasma cells and secrete antibody. Some of the antibodies are carried to the cell surface where they serve as antigen receptor. Secreted antibody circulates in the blood, where it binds to antigen to form antigen-antibody complex which is removed from the system by phogocytes. Thus antibody serves as a major humoral immune component of adaptive immunity. It was first identified in the globulin portion of the serum protein that constituted specific component of the immune function therefore, It is also called *immunoglobulin*. The term 'antibody' however, refers that these are specific *anti-* for *foreign-bodies*. Antibodies are important molecules for biotechnology because it can be easily exploited for generating immunoglobulin molecules of diverse specificity for diagnostic, therapeutic and prophylactic applications. Various techniques are available to produce antibodies with specificity for binding to huge range of molecular structures (antigen).

In order to understand its application a brief and simplified overview of antibody structure, functions and biological properties has been given in this chapter. Interested students are advised to consult other textbook of immunology for more details.

II. HISTORY

1. In 1939, Tiselius and Kabat in his classical experiment on rabbit immunized with ovalbumin, demonstrated that the antibody constitute γ-globulin fraction of serum protein.

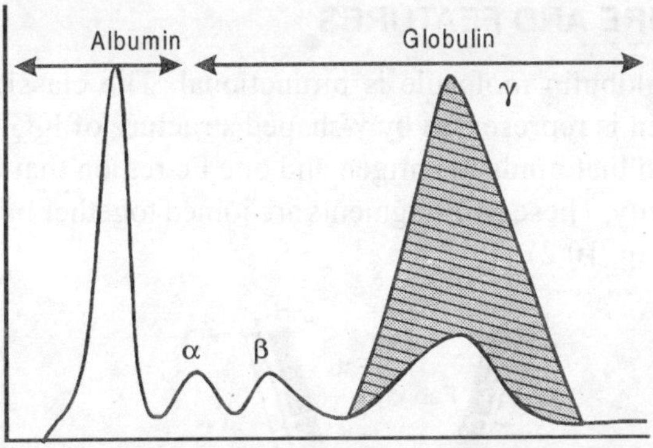

Fig. 10.1: Electrophoretic pattern of rabbit serum. Immunized animal showed high amount of γ-globulin (Shaded area) that was absent when antibody was precipitate with antigen

2. Electrophoretic pattern of serum protein of the immunized animals showed an increased gamma globulin fraction as compared to that of the non-immunized animal's serum (Fig. 10.1). Suggesting an enhanced γ-globulin fraction in the serum upon immunization.
3. When the serum of immunized animal was reacted with antigen, it led to precipitation of antigen-antibody complex resulting in reduced γ-globulin peak upon electrophoresis suggesting that immunological property of serum lies in its γ-globulin fraction.
4. Electrophoretic pattern of *myeloma protein* from patients of B cell tumor (multiple myeloma) suggested further that immunoglobulins are γ-globulin fraction of serum protein. Later on it was found that myeloma proteins are IgG.
5. The discovery of *'Bence Jones'* protein in the urine of multiple myeloma patients provided an opportunity to study the structure of immunoglobulins since it contained homogenous immunoglobulin molecules that was found to be dimer of Kappa (κ) and lambda (λ) light chains.
6. In 1959, a British Scientist, Porter discovered that a proteolytic enzyme **papain** splits immunoglobulin molecule into three fragments. Two of these fragments could bind to antigen, therefore, these were referred as Fab (fragment antigen binding) while the third one could be crystallized and so it was called as Fc (crystallizable fragment). Later it was shown that Fc fragment could not bind to antigen but it mediates a number of biological process when antigen binds to Fab fragment of the intact antibody molecule.
7. Almost at the same time, Edelman in the United States discovered that γ-globulin molecules (molecular weight 1, 50,000 Da) was reduced into four chains when treated with mercaptoethanol (mercaptoethanol breaks disulfide -SS-bond), two larger ones with MW. 53000 Da were called heavy chains and two smaller chains with MW 22,000 Da were called light chain.

III. STRUCTURE AND FEATURES

Each immunoglobulin molecule is bifunctional. The classical structure of immunoglobulin is represented by γ-shaped structure of IgG which contains two Fab portion that binds to antigen and one Fc region that brings about its biological activity. These two fragments are joined together by a more flexible hinge region (Fig. 10.2).

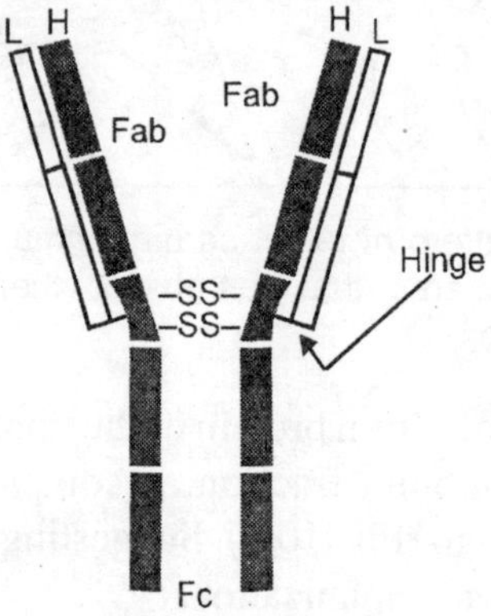

Fig. 10.2: Basic structure of immunoglobulin (IgG) consists of two heavy chains (H) bonded together with disulphide bridge, and two light chains (L) bonded with heavy chain by disulphide bridge. The antibody contains two Fab and a single Fc region

A. Immunoglobulin Classes and Subclasses

Immunoglobulins are a family of proteins consisting of five distinct classes namely IgG, IgA, IgM, IgD and IgE. They differ in size, charge, amino acid composition and carbohydrate contents. Since all the molecules are present in all human beings, these classes are called isotypes (Table 10.1). Each immunoglobulin isotype is made up of two subunits of four polypeptide chains containing two identical light chains (L) and two identical heavy (H) chains joined by disulphide bond (Fig. 10.2).

a. Heavy chain, class and, subclass.

1. The class and subclass of immunoglobulin molecules are determined by its heavy chain type.
2. IgG accounts for 70-75% of total immunoglobulins in serum, it is a four chains molecule with MW 146000 and sedimentation constant 7S. It has four subclasses: IgG1, IgG2, IgG3 and IgG4 with four types of heavy chains (γ), γ1, γ2, γ3, γ4, respectively. It is monomeric antibody molecule (Fig. 10.3a).
3. IgA represents 15-20% of total serum immunoglobulin. It has α heavy chain and two subclasses, IgA1 and IgA2. It is a predominant immunoglobulin of body fluids and secretions. Almost 80% of serum IgA are monomeric while rest of the molecule exists in dimeric form (Fig. 10.3b).

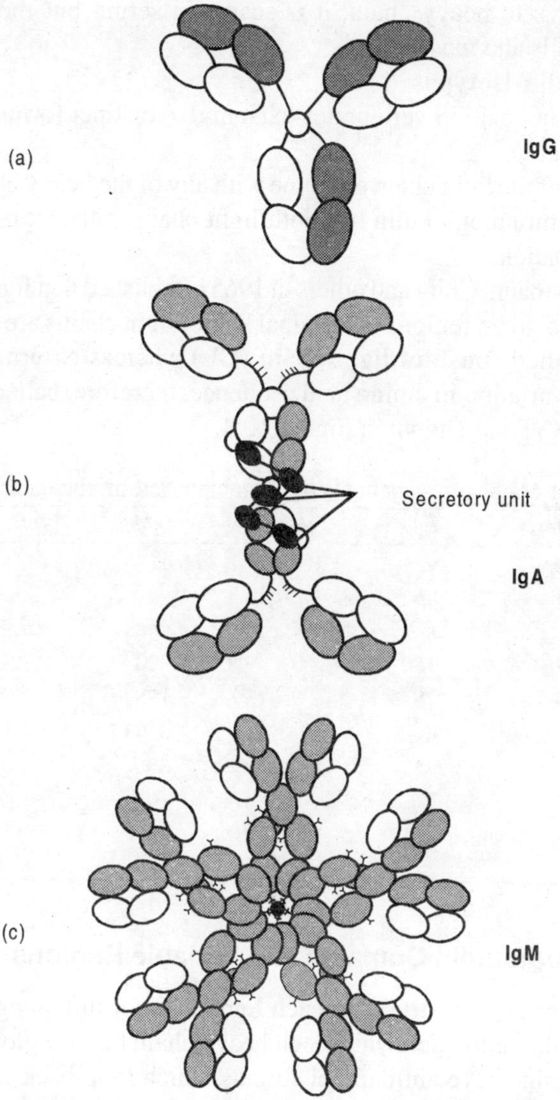

Figs 10.3a to c: Different classes of immunoglobulins (a) IgG is basic unti and (b) IgA contains two units of IgG like structure with additional secretory unit (c) IgM is pentamer of IgG unit. Different classes of immunoglobulins are built up from the same basic structure

4. IgM accounts for almost 10% of serum immunoglobulin, having μ type heavy chain and exists in pentameric form (Fig. 10.3c) with a MW 970,000.
5. IgD forms a major component of the cell membrane of most of the B lymphocytes and about 1% of the serum immunoglobulins. It bears δ heavy chain, MW 180,000 and sedimentation constant 7-8 S.

6. IgE have to heavy chain, it is scarce in serum but mostly present on basophils and mast cells.

b. Light chain isotypes

1. The light chain in vertebrates exist in two distinct forms: kappa (κ) and lambda (λ).
2. Either of the light chain combine with any of the heavy chain, however, a given immunoglobulin has both light chains either κ or λ but never in combination.
3. Hileschmann, Craig and others in 1965 established that light chains consist of two distinct regions C-terminal half of light chains are mostly constant thus called **constant light chain (CL)** whereas N-terminal half shows much variation in amino acid sequence, therefore, called **variable light chain (VL)** as shown in Figure 10.4.

Table 10.1: Major physicochemical characteristics of immunoglobulin classes

Characteristics	*IgG*	*IgA*	*IgM*	*IgD*	*IgE*
Heavy chain	γ (γ_1-γ_4)	α (α_1-α_2)	μ	δ	ε
Light chain	κ/λ	κ/λ	κ/λ	κ/λ	κ/λ
Mol Wt (× 10^3)	150	160	900	180	190
Sed. Const.	7S	7, 9, 11S	19S	7-8S	8S
Serum Con (mg/ml)	6-13	0.6-3	0.3-15	0.03	0.0003
Intravascular dist. (%)	45	42	80	75	51
Carbohydrate content (%)	2-3	12	7-11	9-14	12

B. Immunoglobulin Domains and Variable Regions

1. In terms of protein structure, each light chain of immunoglobulin has two globular domains each while each heavy chain has 4-5 globular domains.
2. The domains have antiparallel strands which loop back on themselves to form β-sheets and these sheets are then rolled back to form barrel like structure (Fig. 10.4). N-terminal domains of both light and heavy chains are called variable regions (V_H and V_L) that vary in amino acid sequence and provide specificity to antibody molecules against antigens.
4. Among variable regions there are three regions of hypervariability to be defined within four or more conserved framework region (FR1-FR4).
5. In three-dimensional structures, the hypervariable region forms loop which combines together to form the principal antigen binding surfaces, therefore, these sequences have been named as **'the complimentarity determining** region or CDRs (Fig. 10.5).

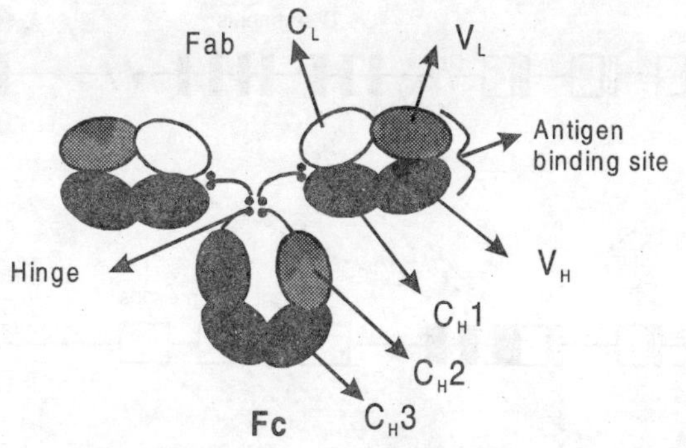

Fig. 10.4: Globular domains of immunoglobulin molecule. The heavy chain domains are shown in dark and light chain domains in lighter shade

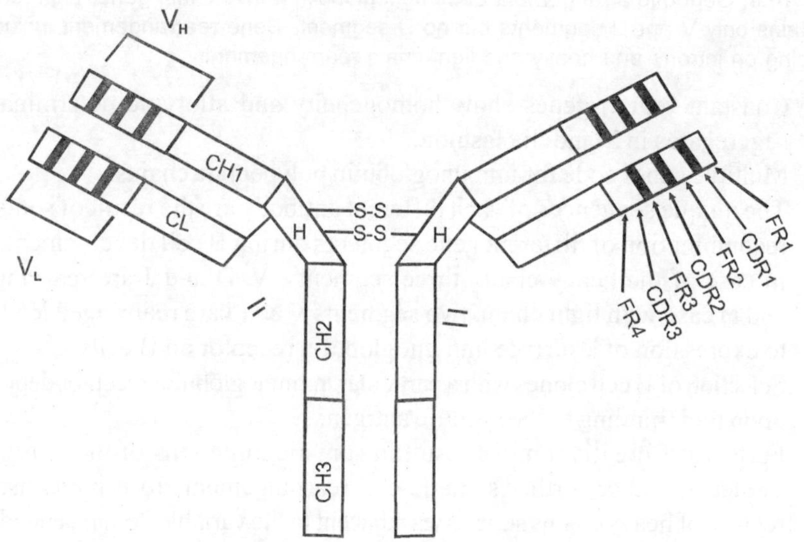

Fig. 10.5: Regions and domians of IgG molecule. Variable regions contain alternate framework (FR) and complimentarity determining regions (CDR)

IV. IMMUNOGLOBULIN GENES AND THEIR REARRANGEMENT

The immunoglobulin gene arrangement has been shown in Figure 10.6.

1. Immunoglobulin molecules are coded by three unlinked gene families, the two light chain genes are located on **chromosomes 2 (κ chains)** and 22 **(λ chains)** while one for heavy chains is located on **chromosome 14**.
2. Variable region genes show diversity in order to provide specificity for different antigens.

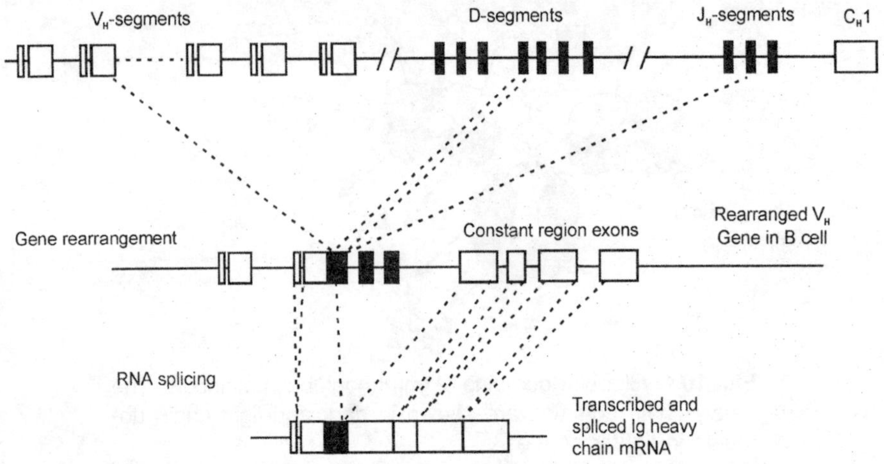

Fig. 10.6: Genomic arrangement of immunoglobulin heavy chain gene. Light chain contains only V and J segments but no D-segment. Gene rearrangement includes splicing on introns and heavy and light chain rearrangement

3. Constant region genes show homogeneity and allotypic determinants seggregates in Mandelia fashion.
4. Multiple genes code for immunoglobulin polypeptide chains.
5. The unique sequences of each different antibody are the result of somatic recombination of different gene segments during B cell development.
6. In case of the heavy chain, three segments, V, D and J are rearranged and in case with light chain, two segments V and J are rearranged leading to expression of a surface immunoglobulin receptor on B cells.
7. Selection of B cell clones with a particular immunoglobulin receptor depends upon their binding to the specific antigen.
8. Further differentiation can result in somatic mutations of the V region sequence and/or further somatic cell rearrangements to bring constant regions of heavy chains subclasses adjacent to the variable domain encoding gene sequence.
9. Post-translational modification of antibody may be critical for its functions, particularly conserved N-linked glycosylation of CH_2 region is important for functions of IgG molecules.
10. Intra- and inter-chain disulphide bridges are important for the overall structure and function of the antibody (for details see textbook of immunology).

V. ANTIGENIC DETERMINANTS OF ANTIBODY MOLECULES

Three antigenic determinants categories are found on immunoglobulin molecules, (A) Isotope (B) Allotypes and (C) Idiotypes:

A. Isotypes: Isotypic determinants are common to all members of a given species hence they cannot be used as molecular markers. They allow the identification of antibody class such as γ for IgG, α for IgA heavy chain, etc. The light chain isotypes are shared by all classes and subclasses of immunoglobulins.

B. **Allotypes.** These are the antigenic determinants inherited from parents in a species that may differ between individuals of a species thus can be used as genetic markers. The loci controlling allotypic markers are codominants, i.e. both are expressed phenotypically in heterozygotes. Some allotypic markers constitute a group on the γ-chain of human IgG called Gm marker, Km and Am on light kappa and lambda chain respectively.

C. **Idiotypes:** Idiotypes are antigenic determinants associated with hyper-variable regions. The antigen combining site in V region of immunoglobulin molecules can also as an antigen and it can induced antibody production when injected in other animals, called anti-idiotype antibody. These antisera contain antibodies specific for several epitopes, called idiotopes. Anti-idiotype antibody behaves as a mirror image of the antigen, therefore, prevents binding of antigen to antibody molecules in some cases.

VI. ANTIBODY PROTEIN FRAGMENTS

a. Proteolytic digestion of antibody can be used to derive different biologically relevant protein fragments of antibody for *in vitro* uses (Fig. 10.7).
   i. Papain digestion splits the antibody molecules into three fragments Fab, Fab and Fc.
   ii. Pepsin digestion cleaves it into $(Fab)_2$ and Fc fragments by the cleavage of c-terminal to disulphide bridge.
   iii. Alternatively mild reducing conditions may be used to separate the two Fab fragments.

b. Protein fragments can also be expressed using recombinant DNA technology. The recombinants products such as **Fv fragments**. contains smallest fragment of variable region domain that still bind to the antigen.

c. Fc portion of the antibody can activate complement even in isolated forms and may be biologically relevant.

d. Alternatively, single chain Fv (ScFv) can be expressed as fusion protein by using artificial linker from C terminus of one domain and N-terminus of the other heavy and light chain genes.

VII. ANTIBODY DIVERSITY

The key effector function of the immune system is to recognize an enormous number of unpredictable diverse and varied antigenic structures which might be encountered during an individual's lifetime. This recognition or targeting, of the effector system is in part, dependent upon the antibody's ability to interact between antigens on the infectious agents as well as the body's effector systems. The effector systems are inherited within the germlines of the individual

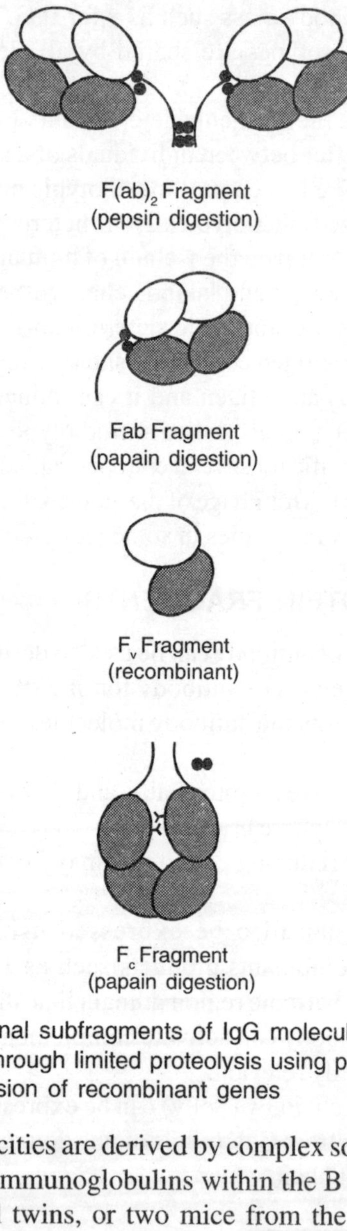

Fig. 10.7: Functional subfragments of IgG molecule. These can be generated either through limited proteolysis using papain and pepsin or through expression of recombinant genes

but the antibody specificities are derived by complex somatic re-arrangements of the genes encoding immunoglobulins within the B lymphocytes. It means that even two identical twins, or two mice from the same strain will have different immunoglobulin sequences expressed at any one time. In addition, antibody diversity could be generated by a number of genetic processes such as combinatorial association, somatic mutations, gene conversion, etc. Existence of a large number of V genes and a small set of D and J segments of germline DNA provides additional opportunity of generating antibody diversity. It seems that such segmentation of germline DNA of antibody gene may have been generated probably due to environmental pressure. It has been estimated that an individual is capable of producing upto 10^9 different antibody molecules. Thus the antibody diversity includes:

- Isotypic variation
- Allotypic variation
- Idiotypic variation

The possible solution to the generation of diversity includes the followings:

- Multiple V gene segments and few D, J gene segments.
- Somating recombination V region forming units.
- Somatic mutation in V region genes.
- Nucleotide addition.
- Gene conversion.

Some animals may have large number of antibody gene pool thus they do not require somatic recombinations mechanisms in generating antibody diversity (e.g. sharks) while other may have few genes thus they undergo a very high level of gene conversion mechanism (e.g. chickens). It is known that mammals may use all the above five mechanisms to generate antibody diversity.

VIII. KINETICS OF ANTIBODY RESPONSE AFTER IMMUNIZATION

The antibody response after immunization can be divided into **primary** and **secondary** response depending upon the exposure of the antigen as discussed in Chapter 8.

1. **Primary response:** Appearance of antibody in the serum following primary immunization generally takes 2-3 weeks depending upon nature of the antigen and genetic background (species) of the animals. First appearance of antibody is called primary response which can be divided into 4 stages (as shown in Fig. 10.8).
   a. **Latent or lag phase:** It is time lag between primary immunization and appearance of antibody, varies between animal species and nature of the antigen and constitute the period between antigen recognition, processing, T cell help (in case with T-dependent antigen), proliferation and differentiation of antigen specific B cell clones.
   b. **Exponential phase:** Antibody concentration in the serum is on the rise during this period.
   c. **Steady state:** The synthesis and metabolism of antibody is balanced during this period.
   d. **Declining phase:** During this phase concentration of antibody in the serum declines, immune response is shut down and effector B cells undergo apoptosis. In the primary immune response, the first antibody that appears is IgM followed by IgG class and decline of IgM level in the serum.
2. **Secondary Response**

   In primary antibody response, effector cells undergo apoptic death but a part of antigen specific cells remains in the circulation as memory cells. In secondary immune response, challenge with the same antigen leads to

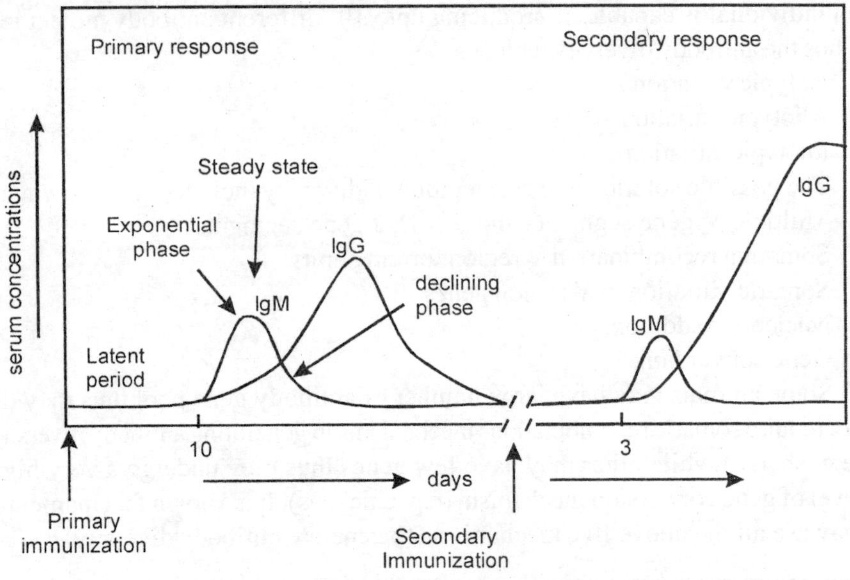

Fig. 10.8: The kinetics of antibody response

proliferation and differentiation of memory cells into the effector cells in antigen primed animals. Secondary immune response is characterized by:

- Considerably shorter lag phase (less than half) than primary response.
- Higher concentration of antibody in the serum.
- Considerably longer steady state (months or even years).
- Class switching of antibody response to IgG, IgA or IgE, etc.
- Increased affinity and avidity of antibody molecules to antigen.

IX. ANTIBODY MEDIATED EFFECTOR FUNCTIONS

As discussed earlier, antibodies, in addition to binding to antigens, participate in a variety of immune functions aimed to eliminate antigen and causing death of the pathogen. In order to invoke immune response, variable region of the antibody interacts with antigen while heavy chains constant region (C_H) is responsible for collaborative functions with other proteins and cells of the immune system leading to **humoral immune response** of the adaptive immunity. Since C_H region is different in different immunoglobulin isotypes, these actions may not be mediated by all classes of antibody. The major effector functions of antibody are as follows:

A. Agglutination

1. Agglutination or clumping of particulate antigens such as micro-organisms is an efficient way of antigen-antibody reaction which results in removal of the complex by phagocytes. IgG and IgM immunoglobulins are potential agglutinating antibodies.

2. Soluble antigens form larger complex when reacted upon by the antibody forming precipitate. Most of the antibody molecules participate in precipitation reaction. This reaction is also employed in the purification of antibody.

B. Opsonization

1. Opsonization (*opsonin means to prepare for eating*) is an important mechanism in bacterial defence which promotes phagocytosis of antigens by macrophage and neutrophils.
2. In the reaction, Fab portion of the antibody binds to the antigen while Fc region of the antibody confers opsonization. The opsonized organisms are recognized by phagocytic cells such as macrophage and PMNs which binds to it by their Fc receptors resulting in phagocytosis of opsonized organisms.
3. Opsonizaiton is a principal effector function of IgG antibodies, therefore, IgG is also called opsonin.

C. Complement activation

1. IgG and IgM antibodies can activate **complement** glycoproteins that result in cascade of reaction leading to perforation of bacterial cell membrane through proteolytic action of complement.
2. The vital components of micro-organisms leak out through the pores causing death of the organisms.
3. The complement protein also acts as opsonin and allows binding of the macrophages and PMNs that posses C 3b receptor resulting phagocytosis of the organisms.
4. Binding of C3b receptor of RBC to antigen antibody complexes allows RBCs to allow the complexes to be delivered to resident macrophages in spleen and liver, that removes the complex without damaging RBCs.

D. Antibody dependent cell mediated cytotoxicity (ADCC)

1. It is one of the important mechanisms through which different cells of the immune system can destroy target cells by their cytotoxic activity (e.g. NK cells, macrophages, neutrophils and eosinophils).
2. In this reaction, antibody binds to target cell through Fab region and Fc region acts as a newly acquired receptor for NK cells and the cytotoxic cells that destroy target cells through their cytotoxic mechanisms other than phagocytosis.
3. IgG molecules play an important role in ADCC.

E. Transplacental passage

1. IgG is the only class of immunoglobulins that crosses placental barrier and provides **passive immunization** to fetus in pregnant females.
2. Placental transfer of IgG is facilitated by expression of **IgG protection receptor (FcRn)** expressed on placental cells.
3. Rapid increase in IgG concentration has been observed in 3rd and 4th trimester of gestation in human.

F. Transcytosis

1. The delivery of antibody to the mucosal surface of the respiratory, gastrointestinal urogenital tracts as well as in milk of lactating females requires transepithelial migration of antibodies, called **transcytosis**.
2. This property lies in the C_H region of IgA in most of the mammals such as mouse and human.

G. Immediate Hypersensitivity

1. IgE, also termed as reaginic antibody participates in allergic reactions and plays an important role in protection against **helminths** through acute **inflammatory reactions**.
2. IgE causes cross linking of mast cells and basophils through its very high affinity receptors (Fc) to these cells with antigen bound to its Fab region.
3. The cells become activated and release a variety of pharmacologic agents such as histamine, heparin, leukotrienes, etc. which triggers **immediate hypersensitivity** response. Such reaction results in **anaphylaxis**.

X. IMMUNOGLOBULIN SUPERFAMILY

1. The term superfamily denotes that the proteins have been derived from a common ancestral gene encoding the basic protein structures.
2. A large number of membrane-bound glycoproteins share **immunoglobulin fold domain** structure therefore considered as a member of immunoglobulin superfamily (Fig. 10.9). Some of them are as follows:
   - Igα/Igβ heterodimer which is a part of B cell receptor.
   - T cell receptor.
   - Accessory glycoproteins on T cells such as CD2, CD4, CD8, CD28 and γ, δ and ε chain of CD3.
   - β_2 macroglobulin and invariant chain associated with MHC I and MHC II molecules.
   - Various cell adhesion molecules such as VCAM-1 ICAM-1, ICAM2 and LFA-3.

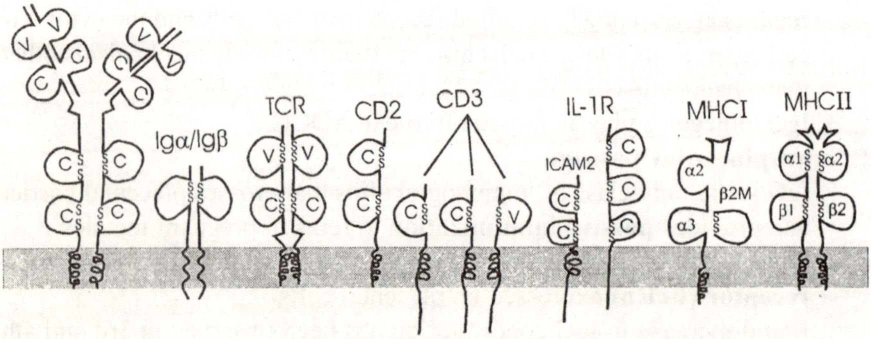

Fig. 10.9: Some members of the immunoglobulin superfamily

- Cytokine receptors, e.g. IL-1R, M-CSFR and C-kit.
- Platelet derived growth factor.

3. The common immunoglobulin fold domain in their light and heavy chains suggests that these proteins have been derived from a common primordial gene through mechanism like gene duplication followed by divergence to give rise to different light and heavy chain combination.
4. Since functions of different proteins of immunoglobulins superfamily are different, it suggests that these genes have evolved independently and do not share genetic linkage or functions.
5. X-ray crystallographic studies of these proteins show that all the members of immunoglobulin superfamily share at least one at more stretches of about 110 amino acids capable of rearranging into antiparallel β-pleated sheet.
6. Since all other members of immunoglobulin superfamily are not capable of binding to antigen, it seems that Ig fold structure may facilitate interactions between membrane glycoproteins.

XI. MONOCLONAL ANTIBODIES

In general, an antigen contains many epitopes that can be recognized by different B cell clones that in turn differentiate into plasma cell and produce antibodies mixture capable of recognizing these epitopes. Therefore, conventional antibody raised in animals is called **polyclonal antibody** due to their origin from several epitope specific clones. Such polyclonal antibody reduces its efficacy in the various immunochemical and immunological diagnostic/therapeutic purposes. In 1975, Kohler and Milstein developed in *in vitro* technique to generate antibody from single clone of antibody producing cells. This technique is called **hybridoma** and the antibody produced by it is called **monoclonal antibody.** Since its development, monoclonal antibody has been widely used in diagnosis and therapy of various diseases as well as in biomedical research. Kohler and Milstein were awarded. 'Nobel Prize in Medicine' in 1984 for this work.

A. Hybridoma and Monoclonal Antibody

Hybridoma technique includes following steps:

1. Animals are immunized with target antigen (need not be pure form).
2. Spleen cells of the animals are removed and fused with myeloma cells (B cell tumor).
3. The hyrbid cells are grown in HAT medium that allows hybrid cells to grow and block DNA synthesis of the fusing partners (unfused myeloma cells).
4. Antibody secereting hybridomas are selected by testing the presence of antibody in their culture supernatant as shown in Figure 10.10.

B. Catalytic Antibodies (Abzymes)

Binding of monoclonal antibodies to antigens involves a weak and non-covalent interaction but with high affinity and specificity. It is similar to that of the

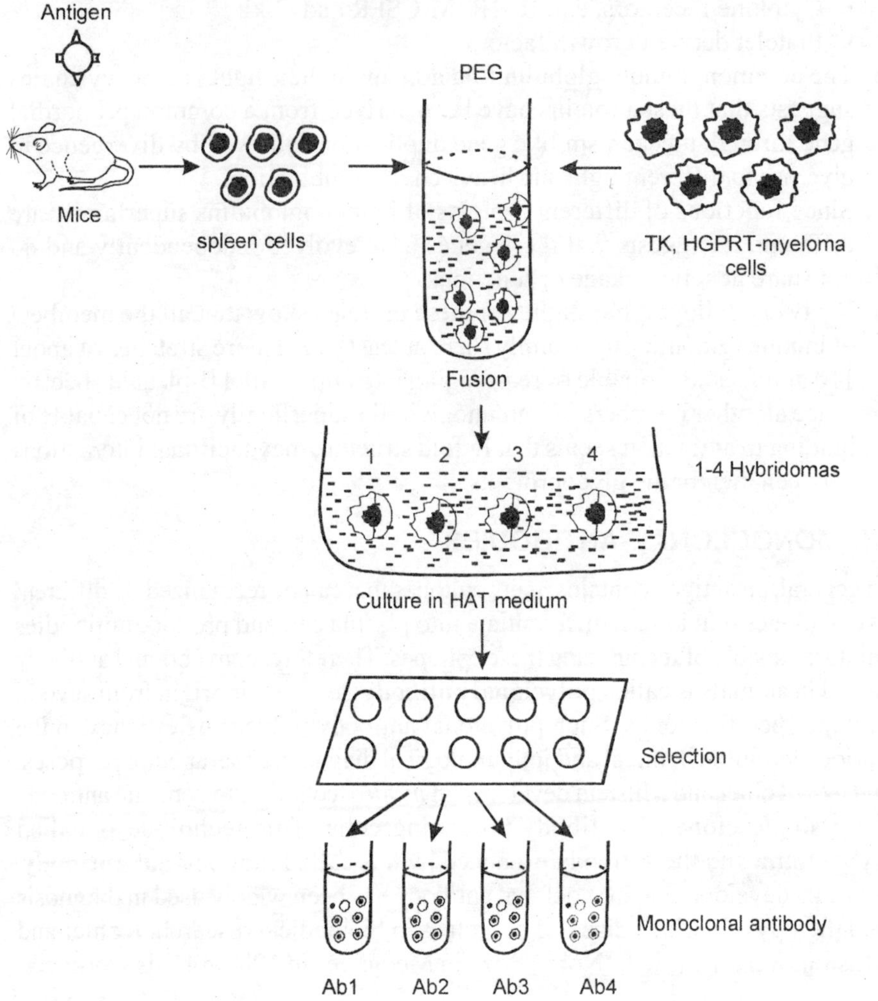

Fig. 10.10: Showing steps in hybridoma technology. Mice is immunized with antigen, spleen cells are separated and fused with nonsecretor myeloma cells. Myeloma cells lack the gene for hypoxanthine guanine ribosyl transferase (HGPRT) and thymidine kinase (TK) that is required for salvage pathway of DNA synthesis. Spleen cells use *de novo* pathway of DNA synthesis but it can also use salvage pathway if hypoxanthine and thymidine is supplied in the culture medium. The fused (hybridoma) cells are selected in HAT (Hypoxanthine, Aminopterin and thymidine). Aminopterin being toxic, blocks *de novo* synthesis of DNA. Thus unfused myeloma cells die off while spleen cells and hybridoma survive using salvage pathway in presence of hypoxanthine and thymidine. Spleen cells being short lived, die off in due course of time while hybridoma survive by virtue of its immortality gained from myeloma fusion partner. Hybridomas are immortal thus can be propagated in tissue culture and provide continuous supply of desired antibody

binding of enzymes to substrate. The only difference between these two reactions is that while the enzyme catalyzes a chemical change in the substrate, antibodies can stabilize the transition state of the antigen molecules. The similarity of both the interactions raises the possibility that whether antibodies can be used as an enzyme. To investigate such possibility a hapten carrier complex was synthesized in such a way that hapten resembled structurally with the transition state of an ester undergoing hydrolysis. The monoclonal antibody raised against such complex accelerated hydrolysis of the ester by almost 1000 fold, suggesting that the monoclonal antibody acted like an enzyme that normally catalyzes hydrolysis of ester. These catalytic antibodies were called abzymes (antibody with enzymatic activity). Such abzymes that can cut peptide bonds at a particular amino acid of a protein can be used in the structural and functional analysis of various proteins as well as in the treatment of different bacterial and viral infections as well as other diseases with a defined target protein. This area is being widely explored by scientist around the globe.

XI. ANTIBODY ENGINEERING

A. Chimaeric and Humanized Antibody

Although monoclonal antibodies are highly specific and it can bind discretely to the desired antigen if so generated, its clinical use is limited due to three major reasons.

i. Monoclonal antibodies are foreign proteins that give serological reactions to human thus these are more toxic.
ii. Often monoclonal antibody could be specific to antigen but may not give desired biological activity due to its class or subclass.
iii. Sometimes antigen binding (Fab) may be required but not the Fc region because it may lead to deleterious effect.
    In order to avoid such obstacles, tailor made recombinant antibody can be generated through recombinant DNA technology or antibody engineering (Fig.10.11).

It is the modular structure of antibody molecules which are composed of a collection of discrete globular domains, encoded by genes with a similar modular structures whereby each domain is coded in a separate exon, makes antibody engineering feasible. It is technically possible through the appropriate cloning strategies, to isolate the genes encoding any antibody made from any immunized species. Through antibody engineering and recombinant DNA technology any V-region of antibody can be expressed in combination with any C-region related for desired properties. Such antibodies are called **Chimaeric antibodies**. Thus variable region from rodent immunoglobulin against selected antigen can be recombined with constant region of human immunoglobulin classes/subclasses to get **humanized** antibody for therapeutic uses with least toxic effect. Fc region or the antibody can be appropriately modified to prevent biological functions like complement activation and Fc receptor binding. Bifunctional antibodies can be generated by recombining two different Fab regions, or antibody can be generated with only Fab region, for antigen binding and for different therapeutic purposes.

After manipulation of the antibody genes, they can be expressed in a number of different, eukaryotic and prokaryotic expression systems, including plant cells, bacteria and viruses. In case with mammalian system, the genes can be expressed in Chinese Hamster Ovary (CHO) cells or myeloma cells where it can be processed and glycosylated in a manner similar to normal B cells. Currently large scale production of recombinant monoclonal antibody is done using either B-lymphoid cell lines or CHO cells. The problem encountered in this technique is that the cell must be able to assemble different parts (domains) of the molecules correctly.

B. Combinatorial and Phage Display Library

The development of molecular biology and recombinant DNA technology has enabled scientists to done and express any gene. In **phage display**, genes coding variable region of antibody (Fab or Fv fragment) are cloned into phage vectors. The modified phage vectors are then used to transform bacteria. During assembly of the phage particle, immunoglobulin Fab/Fv regions are expressed on the phage that are related by panning technique (binding to immobilized antigen on solid surface) and isolated phages are allowed to infect bacteria. If the bacterium is infected by only one phage type, all of the newly assembled phage particle will carry some antibody DNA and display similar Fab or Fv fragments. This way a phage display library can be created. The random coming of libraries of different single light and a single heavy chain variable region genes can be used to create *a combinatorial phage display library*. (Fig. 10.11).

A major disadvantage of phage display libraries is that the only antigen binding property or antibody is tested that may not reflect the crucial biological function may not be therefore final application. Secondly, it is easy to screen the library with purified homogenous antigen preparation but very difficult in case with complex mixture of antigen. However, in case with hybridoma the screening is based on limiting dilution of hybrid clones, therefore, purified antigen is not required for raising monoclonal antibody.

C. Applications of Recombinant and Monoclonal Antibodies

The development of hybridoma and latest recombinant DNA technology has provided enormous opportunity to solve many exclusive and unsolved issues in biology and medicine. Monoclonal and engineered antibody have been applied in the clinical diagnosis and treatment of an array of human diseases such as infections and autoimmune diseases cancer and in immunomodulation in

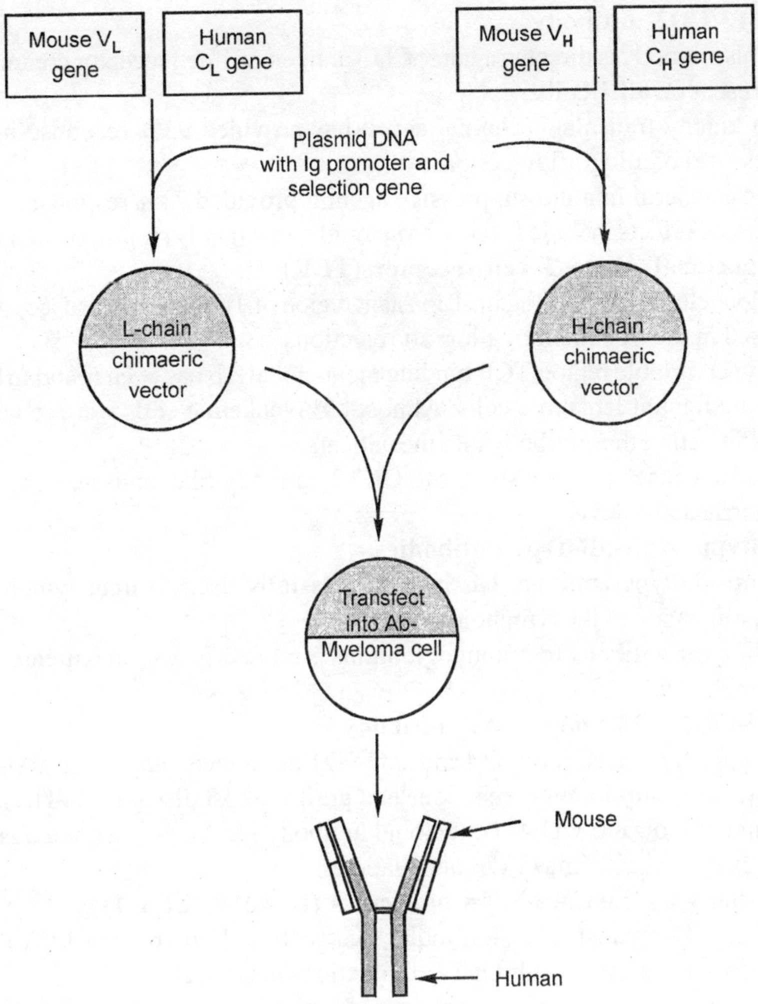

Chimaeric Humanized Antibody

Fig. 10.11: Schematic diagram showing production of chimaeric mouse-human monoclonal antibody by genetic engineering. The promoter, leader and variable region sequences from mouse antibody gene and constant region exons from human antibody gene is cloned and expressed in myeloma cells resulting in production of humanized antibody.

transplantation and immunosuppression. These applications can be broadly divided into following categories.

1. Cell surface antigenic targets

Monoclonal antibody against various cell surface targets has been widely used for immunosuppression in transplantation, treatment and diagnosis of leukemia and in understanding cell interactions. Some of the important cell surface targets are as follows:

a. OKT3/CD3 antibody

i. This antibody is directed against CD 3 antigen of T cell receptor commonly present on all T cells.
ii. In kidney transplantation patient, it has provided 93% response in the reversal of allograft rejections.
iii. As a general immunosuppressive agent it provided 73% response.
iv. Its side effects include fever, diarrhea and occasionally respiratory distress.

2. Monoclonal Ab to T cell receptor (TCR)

i. Monoclonal antibody against constant region of TCR has been successfully used in the treatment of allograft rejection.
ii. Hypervariable region TCR binding monoclonal Ab has been found to bind to malignant leukemic cells. Almost 80% leukemic cells were removed by injecting this antibody into the patients.
iii. It also causes toxicity similar to OKT3 antibody like, nausea, vomiting, diarrhea and fever.

a. Idiotypic Anti-idiotype antibodies

i. Anti-idiotype antibody has been successfully used to treat lymphoma (malignancy of the lymphoid system).
ii. Idiotypic antibody to immunoglobulins is more effective in patients with lymphoma.

b. Anti-CD4/CD8 monoclonal antibody

i. In animal models, anti-CD 4 and anti-CD 8 antibodies suppressed allograft rejection, autoimmune reactions and graft vs. host disease (GVHD).
ii. Injection of anti-CD 4 monoclonal antibody has been effective against arthritis, myasthenia gravis and diabetes.

c. Antibody against adhesion molecules (ICAM-l, LFA-1)

i. Antibody against adhesion molecules such as ICAM-l and LFA-l has been widely used to inhibit T cell functions in GVHD.
ii. These antibodies are helpful in understanding cell cell interactions, lymphocytes migration and homing.

d. Flowcytometry

In flowcytometry, monoclonal antibodies against cell surface targets have been used to enumerate various cell types and their changes in disease conditions such as HIV, (CD4 cell cout), leukemia, lymphoma, etc.

3. Genetically engineered antibodies

As described earlier, genetically engineered antibody is of great value in the treatment of human disease as it has following advantages over monoclonal antibody:

i. Antibody can be generated with defined biological functions such as Fv fragments for antigen binding region.
ii. Mouse-human chimaeric antibody can be produced that may increase its efficacy and decrease toxicity.

iii. Human Fab (Fv) have a longer half-life as compared to mouse monoclonal antibody. Thus they may providing long-lasting response.
iv. The effector region can easily be conjugated with radionucleides, toxins etc.
v. Genetically engineered antibody has been used against cancer, autoimmune disease and in GVHD.

4. Human monoclonal antibodies

Human monoclonal antibodies have been produced by bypassing hybridoma technology by cloning V-region genes (as shown in Fig. 10.12) using phage display technique. Such antibodies can be potentially used for diagnosis and treatment of various diseases.

5. Bifunctional antibodies

i. Antibodies with two distinct binding activities have been generated for targeted delivery of toxin, radiation and cytotoxic drug for treatment of various diseases.
ii. In this strategy, one arm of Fab is specific for cell surface target to facilitate binding and other is labelled with toxin, or radionucleide for destruction of the bound and neighboring cells.

6. Radiolabelled antibodies

i. Radiolabelled monoclonal antibody has been used for treatment of cancer particularly metastatic and deeply situated tumours that otherwise cannot be treated by conventional radiotherapy.
ii. It has been successfully used in radio imaging techniques for diagnosis of myocardial infarction and cancer such as brain tumours and metastasis.
iii. Such antibodies can be radiolabelled with different types of particles such as α, β, γ emitters as per requirement.
iv. Radiolabelled monoclonal antibodies can be effectively used in radioimmunoassay for diagnosis and monitoring of various diseases, pregnancy, etc.

7. Immunotoxins

Immunotoxins are composed of tumor specific monoclonal antibody conjugated with toxins. Tumor specific part leads to binding of antibody to tumor cell surface that follows its internalization in the endosome where toxin is released in the cytoplasm and inhibits protein synthesis.

i. In this strategy, most commonly used toxins are Ricin, Shigella and Diphtheria toxins.
ii. These toxins consist of two functional chains, one toxin chain and one or more cellular binding chains.
iii. An immunotoxin is prepared in such a way that cellular binding chain is replaced by monoclonal antibody specific for tumour antigens while toxin chain remains as such to impart its toxic effect.

8. Catalytic antibodies (Abzymes)

i. Abzymes can cut peptide bond between amino acids in a similar way as the restriction endonucleases cleave nucleotide at a specific site.

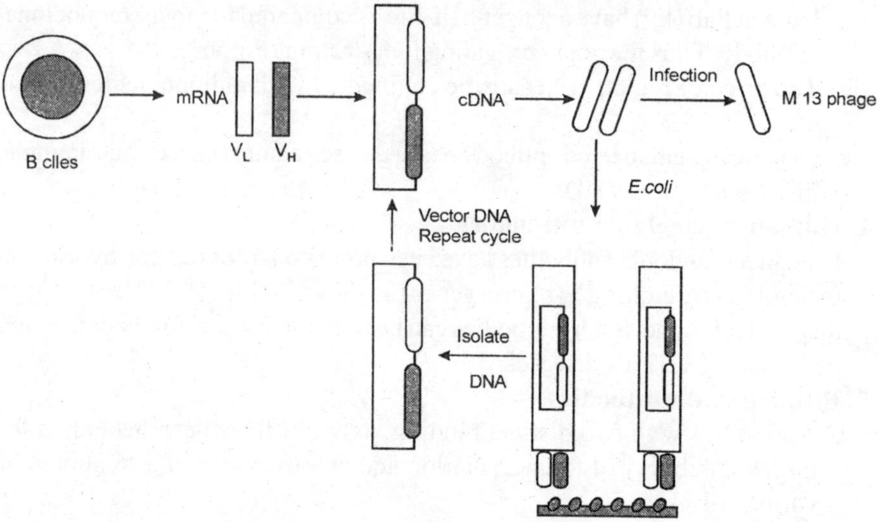

Fig. 10.12: Production of pahge-display library. Antibody V_H/V_L genes are first amplified from mRNA of B cells by reverse transcriptions and PCR. V_H/V_L are joined by spacer to provide gene for Fv fragment. *E. coli* is then transfected with phagemid vector containing a leader sequence. The phage replicates and displays Fv on its tip. These are then selected by panning with antigen coated plate. The Fv DNA can be isolated and cycle can be repeated to increase the yield of high affinity phage

ii. A battery of abzymes would be an invaluable tool for structural and functional analysis of proteins.

iii. Abzymes with an ability to dissolve blood clots can be used for treatment of hypertension.

iv. Abzymes with proteolytic activity may be used to cleave viral proteins to blocks its infectivity. There has been tremendous development in the area of antibody technology and antibody engineering, however, much more is to be achieved through application of these molecules in the area of pathogenesis of diseases, vaccine development and other therapeutic uses.

STUDY QUESTIONS

1. Describe structure of antibody molecule with a suitable diagram. What do you understand by antibody classes and subclasses?
2. Describe briefly uses of monoclonal antibodies in serodiagnosis and therapy giving suitable examples.
3. Write short notes on (a) Catalytic antibodies (Abzymes) (b) Combinatorial library (c) Immunotoxin (d) Bifunctional antibody.

MCQs

Select one most appropriate answer given for each questions below:

1. **The antigens binding region of an antibody molecule is:**
   A. Fab
   B. Fc
   C. Hinge region
   D. Heavy chain only
2. **Fc region of antibody is related to:**
   A. Antigen binding
   B. Antigen specificity
   C. Biological activity
   D. Internalization of antibody in the cell
3. **Immunotoxins are:**
   A. Antitoxin antibody
   B. Antibody specific for toxin binding
   C. Antibody toxin conjugate
   D. Cytotoxic toxins
4. **Pepsin cleaves antibody, molecules into:**
   A. Fab, Fab and Fc fragments
   B. $(Fab)_2$ and Fc fragments
   C. Heavy and light chain fragments
   D. Framework and CDR fragments
5. **Humanized monoclonal antibodies contain**
   A. Mouse heavy chain and human light chain
   B. Mouse variable region and human constant region
   C. Human Fab region and mouse Fc region
   D. Mouse light chain and human heavy chain
6. **Secretory component is present on:**
   A. IgG
   B. IgA
   C. IgM
   D. IgE
7. **The hinge region of an IgG heavy chain is located:**
   A. Within the CH1 intrachain loops
   B. Between CH1 and CH2
   C. Between CH2 and CH3
   D. Between CH3 and CH4

8. **The class specific antigenic determinants of Igs are associated with:**
   A. L-chians
   B. J-chains
   C. Variable region
   D. H-chains

9. **The idiotype of an antibody molecule is determined by the amino acid sequence of the:**
   A. Variable regions of H- and L-chains
   B. Constant region of the L-chain
   C. Variable region of the L-chain
   D. Constant region of the H-chain

10. **Antibodies to human Ig will react with:**
    A. Human IgM
    B. K–light chains
    C. Human IgG
    D. All the above

11. **An individual was found to be heterologous for IgG1 allotypes 3 and 12. The different possible IgG1 antibodies produced by this individual may have:**
    A. Two H chains or allotype 12
    B. Two L chains or K or γ
    C. Two H chains of allotype 3
    D. All the above

---

1 A 2 C 3 C 4 B 5 B 6 B 7 B 8 D
9 A 10 D 11 D

11

Antigen Antibody Interactions and Its Applications in Immunodiagnosis

I. INTRODUCTION

It has been shown by X-ray crystallographic studies that antigen binds to the hypervariable loops of immunoglobulin variable region that also determines the specificity of immunoglobulins (antibody molecules). The antigen-antibody interaction can be prevented by high salt concentrations, extreme pH and detergents suggesting that it is noncovalent interaction. The noncovalent forces in such interactions involve electrostatic forces, hydrogen bonds, Van der Waals forces and hydrophobic forces—in the primary interaction between antigen and antibody. Thus such interaction requires a close fit between an epitope and antibody like lock and key. The secondary interaction between antibody and multivalent antigen may result in a variety of consequences including precipitation of the soluble antigen, agglutination of particulate antigens and activation of complement. Such property of antigen-antibody interaction serves as a basis of various immunoassay techniques that are used in the diagnosis and monitoring of disease. This chapter deals with basis of antigen-antibody interactions and its application in immunoassay techniques.

II. PRIMARY INTERACTIONS AND ITS CHARACTERISTICS

Antigen-antibody reaction involves non-covalent interaction and the binding of antigen to antibody is reversible. The following intermolecular forces are associated with this interaction.

A. Physicochemical Nature

1. **Electrostatic bonds:** It involves interaction of ionic change on two groups such as (+ ve) charge of lysin NH_4^+ in the antibody molecule with (-ve) charge on aspartate residue of antigen (COO^-).
2. **Hydrogen bonds:** When antigen-antibody is in close contact, a weak hydrogen bond is formed between hydrophilic groups of the two molecules.
3. **Hydrophobic interactions:** Upto 50% of the energy of the total antigen-antibody bond is contributed by hydrophobic interaction. In this reaction, hydrophobic groups, such as side chains of amino acids (Leucine, valine and phenylalanine) tend to associate due to Van der Waals bonds.
4. **Van der Waals bonds:** These forces depend upon the electron clouds surrounding the antigen-antibody molecules.

All these interactions depend upon the close proximity of the antigen and antibody molecules.

B. Antibody Affinity

X-ray crystallographic studies of antibody V domains show that hypervariable regions are clustered at the end of Fab arms, i.e. particular residues in these regions interact with the antigen. Antibodies form multiple non-covalent bonds with antigens. **The strength of bond between Ag and an antibody is known as antibody affinity.** It is expressed as the sum of the attractive and repulsive forces. It may be calculated using law of mass action:

$$[Ab] + [Ag] \underset{\text{K back}}{\overset{\text{K foreward}}{\rightleftharpoons}} [Ab\ Ag]$$

$$Ka = 1/kd = \frac{[Ag \cdot Ab]}{[Ag] \cdot [Ab]}$$

Ka = Affinity constant; Kd = dissociation constant at equilibrium of the antigen-antibody reaction.

Two antibodies can have a similar affinity for antigen measured at equilibrium but one may have much slower on-rate ($K_{forward}$) as well as a proportionally slower off-rate (K_{back}).

C. Antibody Avidity

Antibodies are potentially multivalent as far as their reaction with antigen is concerned since each antibody molecules consists of four polypeptide chains with two antigen binding sites. The strength with which a multivalent antibody binds to a multivalent antigen is termed as avidity. Thus the avidity of an antibody for its antigen is a sum of the affinities of the individual antigen combining sites and even greater than that. Like affinity, avidity of an antibody for antigen is also related to the ratio of the rates of the foreward reaction for the formation of antigen-antibody complex to back reaction for decay of the complex as in case with antibody affinity.

D. Antibody Specificity and Cross-Reactivity

The specificity of antibody is another important property which should not be confused with affinity and avidity.

1. Antigen—antibody can show high level of specificity to a particular antigens. For example, antibody to cowpox virus will bind to cowpox virus only but not with poliovirus.
2. Antibodies recognize the overall conformation of antigen, i.e. they recognize overall shape of an epitope rather than a particular chemical residue.
3. It is highly likely that an antibody may have a spectrum of affinity towards antigens of similar topography (conformation) and may react to different antigens (cross-reactivity). For example, antibody to cowpox virus provides protection against smallpox.
4. Polyclonal antisera obtained from immunized animals may contain a heterogeneous antibody that shows more cross-reactivity to different antigens.

III. SECONDARY INTERACTIONS

Secondary interaction of multivalent antigen with divalent antibody leads to (i) agglutination reaction when the antigen is particulate such as cells (bacteria, RBC) and (ii) precipitation of antigen-antibody complex in case with soluble antigens.

A. Agglutination Reaction

i. The reaction of antibody with multivalent antigen leads to cross-linking of various antigen particles by the antibody.
ii. Such cross-linking causes clumping or agglutination of cells bearing antigens, e.g. RBC treated with blood group antibody or bacteria treated with antibody against its cell surface antigen molecules.
iii. Agglutination reaction may be used in determining antibody titre which is a measurement of the highest dilution of antibody in the sample that can agglutinate the antigen bearing cells.
iv. In clinical practice, it is used to determine blood group of Rh-type of patients. In this test, patient's blood is treated with anti-blood group (A, B, O) and anti-Rh antibody. The specific reaction of the blood group antigen with antibody leads to formation of visible clump of RBC. Such agglutination reaction is called direct agglutination. Since RBC is used

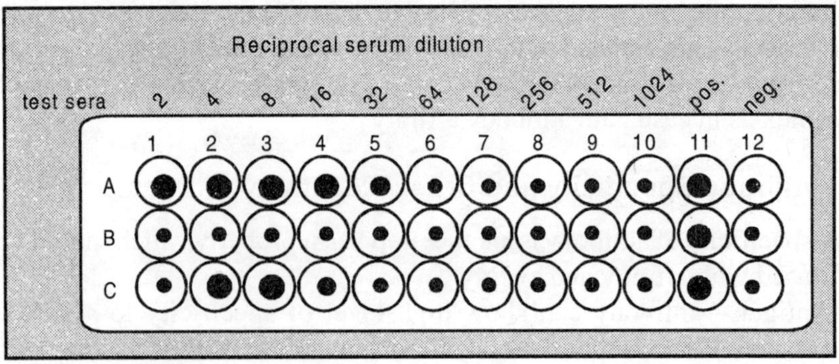

Fig. 11.1: Haemagglutination test against RBC. Fixed number of RBC is added to each well of the haemagglutination plate. Serial dilution of different antibodies along with positive control (row 11) and negative control (row 12). Agglutination is seen in wells A1–5 and C 1–4 as well as in positive control wells

for agglutination reaction in the case, it is also called hemagglutination (Fig. 11.1).

v. In case with soluble antigen, agglutination reaction can be carried out by binding antigen to solid support such as latex particle. For example, thyroglobulin can be attached to latex particle and when anti-thyroglobulin antibody is applied, there is agglutination of the thyroglobulin coated latex particles. Such agglutination reaction is called **direct** or **passive agglutination reaction.**

B. Precipitation Reaction

1. In contrast to agglutination reaction, precipitation reaction takes place in solutions.
2. In this interaction, bivalent antibody is linked to multivalent antigen, forming *lattice.* Since the antigen-antibody complex becomes heavier, it precipitates out.
3. A precipitin curve can be obtained using fixed quantity of antibody with varying concentration of antigen as shown in Figure 11.2. Such property of antigen-antibody interaction can also be used to determine antibody titre.
4. *Precipitation reaction can also be performed in semisolid medium such as agar gel,* in which carbohydrate matrix of the agar functions as container of buffer solution that fills the interstitial spaces of the agar and precipitation of antigen-antibody complex can be achieved. In this strategy, antibody is either mixed in the agar or put on the punched wells of the agar and antigen is applied in another well. Both the molecules diffuse towards each other and form precipitin band. Variation of agar gel technique has been used for determining antibody titre, quantitation, specificity, cross-reactivity and in other clinical applications as discussed later.

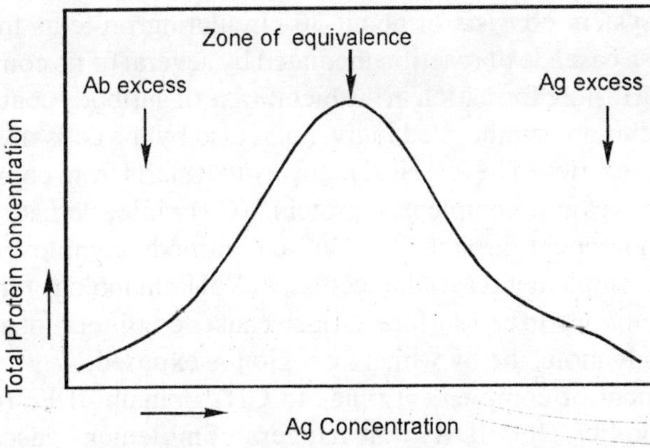

Fig. 11.2: Precipitin curve of Ag Ab reaction. Fixed amount of Ab is allowed to react with increasing concentration of Ag. At zone of equivalence all Ag and Ab are used up in the reaction giving peak of precipitation curve. At this stage neigther Ag nor Ab is detected in the supernatant. At Ab excess, all Ag are used up while Ab is detected and in case with Ag excess all Ab are used up and free Ag is detected in the supernatant

IV. BIOLOGICAL CONSEQUENCES OF THE ANTIGEN—ANTIBODY INTERACTION

Antigen-antibody interaction in the host leads to various biological reactions that are aimed to destruction and removal of antigen bearing parasites/microbes from the body. These consequences can be classified into three categories, (i) opsonization and phagocytosis, (ii) complement activation and (iii) neutralization of drugs and toxins.

A. Opsonization

After binding to particulate antigens such as bacteria, antibodies may unfold. Such unfolding exposes Fc region of the antibody molecules to which a large number of Fc receptor bearing cells can bind and mediate effector functions.

i. Three subtypes of Fc γR of IgG are present on different cell types having differential affinity to Fc region of IgG molecules.
ii. Fc γ RI is a high affinity receptor present on monocytes and macrophages while Fc γ R II and Fcγ R III are low affinity receptors present on phagocytic cells, B cells and NK cells respectively.
iii. Opsonization of bacteria leads to binding of phagocytic cells to unfolded Fc region of antibody and its subsequent phagocytosis.

B. Complement Activation

Fixation and activation of complements is one of the important consequences of antigen-antibody interactions.

Complement system consists of about 30 circulating proteins that when activated leads to a cascade of reaction mediated by several of its components ultimately leading to pore formation in the membrane of antibody coated target cells. These proteins are synthesized in liver and also by the cells involved in the inflammatory reaction. The classical pathway that starts from complement protein 1 (C1) involves upto complement protein 9 (C9) while alternate pathway starts from C3 component through C9. Antigen-antibody complexes are the major activator of complement cascade (see textbook of immunology for details).

i. Binding of antibody to cell surface antigen causes configurational change in the antibody molecule by which Fc region is exposed.
ii. C1q component of complement binds to CH2 domain of Fc region of bound antibodies (IgG/IgM) that triggers complement cascade and sequential activation of C1-C8 complement proteins.
iii. The terminal complement component C9 (poreforming unit) which polymerizes and forms transmembrane channel in the target cells.
iv. Ion passes through the channels and water enters the target cell resulting in swelling of the cell membrane, lysis and cell death.

C. Neutralization

i. Antibody binding to toxic chemicals and toxins leads to neutralization of its toxic activity.
ii. IgG is a very efficient antibody in neutralizing snake or scorpion venoms, tetanus and botulinus toxins mostly by blocking the tissue/protein binding part of the toxins, thus it is most commonly used in passive immunization.
iii. IgG antibody immobilizes bacteria and neutralizes viruses.

V. APPLICATIONS OF ANTIGEN-ANTIBODY REACTIONS

Antigen-antibody interactions have been used in various immunological techniques such as precipitation reactions, immunoassays including enzyme-linked immunosorbent assay (ELISA), radioimmunoassay, immunofluorescence, flowcytometry and various functional assays of B and T lymphocytes, NK cells, etc. Some of the important basic immunoassay techniques are given below.

PRECIPITATION REACTIONS (GEL TECHNIQUES)

In this assay, antigen-antibody interaction is allowed to take place in semisolid agar gel placed on the glass plates or microscope slides. Precipitin band can be stained with dyes and visualized by naked eyes.

Some of the gel techniques are as below:

A. Double Immunodiffusion

1. Developed by Ouchterlony, it is used to determine relative concentration

of antibody/antigen, to compare quality of antigen (purity) as well as to determine specificity/cross-reactivity of antibody molecules.

2. Antigen and antibody (diluted) are placed in the wells and allowed to diffuse towards each other, a precipitin line is formed at the zone of equivalence.

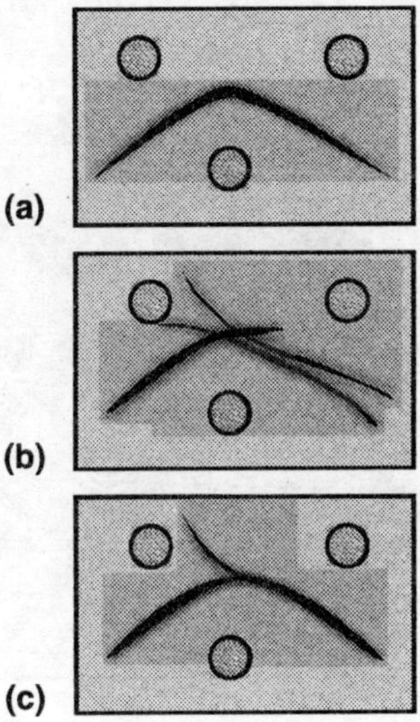

Fig. 11.3: Double immunodiffusion showing (a) complete identity (b) non-identity (c) partial identity

3. Patterns of precipitin line suggest complete identity, partial identity or non-identity of the antibody with antigens (Fig. 11.3).

B. Radial Immunodiffusion

1. It represents a variation of double immunodiffusion technique in which antibody is distributed uniformly in the agar surface while varying concentrations of antigen is placed in the well.
2. Antigen-antibody forms a precipitin ring around the wells (Fig. 11.4).
3. The distance of precipitin ring from the centre of the well is directly proportional to the antigen concentration.
4. This technique is used to quantitate concentration of different immunoglobulins in serum.

C. Immunoelectrophoresis

1. In immunoelectrophoresis, a mixture of proteins is separated on the basis of their mobility in an electrical field, by electrophoresis in gel.
2. Antibody is placed in a trough cut on the gel and is allowed to diffuse.
3. Antibody forms precipitin band with the separated antigens (Fig. 11.5) at the respective zone of equivalence.

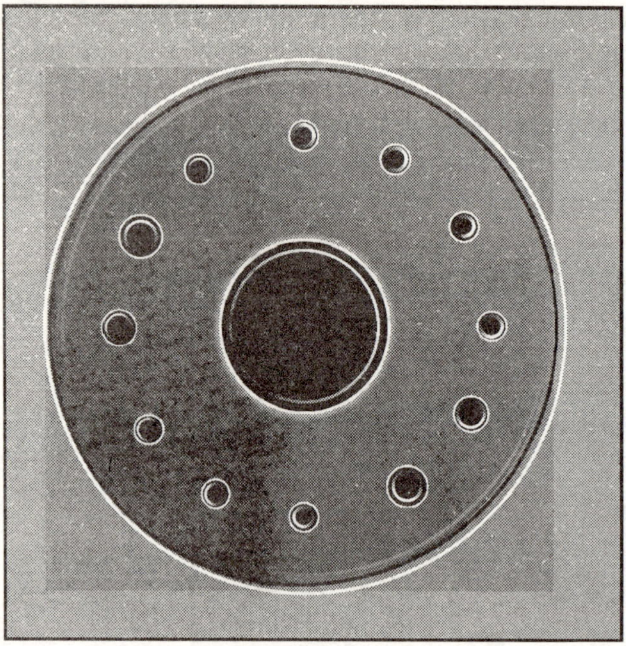

Fig. 11.4: Radial immunodiffusion for determination of IgG in serum. Antibody has been added in agar while diluted serum (Ag) is added to the wells. Precipitatin ring is visible around the wells

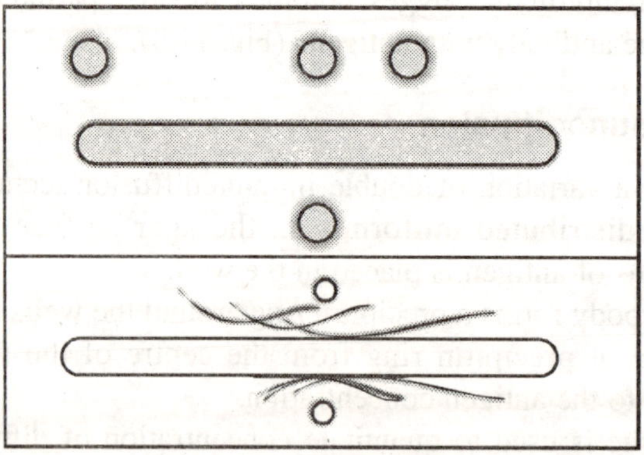

Fig. 11.5: Immunoelectrophoresis. Precipitation band is seen with separated antigens

4. The clinical application of this technique has been in the detection of presence absence or abundance of a particular serum protein in patients as compared to the normal individual.
5. **Burtons agammaglobulinemia,** which is characterized by deficiency of immunoglobulin, was discovered by this technique.

D. Rocket Electrophoresis

1. Rocket electrophoresis is an adaptation of the radial immunodiffusion technique used for quantitative measurement of many proteins including immunoglobulins.
2. The pH of the gel is chosen in such a way that the antibody remains immobile while antigens being negatively charged, move towards anode through the immobilized antibody in the gel.

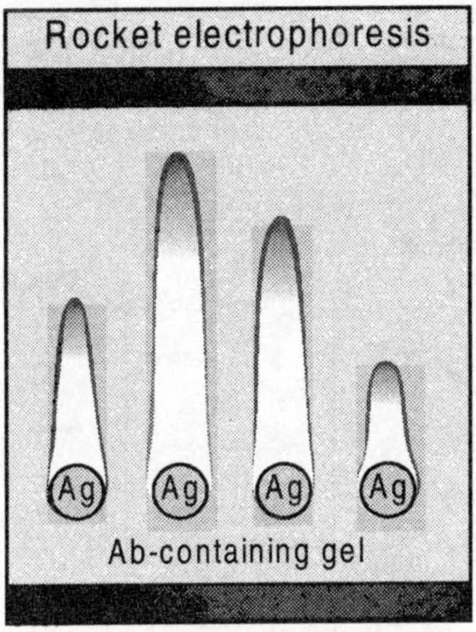

Fig. 11.6: Rocket electrophoresis

3. As the equivalence zone is reached, antigen-antibody reaction stops and an elongated precipitate is formed in the form of a rocket (Fig. 11.6).
4. The length of the rocket is proportional to the concentration of antigen in the wells.

E. Western Blot (Immunoblots)

1. In western blot, antigen is first separated from antigen mixture by using sodium dodecyl sulphate (SDS) gels (Fig. 11.7).
2. The separated material is then transferred on to the nitrocellulose membrane (protein binding sheet) by using electroblotting technique.
3. The transferred antigen may be reacted upon by labelled antibody and visualized.
4. Western blot technique can be used for identification of a large number of proteins in the sample, e.g. for the diagnosis of various infections such as HIV.

IMMUNOASSAY TECHNIQUES

Immunoassay techniques utilizes antigen-antibody interaction to detect antigens present in serum, body fluids or on cells. Cell surface receptor, proteins as well

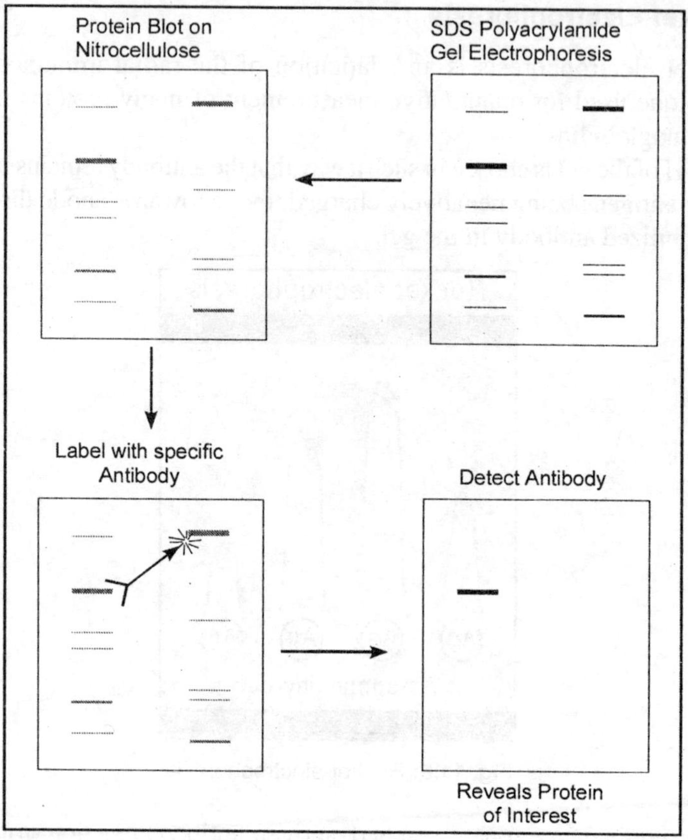

Fig. 11.7: Protein is separated by SDS-PAGE and blotted on the nitrocellulose paper. It is then incubated with enzyme labelled antibody. Colour reaction develops by incubating the complex with chromogenic substrate

as intracellular components in live cells and tissues can be detected by this technique for diagnosis, and monitoring of diseases or to assess immune reactions. Immunoassay techniques are highly sensitive compared to gel techniques. Following are some of the most commonly used immunoassay techniques.

A. RADIOIMMUNOASSAY (RIA)

In this assay, free antigens such as hormones, growth factors, etc. càn be measured using radionucleide bound to antigen/antibody for quantitative measurement of the protein. This technique is highly sensitive, therefore, it can be used to assay very small quantity of anylate (proteins) in the sample.

1. **Principle:** The assay is based on competitive binding of radiolabelled antigen to antibody in presence of unlabelled antigen. The quantity of free radiolabelled antigen is directly proportional to quantity of unlabelled antigen in the mixture.

2. **Procedure:**
 i. Mixture of sample and known amount of radiolabelled antigen is prepared, added to fixed quantity of antibody and incubated.
 ii. Antigen-antibody complex is precipitated with polyethylene glycol or by adding secondary antibody.
 iii. Radioactivity count in precipitate is taken which indicates amount of radio- labelled antigen incorporated in the complex (Fig. 11.8a).
 vi. A standard curve is plotted by counting radioactivity (% binding) incorporated in the complex against concentration of standard (unlabelled antigen) and quantity of anylate is extrapolated from the standard curve (Fig. 11.8b).
 v. This technique is widely used for assay of thyroid hormone, estrogen progesterone, human chorionic gonadotrophin, etc. in clinical samples.

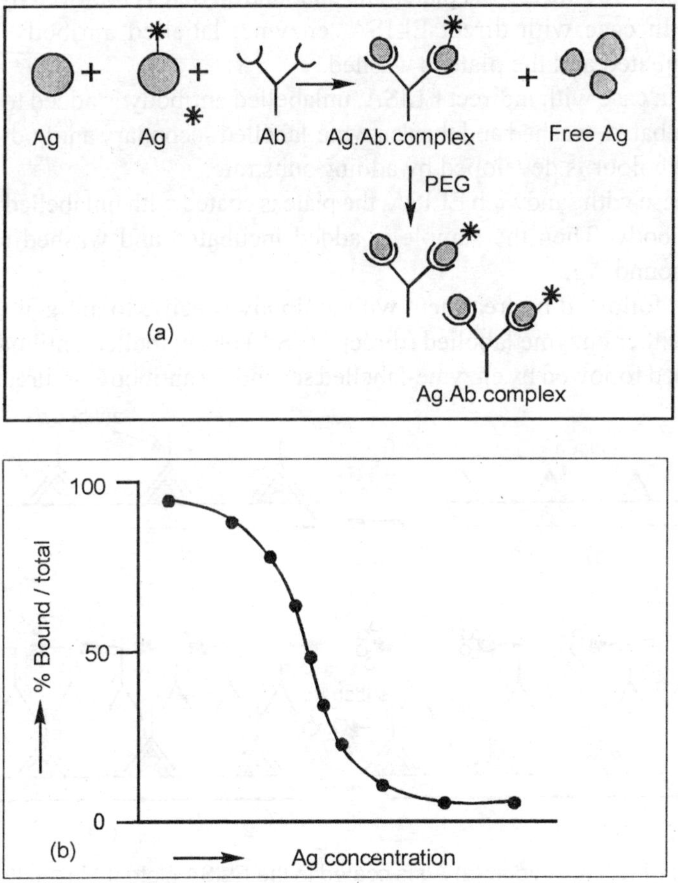

Fig. 11.8: Diagrammatic representation of radiommunoassay (a) Sample (Ag) and labelled antigen (Ag*) are incubated with fixed amount of antibody (Ab) and Ag Ab complex is precipitated. (b) A standard curve is plotted against bound/total versus, concentration of standard to obtain concentration of anylate in the sample

B. ENZYME-LINKED IMMUNOSORBENT ASSAY (ELISA)

It is also called solid phase assay as against RIA which is done in liquid. In case with ELISA or enzyme immunoassay (EIA), the antigen is bound to solid support such as plastic ELISA plates or microtitre plate and reacted with enzyme-labelled antibody. Colour of the reactions developed by incubating it with chromogenic substrate.

1. **Principle:** Antigen bound to solid support is treated with enzyme labelled antibody and bound enzyme is allowed to react with chromogenic substrate. The optical density of the reaction mixture is proportional to bound antibody/antigen concentration.
2. **Procedure**
   i. The anylate (antigen antibody which ever is to be assayed) is coated to the wells of the ELISA plate. The unbound anylate is removed by washing.
   ii. (a) In case with direct ELISA, enzyme, labelled antibody is added, incubated and the plate is washed.
   (b) In case with indirect ELISA, unlabelled antibody is added to the plate, incubated, washed and then enzyme-labelled secondary antibody is added, and colour is developed by adding substrate.
   iii. In case with sandwich ELISA, the plate is coated with unlabelled capturing antibody. Then the sample is added incubated and washed to remove unbound Ag.
   iv. It is followed by treatment with antibody specific to antigen that could be either enzyme labelled (direct ELISA) or unlabelled antibody may be added followed by enzyme-labelled secondary antibody (indirect ELISA).

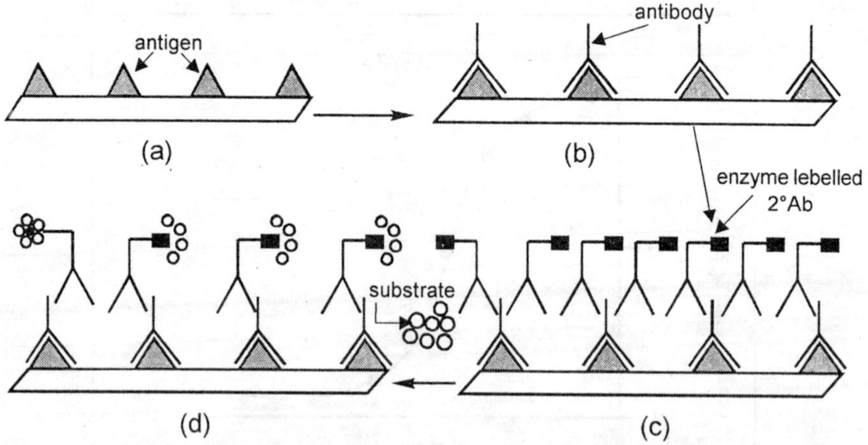

Fig. 11.9: ELISA technique (a) Ag is coated to the ELISA plate (b) incubated with the Ab (c) An enzyme labelled secondary Ab is added that binds to the primary Ab bound to Ag (d) Substrate is added and incubated to develop colour

   v. The plate is washed again to remove unbound antibody and colour is developed by adding chromogenic substrate followed by incubation.

vi. Optical density is taken by spectrophotometer and standard curve is plotted against OD vs concentration of standard. Concentration of the anylate is extrapolated from the standard curve (Fig. 11.9).

C. IMMUNOFLUORESCENCE

Immunofluorescence assay can be performed on live cell in cell suspension or in cryostat tissue sections. This technique is conveniently applied to observe distribution of proteins/receptors on the cell surface and also enumeration of different types of cells, e.g. T cells, B cells, or T cell subsets in peripheral blood or tissues. There has been tremendous development in immunofluorescence techniques to make it more sensitive and quantitative (Flow cytometry), dynamic (Confocal microscopy) and more invasive (Immunoelectron microscopy). Students may refer to different books on these techniques for further details.

1. **Principle:** The antigen present in cells and tissues can be visualized by treating cells/tissues with fluorochrome-labelled antibodies (monoclonal antibody is preferred) and the fluorescing cells can be counted under fluorescence microscope.
2. **Procedure**
   1. In *direct immunofluorescence,* cells are treated with flurochrome-labelled monoclonal antibody, incubated, washed and fixed with 1% paraformaldehyde. Then the fluorescence is observed under fluorescence microscope.

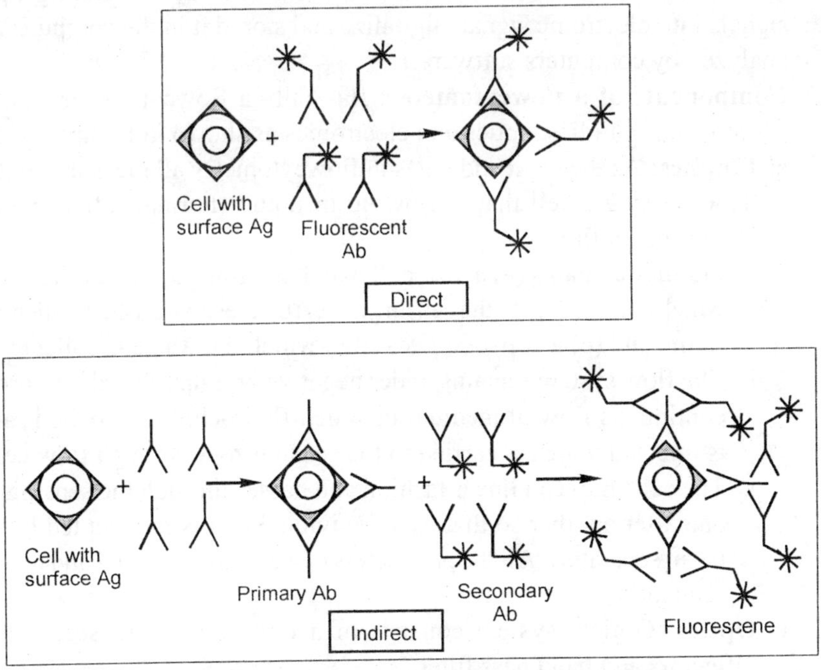

Fig. 11.10: Direct and indirect immunofluorescence
Note amplification of fluorescence signal in indirect labelling

2. In *indirect immunofluorescence* (most preferred), the cells are first treated with unlabelled specific monoclonal antibody followed by fluorochrome-labelled polyclonal secondary antibody and visualized under microscope after washing and fixation (Fig. 11.10).

3. **Applications**
   1. Diagnosis of leukemia and lymphoma
   2. Monitoring HIV infection by CD_4^+ T cell counts.
   3. Diagnosis of autoimmune diseases (antinuclear antibody)
   4. Lymphocyte and T-cell subset analysis in case with infections, immunodeficiency, and other diseases, etc.

D. FLOWCYTOMETRY

Flowcytometry is a highly sensitive semiautomated technique to detect and measure various receptors, intracellular proteins (cytokines), DNA, etc. in cells. It can simultaneously detect many different components of a large number (10,000-100,000) of cells in a very short time. With an additional cell sorting attachment it can sort out homogenous cells from heterogenous samples. Therefore, flowcytometry has a wide application in biology and medicine.

1. **Principle:** The technique relies on measuring scattered and fluorescent optical signals from fluorochrome-labelled cells by converting optical signals into electronic signal, digitalize and stored it in the computer to be analyzed by computers software.
2. **Components of a flowcytometer**: Basically a flowcytometer has three components: fluidics, optics and electronics as shown in Figure 11.11.
   a. **Fluidics:** As the name indicates in flowcytometry all measurements are done on single cell that is flowing in liquid (normal saline) medium called sheath fluid.
      i. Fluidics contains (a) a *flow* cell which is a thin capillary tube through which cells passes to the sensing zone. (b) a reservoir filled with normal saline and (c) a collecting vessel in which the waste is collected.
      ii. The flow system remains under negative pressure in order to provide continuous flow of medium in which fluorochrome-labelled sample is injected which is carried to the sensing zone through flow cell.
      iii. The cells move in linear fashion and passes through the sensing zone one after another so that each individual cell is intercepted by laser beam separately. All the parameters of a cell are observed and recorded sequentially.
   b. **Optics**: Optical system consists of a light source, lenses, dichroic reflectors and band pass filter.
      i. *Laser* provides the best light source as it has constant intensity, parallel beam and it can be focussed to a very small point (cells in this case).

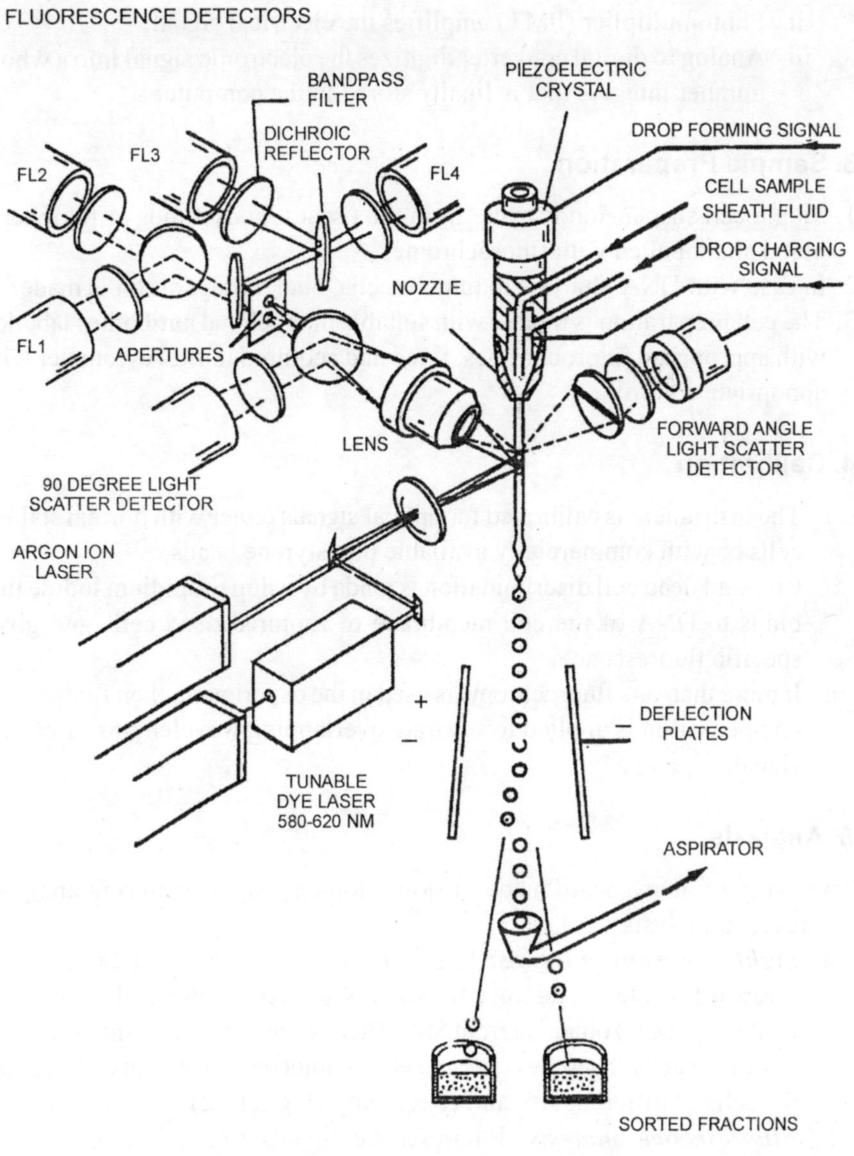

Fig. 11.11: Schematic diagram of flowcytometer

ii. *Lenses* are used to focus the scattered light to optical sensors.

iii. Dichroic reflector and band pass filter are used to sort and collect optical signals based on its wavelength.

c. **Electronics:** Electronics consists of a photocathode tube, photomultiplier (PMT) and an analog to digital converter.

i. Photocathode converts optical signals (photons) coming from the cell into electrical (electrons) signals.

ii. Photomultiplier (PMT) amplifies the electrical signal.
iii. Analog to digital converter digitizes the electronic signal into a whole number integers that is finally stored in the computer.

3. Sample Preparation

1. Single cell suspension is made from the tissues, body fluids or peripheral blood and labelled with fluorochrome.
2. In case with DNA analysis in tumour cells, nuclei preparation is made.
3. The cell preparation is treated with suitable monoclonal antibodies labelled with appropriate fluorochromes, fixed and acquired in flowcytometer with appopriate controls.

4. Calibration

i. The instrument is calibrated for optical signals either with normal stained cells or with commercially available polystyrene beads.
ii. Live and dead cell discrimination is made by using propidium iodide that binds to DNA of the cell membrane of ruptured dead cells and gives specific fluorescence.
iii. If more than one fluorochrome is used in the experiment, then fluorescene compensation is applied to separate overlapping wavelengths of optical signal.

5. Analysis

Two types of analysis are made in flowcytometry, light scattering analysis, fluorescene analysis.

a. ***Light scattering analysis:*** Foreward scattering (FSC) obtained from foreward scatter detector corresponds to size of the cell while side scattering (SCC) obtained from SSC detector corresponds to the complexicity of the cells. These two parameters combined together can differentiate the cells of different size and complexity (Fig. 11.12).
b. ***Fluorescence analysis:*** Fluorescence signals from different cellular constituents as detected by fluorescence detector is analyzed after the light scattering analysis. In this case, the desired cell is electronically gated on the basis of light scattering analysis (Fig. 11.13a) and further fluorescence analysis is done on the gated populations for various parameters (Fig. 11.13b).

6. Data Display

The collected data can be displayed basically in the following four formats.

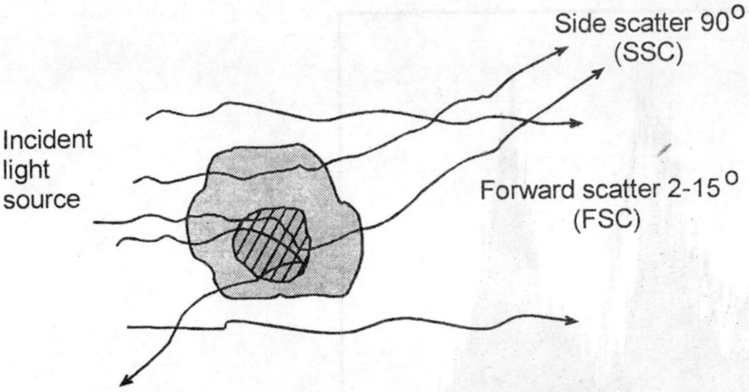

Fig. 11.12: Ligth scattering depends upon shape and internal architecture of the cell

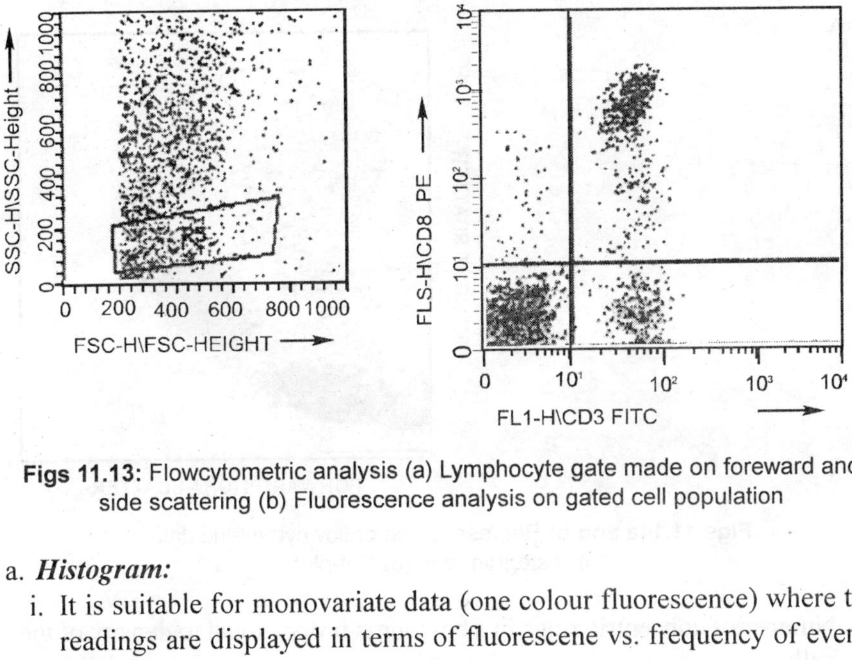

Figs 11.13: Flowcytometric analysis (a) Lymphocyte gate made on foreward and side scattering (b) Fluorescence analysis on gated cell population

a. ***Histogram:***
   i. It is suitable for monovariate data (one colour fluorescence) where the readings are displayed in terms of fluorescene vs. frequency of events (Fig. 11.14a).
   ii. It is commonly used for DNA analysis of tumour cells.
   iii. Quantitation of the parameters can be made in histogram plot.

b. ***Dot display:***
   i. It is most commonly used for multivariate (multiparametric) analysis where more than one parameters of the cells are measured.
   ii. In this format, each event is represented by a dot.
   iii. Quantitative analysis of different parameters can be performed as shown in Figure 11.14b.

c. ***Contour plot:***
   i. Each event is represented in the form of a contour.

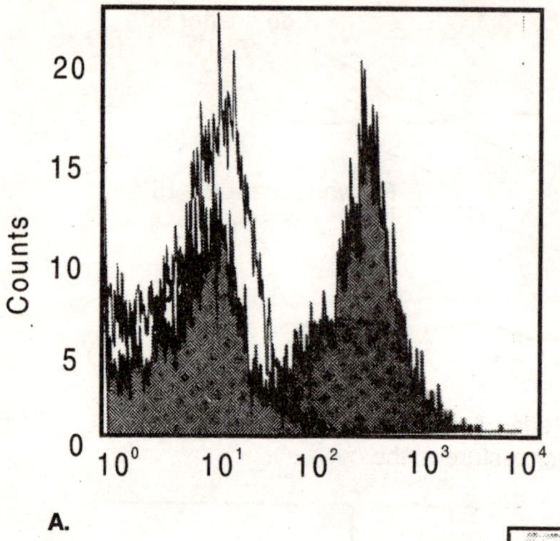

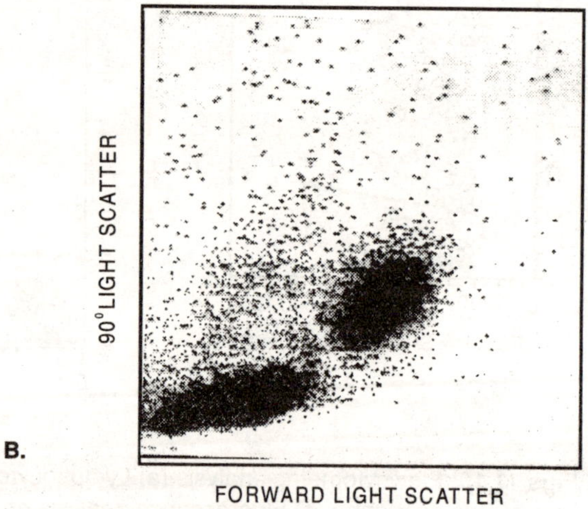

Figs 11.14a and b: Representation of flowcytometric data (a) Histogram and (b) Dot-plot

ii. Number of concentric rings in a contour is proportional to density of the cell.
iii. Quantitative analysis of parameters can be made.
iv. A rare cell subset can be identified easily even if it is present in a very low frequency (Fig. 11.14c).

d. ***Isometric plots:***
i. It represents hills and valley and shows frequency of distribution of a particular parameter.
ii. Many parameters can be visualized (Fig. 11.14d) in this format.
iii. Isometric plots are not suitable for quantitative measurements.

7. Applications

Followings are the potential applications of flowcytometry.

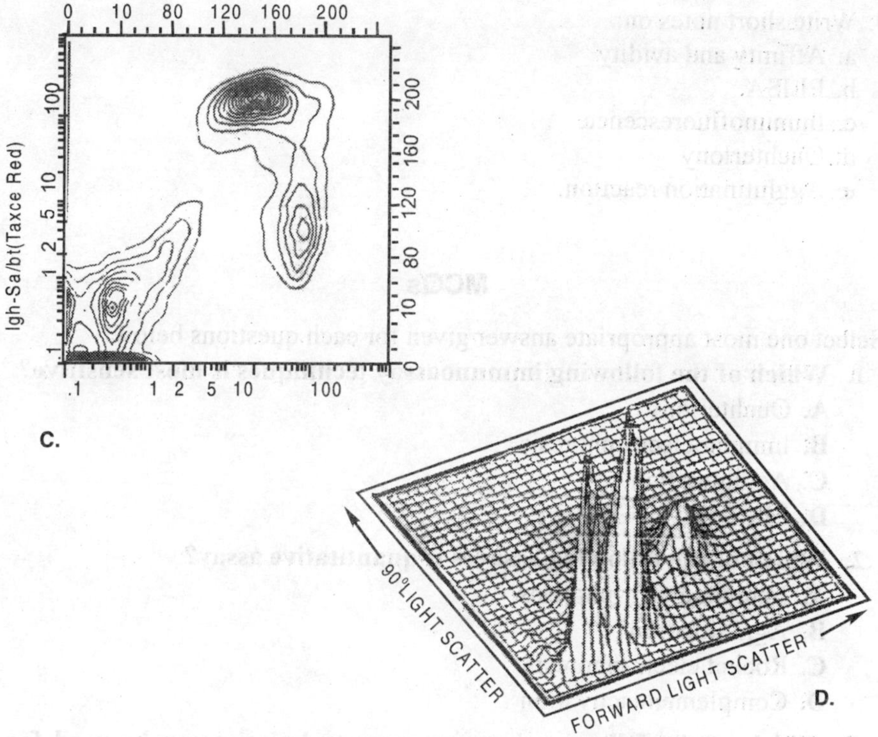

Figs 11.14c and d: Representation of flowcytometric data
(c) Contour plot and (d) Isometric plot

1. DNA ploidy analysis of tumor cells, apoptosis, cell cycle, etc.
2. Diagnosis of leukemia and lymphoma.
3. Monitoring of HIV by CD_4^+ T-cell subset analysis.
4. Analysis of lymphocyte subsets, receptors and intracellular proteins such as cytokines.
5. Reporter gene assay in molecular biology.
6. Cell activation and signal transduction marker assays.
7. Chromosomal analysis and karyotyping
8. Detection of minimal residual disease in patients undergoing cancer therapy.

STUDY QUESTIONS

1. What are the physical characteristics of the antigen-antibody interactions? Describe in brief the biological consequences of antigen-antibody interactions.
2. Name various immunoassay techniques based on antigen-antibody interaction and describe one of them in details.
3. What is flowcytometry? Describe various components of a flowcytometer and their uses.

4. Write short notes on:
   a. Affinity and avidity
   b. ELISA
   c. Immunofluorescence
   d. Ouchterlony
   e. Agglutination reaction.

MCQs

Select one most appropriate answer given for each questions below:

1. **Which of the following immunoassay techniques is most sensitive?**
   A. Ouchterlony
   B. Immunoelectrophoresis
   C. Agglutination
   D. ELISA
2. **Which of the following assayes is quantitative assay?**
   A. Immunoelectrophoresis
   B. Agglutination
   C. Rocket electrophoresis
   D. Complement activation
3. **Which of the following immunoassay techniques can be used for diagnosis of HIV?**
   A. Immunoelectrophoresis
   B. ELISA
   C. Agglutination
   D. Complement fixation
4. **Which of the following techniques can be used for monitoring HIV infection?**
   A. Agglutination
   B. Complement fixation
   C. Flowcytometry
   D. EIA
5. **The antibodies that fixes complements are:**
   A. IgG and IgM
   B. IgG and IgA
   C. IgA and IgM
   D. IgD and IgE
6. **In RIA:**
   A. Radio-labelled Ab is immobilized on the solid support.
   B. Radio-labelled Ag is immobilized on the solid phase

C. Free Ag is precipitated out
D. Ag Ab complex is precipitated.

7. **The forces involved in antigen-antibody interaction:**
   A. Depend upon the presence of complement
   B. Involve covalent interactions
   C. Operate between antigen in solution and antibody within the cytoplasm of plasma cells
   D. Involve electrostatic interaction
8. **The specificity of the Ag-antibody reaction depends upon:**
   A. Molecular weight of the antigen
   B. Charge of the antibody
   C. Ig isotype
   D. Confirmation of the reactants

---

1 D 2 C 3 B 4 C 5 A 6 D 7 D 8 D

12

Biology of Lymphocytes and Generation of Humoral Immunity

I. INTRODUCTION

In chicken, it was observed that if one of its organ *bursa of Fabricius* which is present near cloaca, is surgically removed, the chicken fails to develop antibody response. It was later shown that bursa of fabricius is important for development and maturation of antibody producing lymphocytes, thus such cells were named as Bursa of Fabricius-derived lymphocytes (in short, B lymphocytes) or B cells. In human and mice, B lymphocytes develop and mature in bone marrow throughout the life. These are a heterogenous group of short-living cells that differentiate into antibody producing plasma cells upon antigenic stimulation and can be easily recognized by its membrane bound immunoglobulin molecules. These cells are originally produced in large numbers but only few of them survive in the peripheral circulation. Once mature it acquires all the cardinal features of adaptive immune response such as specificity, memory, self-non-self discrimination, and diversity. A mature B cell produces only one heavy chain and one light chain of surface immunoglobulin molecule, thus it bears an antigen receptor of single specificity. Interaction of mature B cells with antigen, T_h-cells and macrophage induces activation and differentiation of B cell clone to produce a large number of plasma cells capable of producing antibody molecules specific to same antigen. Thus B cells generate humoral immune response.

II. ONTOGENY OF B LYMPHOCYTES

B cell development and maturation is a sequential process which begins from the haematopoietic stem cells in the bone marrow and foetal liver by antigen independent mechanisms and terminates in the peripheral circulation where mature B cells encounter antigens and differentiate into antibody secreting

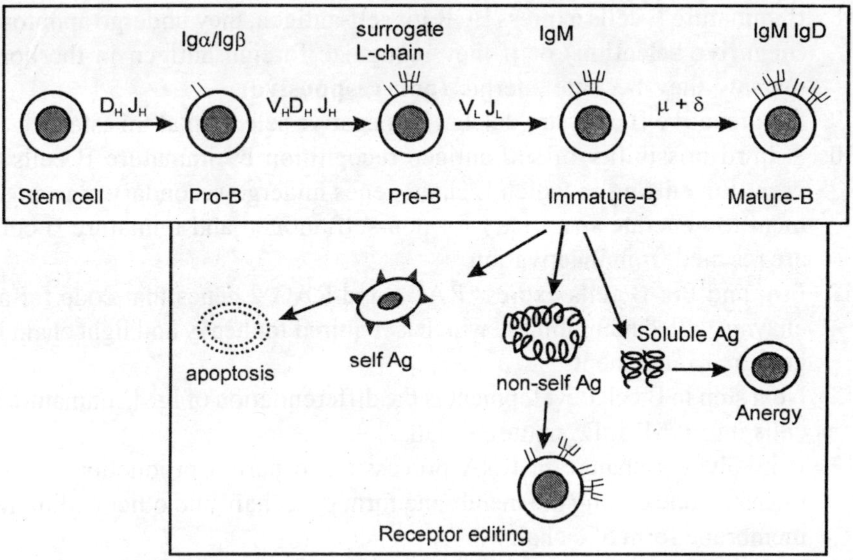

Fig. 12.1: Stages in B cell ontogeny in the bone marrow

plasma cells with short life span and memory cells with long life. The key feature of B cell development is an orderly sequence of Ig-gene rearrangement (Fig. 12.1).

1. The first step in B cell ontogeny is the differentiation of lymphoid stem cells into a distinct B cell progenitors (pro-B cells).
2. Pro-B cells interact with stromal cells in the bone marrow through c-kit receptor, rearrange Ig heavy chain D_H gene to J_H segment, express IL-7R and differentiate into pre-B cells.
3. Cytokine IL-7 secreted by stromal cells of bone marrow activates pre B cells to synthesize μ chain of immunoglobulin by rearranging VDJ segment of heavy chain gene and progress to IgM expressing immature B cell stage. If pre-B cells fail to rearrange Ig heavy chain gene, they undergo apoptotic death.
4. Pre-B cells express surrogate light chains, μ chain and two other closely associated transmembrane molecules known as Igα and Igβ, which are disulphide-linked to each other and is known as B cell receptor (BCR).
5. Development of pre-B cells into immature B cells requires a productive Ig-light chain gene rearrangement that first begins with κ-chain followed by λ-chain gene. In case with non-productive L chain gene—rearrangement the cells undergo apoptosis.
6. Successful rearrangement of κ and λ-chain genes drives pre-B cells into immature B cell stage which acquires antigenic specificity determined by functional heavy chain VDJ sequence and light chain VJ sequence.
7. Immature B cells express membrane IgM molecules on its surface.

8. If immature B cells express BCR for self-antigen, they undergo apoptosis (**negative selection**) or if they recognize foreign antigen in the bone marrow they become anergic (**non-responsive**).
9. Autoreactive B cells are deleted by negative selection at this stage.
10. A third possibility of self-antigen recognition by immature B cells is **receptor editing** in which L chain genes undergo secondary rearrangement to generate specificity for non-self-antigen and immature B cells are rescued from inactivation.
11. Pro- and Pre-B cells express RAG1 and RAG 2 genes that code for an enzyme called recombinase which is required for heavy and light chain Ig gene rearrangement.
12. Next step in B cell development is the differentiation of IgM^+ immature B cells into IgM^+ IgD^+ mature B cells.
13. It involves a change in RNA processing to permit production of two mRNAs, one coding the membrane form of μ chain and other coding for membrane form of δ chain.
14. In δ chain knock out mice, it has been shown that IgD is not essential either for B cell development or for antigen responsiveness but it may be responsible for silencing autoreactive signal.
15. After acquiring a complete IgM and IgD on its surface the mature B cells become Ag specific and migrate out of the bone marrow to peripheral lymphoid organs and peripheral circulation.
16. Development of B-lymphocytes beyond pre-B cell stage requires a kinase associated with the pre-BCR called B-cell tyrosine kinase (btk). In humans, mutations in btk gene results in **X-linked agammaglobulenemia,** characterized by failure of B-cell maturation.

III. CELL SURFACE MARKERS

The developmental stages of B cells are characterized by changing pattern of expression of cell surface molecules.

1. Pro-B cells express common leukocyte marker, CD45R which is a form of protein tyrosine phosphatase, a growth promoting ligand c-kit and CD 19 and $Ig\alpha/Ig\beta$ chains.
2. Pre-B cells loose c-kit and begin to express CD 25 (α-chain of IL-2R).
3. Immature B cells loose CD25 and cease to express CD 25 as well but express IgM.
4. Mature B cells express large number of proteins which can be grouped under (a) Signal transduction receptor (BCR) (b) BCR associated co-receptor molecules and (c) Molecules responsible for T cell, B cell and antigen presenting cell interactions (Fig. 12.2).
5. **B cell receptor complex (BCR):** BCR consists of Ag binding membrane immunoglobulin molecules (m Igs) and signal transducing Ig-α/β hetero-

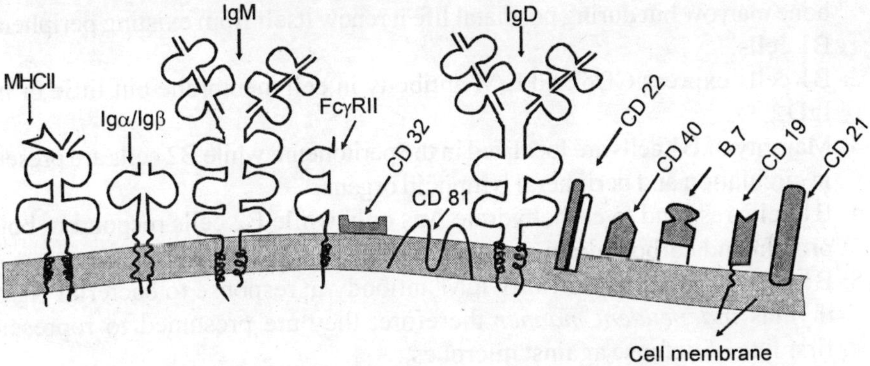

Fig. 12.2: Receptors present on mature B cells

dimers. The Igα chain consists of a long cytoplasmic tail containing 61 amino acids and cytoplasmic tail of Ig-β chain consists of 48 amino acids. The cytoplasmic tail of Igα/β chain also consists of 18 residue motif known as **immunoreceptor tyrosine-based activation motifs (ITAMs)** which phosphorylate its tyrosine residue during Ag mediated cross-linking of BCR. It is the early event of signal transduction pathway leading to activation and differentiation of mature B cells into effector plasma cells and memory cells.

6. **B cell coreceptors** are other molecules that play an important role in Ag cross-linking and signal transduction pathway of B cells. It consists of a common CD 19 molecule which is present in all the stages of B cell development. Monoclonal Ab to CD19 is commonly used for immuno-phenotyping of B cells. Other B cell co-receptor molecules are CD21 which is a receptor for complement component C3d which binds to microbial pathogen. It also acts as a receptor for *Epstein-Bar Virus* (EBV) which is implicated in B cell lymphoma. CD 32 coreceptors of B cell regulates negative feedback mechanism of antibody response, is the low affinity receptor for Fc region of IgG (Fc γ RII) while CD81 coreceptor is involved in signal transduction.
7. **Receptors involved in T and B cooperation:** B cells are also antigen presenting cells therefore, it possess certain molecules also present an other APCs. These include MHC II molecules, co-stimulatory molecules such as B7 (CD 80/86) and CD40. Mutation in CD40 Ligand (CD154) leads to X-linked hyper-IgM syndrome.

IV. B CELL SUBSETS

Classical B cells that predominate in the blood, lymph and secondary lymphoid organs are called B2 cells in contrast to B1 cell subset that forms about 5% of the total B cell population. Followings are the distinguishing features of B1 cells as opposed to B2 cells.

1. Like B2 cells, during fetal life B1 cells originate from the stem cells of bone marrow but during postnatal life it renew itself from existing peripheral B1 cells.
2. B1 cells express CD5 and IgM antibody in cell membrane but little or no IgD.
3. Majority of B1 cells are localized in the peritoneum while B2 cells are present in circulation and peripheral lymphoid organs.
4. B1 cells respond to carbohydrate Ags only while B2 cells respond to both protein and carbohydrate antigens.
5. B1 cells produce low affinity IgM antibody in response to bacterial Ag in thymus *independent manner* therefore, they are presumed to represent first line of defense against microbes.

V. B CELL ACTIVATION, PROLIFERATION AND DIFFERENTIATION

Once the mature B cells migrate to peripheral lymphoid organ and bloodstream, they require further activation signal by antigen-BCR interaction in order to proliferate and differentiated into Ab synthesizing plasma cells and memory cells. The signal transduction mechanism in B cell activation is similar to that of T cells and has been represented in Figure 12.3. Details have been discussed in Chapter 13.

1. B cell signal transduction pathway begins with antigen cross-linking to BCR resulting in phosphorylation of ITAMs.
2. Phosphorylated ITAMs form docking site for Src family of tyrosine kinases activating Syk kinases.
3. Phosphorylation of BCR generates second messengers namely inositol triphosphate (IP3) and diacyl glycerol (DAG). IP3 causes opening of Ca^{++} channel in the cell membrane, resulting in influx of Ca^{++} in the cell as well as release of Ca^{++} from endoplasmic reticulum, while DAG activates protein kinase C (PKC).

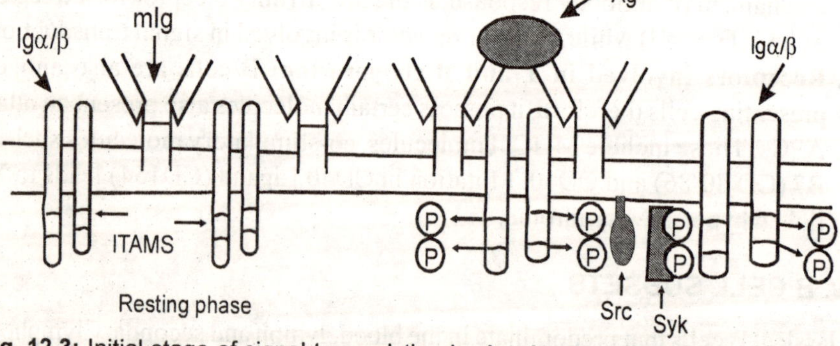

Fig. 12.3: Initial stage of signal transudction by Ag. Ab cross-linking-leads to phosphorylation of ITAMs which when phosphorylated, forms docking site for members of Src family of tyrosine kinases (Lck, Fyn, etc)

4. There is also activation of G-protein which is mediated by signals from BCR.
5. Such activation leads to expression of transcription factors that follows transcription of specific genes.
6. Signal transduction drives the cells into cell cycle.
7. Once activated, B cells undergo *affinity maturation* through a complex process of *somatic hypermutation* of Ig genes and differentiates into antibody secreting plasma cells and memory cells in the germinal centres of peripheral lymphoid organ.

A. Immunoglobulin Synthesis and Class Switching

Activation of B cells proceeds by two different routes depending upon nature of antigens. Bacterial lipopolysaccharide (LPS) activates B cells in **T cell independent** manner to produce Ab without co-operation of CD_4^+ T cells. Hence, a large number of cells can be activated by LPS and it is also called polyclonal B cell mitogen. However, mostly protein Ags requires physical contact of CD_4^+ T cells to B cells in antibody synthesis. Therefore, such antigens are called **T cell dependent antigens.**

1. In the early phase of activation, IgM class of antibody is produced by activated B cells.
2. In the later stage, T and B cell interaction leads to class switching of IgM Ab to IgG, IgA or IgE.
3. Activated B cells express receptors for IL-2, IL-4 and IL-5 which can bind to respective cytokines produced by activated $CD4^+$ T cells.
4. Binding of IFN-γ and IL-2 to its receptors leads to production of IgG while IL-4 and IL-5 binding on B cells leads to Ig class switching from IgG to IgE.
5. The signals produced by cytokine-receptor interaction also leads to differentiation of B cells into memory cells and plasma cells.

B. Affinity Maturation and Generation of Memory Cells

1. In response to T-dependent antigens, B cells migrate to **germinal centres**, where they down regulate expression of MIG and undergo rapid cell division.
2. During this huge expansion of B cells undergo somatic **hypermutation** of Ig variable region genes to produce **higher affinity** immunoglobulins on its surface and cell division stops. At this stage, B cells require minimal amount of Ag exposure to be activated.
3. B cells move to the lighter zone of the germinal centre and interacts with CD_4^+ T cells and follicular dendritic cells.
4. At this stage B cells bearing high affinity membrane Ig passes through antigen-mediated selection a phenomenon called '**affinity maturation**'.

These cells called plasma cells may undergo Ig class switching and exit lymphoid organs to produce antibody at the relevant site.

5. Another set of B cells differentiate into **memory B cells** and leave lymphoid organ, migrate to tissue where they can live for longer period waiting for re-exposure of the same antigen to repeat the cycle.
6. Memory B cells are long lived, non-proliferating cells expressing higher levels of CD44 and surface Ig/IgA or IgE instead of IgM and IgD. Upon reexposure of the same Ag these cells differentiate into Ab producing plasma cells and memory cells again.

VI. REGULATION OF B CELL DEVELOPMENT

1. B cell development is regulated by a number of *transcription factors* such as NF-kappaB (KB), BSAP, Ets-l, c-Jun, I Karos, etc. that interact with the promoter region of the gene.
2. Among these, BSAP (B cell-specific activation protein) seems to be master regulator of B cell development and differentiation.
3. Binding of BSAP to 3 α enhancer appears to influence B cell development by preventing binding of other transcription factors.
4. In case with low BSAP levels NF-αP binds to enhancer 3α resulting in the increased transcription of the immunoglobulin heavy chain gene and formation of plasma cells.

STUDY QUESTIONS

1. With suitable diagram illustrate different stages of B cell development.
2. Write short notes on
   a. Affinity maturation
   b. B cell receptor
   c. B1 cells.

MCQs

Select one most appropriate answer given for each questions below:

1. B cells may be enumerated by immunofluorescence using:

A. FITC, CD19 monoclonal antibody
B. FITC, IgM monoclonal antibody
C. FITC, FcγRII monoclonal antibody
D. FITC, CD 25 monoclonal antibody

2. Stem cells express:

A. c Kit
B. IgM and IgD
C. IgM
D. IgD

3. Expression of high affinity receptors to self Ag during B cell development leads to:

A. Clonal selection
B. Clonal deletion
C. Clonal anergy
D. Affinity maturation

4. Surrogate light chain is expressed in:

A. Pro-B cells
B. Pre-B cells
C. Immature B cells
B. Mature B cells

5. Rearrangement of μ and δ chain drives B cells to:

A. Pro-B to Pre-B stage
B. Pre-B to immature B stage
C. Immature B to mature B stage
D. Apoptosis of B cells

6. Affinity maturation of B cells takes place in:

A. Bone marrow
B. Germinal centre
C. Peripheral circulation
D. Afferent lymphatics

7. Which of the following is expressed on the surface of the mature B lymphocytes:

A. CD40
B. CD32
C. IgM and IgD
D. All of the above

8. Immature B lymphocytes:

A. Have rearranged only D and J gene segment
B. Are progenitors of T and B cells
C. Express both IgM and IgO on their surface
D. Are at the stage of development where contact with antigen may lead to anergy.

9. Which of the following is not found on a memory B cell:

A. Igα and Igβ
B. γ heavy chians
C. Epsilon (ε) heavy chain
D. Surrogate ligand.

10. Antigen binding to the BCR:

A. Tranduces a signal the Ag binding chains
B. Invariably leads to B cell activation
C. Tranduces a signal through the Igα and Igβ molecules
D. Leads to synthesis of cytokines.

---

1 A 2 A 3 B 4 B 5 C 6 B 7 D 8 D
9 D 10 C

13

Biology of T lymphocytes and Generation of Cellular Immunity

I. INTRODUCTION

T lymphocytes play a central role in the cellular arm of adaptive immune response **exclusively** to protein antigens. Progenitors of these cells develop in the bone marrow stem cells, migrate to thymus mature and acquire specificities for antigens (called **T cell repertoire**) under the influence of **thymic micro-environment**. Therefore, these cells are named as thymus-dependent (T) lymphocytes or T cells. Unlike B cells, T cells recognize antigens in association of MHC molecules therefore, these cells are also called MHC-restricted cells. The two major subsets of T cells, CD_4^+ T-helper cells and CD_8^+ cytotoxic T cells are restricted to MHC II and MHC I molecules respectively, together they form a strong resistance force to exogenous bacteria and parasites as well endogenous virus infected and tumor cells. Similar to B cells, T cells possess antigen receptors (TCR) on their surface with more advanced features than BCR. TCR belongs to immunoglobulin superfamily. Therefore, it seems to have evolved from the immunoglobulin genes only, but have acquired more diversity in order to counteract a plethora of antigens present in the nature. The complexity of antigen recognition by T cells, such as MHC restriction and costimulation has been acquired probably to prevent the host from its own destruction by this powerful army of the immune system.

T cells play a crucial role in the generation of both cellular and humoral arms of the immune system. CD_4^+ T cell subsets provide help to B cells in antibody production as well as in amplification of immune functions by releasing a number of cytokines, on the other hand CD_8^+ T cell subsets are cytotoxic to virus infected cells and tumor cells. On the basis of T cell receptors (TCR) two types of T cells have been identified, αβ **TCR bearing versatile T cells** that are commonly present in peripheral circulation and γδ **TCR bearing T**

cells which have restricted distribution in peripheral tissues. This chapter refers to mainly characteristics and functions of $\alpha\beta^+$ T cells.

II. ONTOGENY OF T LYMPHOCYTES

Similar to B cells, T cell precursors develop from bone marrow stem cells however, unlike B cells further development and acquisition of T cell repertoire take place in the **thymus**, a lymphoid organ located in the upper thoracic region. It has been observed in human with **DiGeorge syndrome** and in **nude mice** (where thymus fails to develop) that their peripheral circulation has normal number of B cells but T cells are absent, suggesting that the thymic microenvironment plays an important role in the development and maturation of T cells.

The stages of T cell development has been shown in Figure 13.1.

1. In the early stage, T cells arriving thymus possess c-kit, the receptor for stem cell growth factor, CD44, an adhesion molecule required for homing of precursor cells into the thymus and CD25, α-chain of IL-2R. However, TCR genes remain unarranged and is not expressed.
2. These cells do not express either CD4 or CD8 molecules therefore these are called **double negative** cells and early thymocytes.
3. Most double negative thymocytes further express RAG1 and RAG2 recombinase gene which is required for the rearrangement of TCR α, β and γ, δ chains.

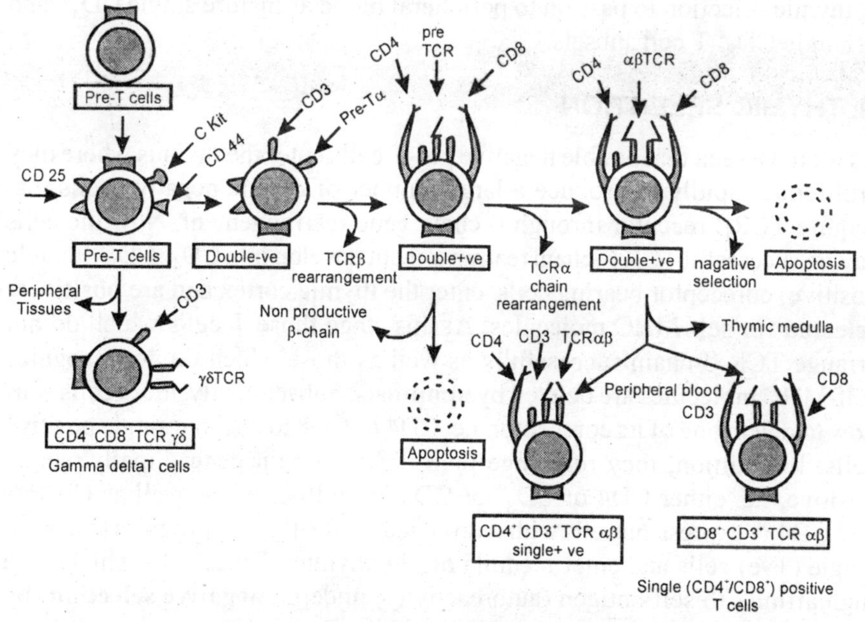

Fig. 13.1: Stages of T cell development

4. About 5% of thymocytes rearrange δ/γ TCR chains and differentiate into CD_3^+ γδ$^+$ T cell subsets while majority of thymocytes down regulate c-kit, stop proliferation and begin to rearrange TCR β gene and express TCR β-chain. Those cells which fail to rearrange TCR β chain undergo apoptosis and die off.
5. Subsequently, TCR β chain combine with TCR α chain and associate with CD3 molecule to form pre-TCR (CD_3^+ TCR αβ) complex.
6. Since CD3 molecules have a long cytoplasmic tail they can transmit intracellular signal upon binding to some of the ligands. Ligands present on the thymic epithelium lead to activation and proliferation of these cells and induce its progression to CD_4^+ CD_8^+ (double positive) stage.
7. Double positive cells now express both CD_4 and CD_8 co-receptors along with CD_3 TCRαβ complex. Though rearrangement of TCR β chain is complete and arrested, α-chain gene rearrangement does not take place probably because RAG2 proteins are degraded in the rapidly proliferating cells.
8. When sufficient number of double positive cells are attained, they stop dividing and accumulate RAG-2 protein, increase diversity of α-chain by differential re-arrangement of α-chain gene unique to each clone of thymocytes. In this way, **a diversity in αβ TCR is generated.**
9. The acquisition of fully developed αβ TCR in the double positive cells, follows a rigorous selection process in the thymus in order to generate single positive CD3. TCR α/β CD_4^+ and CD3. TCR α/β CD_8^+ T cell subsets. These cells undergo additional selection process in the thymus where almost 98% of them die off and only 2% of thymocytes successfully completes thymic selection to pass on to peripheral blood as mature single CD_4^+ and single CD_8^+ T cell subsets.

III. THYMIC SELECTION

As we have seen that double negative pro-T cells enter the thymus where they proliferate rapidly to produce a large number of double negative cells that acquire TCRβ receptor through β chain gene rearrangement. Now the cells having successful TCR β-chain rearrangement develop into CD_4^+ CD_8^+ (double positive) coreceptor bearing cells, enter the thymic cortex and are **positively selected** for self-MHC molecules. At this stage those T cells which do not arrange TCR β-chain successfully as well as those which do not recognize self-MHC molecules are deleted by **apoptosis**. Subsequently, these cells start now to delete one of its coreceptor, i.e. CD4 or CD8 to become single positive cells. In addition, they rearrange their TCR α-chain gene. Finally, single positive, i.e. either CD4 or CD_8^+ or $CD4^{high}$ and $CD8^{low}$ as well as $CD8^{high}$ and $CD4^{low}$ expressing cells with rearranged TCR αβ chain progress to mature single (+ve) cells and enter medulla of the thymus. Those cells which have high affinity to self-antigen (autoreactivity) undergo **negative selection** (by apoptosis) and cells having low or no affinity for self-antigen pass through the thymic selection. It is apparent therefore, that thymic selection is basically a self-editing process to exclude useless T cells, select useful T cells and eliminate dangerous T cells (autoreactive cells) through negative selection. Thus perfect T cells with high affinity for self-MHC but low affinity for self-antigens are generated through thymic selection. It is presumed that factors present in thymic **microenvironment** influence this self-editing process of positive and negative selection of thymocytes.

IV. T CELL RECEPTOR AND CORECEPTORS

T cells recognize antigens along with the MHC molecules, get activated, produce cytokines that again stimulate T cell in autocrine manner and the activated lymphocytes migrate and home to the effector site. All these phenomena are based on receptor-ligand interaction. Thus, early T cell expresses a number of receptors to interact with these ligands. T cell receptors can be classified as follows:

- Antigen receptors
- Signal transduction receptors
- Ligand for costimulatory molecules
- Coreceptor molecules to interact with MHC I and II
- Adhesion molecules
- Molecules to interact with B cells and APCs
- Homing receptors

A. Antigen Receptors (TCR)

1. Each T cell expreses antigen recognizing T cell receptors (TCR) which are member of immunoglobulin superfamily. TCR contains α, β chain (**heterodimer** molecule) linked with disulfide bond similar to the immunoglobulin molecule.
2. Each chain of TCR is made up of variable and constant regions which fold like immunoglobulin domains.
3. Unlike BCR that undergoes somatic hypermutation after binding with Ag, TCR does not change rather its diversity is generated during maturation of T cells by β and α chain gene rearrangement.
4. TCR recognize only small part of proteolytically cleaved linear peptide of Ag in association with MHC molecules.
5. Unlike BCR (which is present in soluble form), TCRs are not cleaved from the cell surface.
6. Different region of TCR interact with different part of antigen, thereby TCR-antigen interaction is highly specific.

B. Signal Transduction Molecule

1. TCR has a very short cytoplasmic tail. It can bind to antigens but cannot transmit intracellular signal. Thus T cells have another signal transduction molecule called, CD3.

2. CD3 is a unique molecule expressed only on the surface of T cells, therefore, it is commonly used as a phenotypic marker of T cells.
3. CD3 is made up of three distinct peptides, **25 kDa γ chain, 20 kDa δ chain** and **20 kDa ε chain**.
4. It remains covalently associated with TCR and another 16 kDa ζ (zeta) chain forming **T cell receptor complex**.
5. Each chain of CD3 contains ITAM motif which is essential component of signal transduction.

C. Costimulatory Ligands

1. We have learned that antigen presenting cells express costimulatory molecules B7.1 and B7.2 on its surface which interacts with CD 28 molecules on T cells to provide costimulatory signal.
2. T cells express CD 28 ligand for B-7 costimulatory molecule of APC which is required for activation of T cells upon TCR-Ag interaction.
3. If TCR-Ag interaction takes place in absence of B-7-CD28 interaction, T cells become anergic and do not respond to Ag.
4. In addition to CD28, T cells also express CTLA-4 **(cytotoxic T lymphocyte associated antigen 4)** which is antagonist of CD28. Interaction of B7 with CTLA-4 provides negative signal to T-cells and inhibits T cell activation even in case with TCR-Ag interaction.

D. T Cell Coreceptor Molecules

1. In addition to TCR-Ag interaction, T cells require interaction with MHC molecules present on APC along with the antigen.
2. T-helper cells bear CD4 coreceptor to interact to MHC II molecules on APC. CD4 coreceptor is also used as a marker for T-helper cells.
3. T-cytotoxic cells possess CD8 coreceptor which interacts with MHC I on APC. Thus CD8 is also used as a maker for cytotoxic T cell subset.
4. CD4 and CD8 molecules also act as adhesion molecules for MHC II and MHC I complex respectively, to provide strong association between APC and T cells in case with antigen presentation.
5. Thus CD4 and CD8 receptors are the fundamental component of MHC restriction which also directs immune response either towards Ab response/ inflammatory response or towards cytotoxicity.

E. Adhesion Molecules

1. During immune response the immune cells produce different chemoattractants (chemokines) to attract different immune cells to the site of reaction. Such chemokines further induce expression of adhesion molecules on T cells.
2. T cells express LFA-1 (leukocyte functional antigen) and VLA-4 which binds to ICAM-l and VCAM 1 respectively expressed on endothelial cells for attachment of T cells and its subsequent homing to the site of action.

3. These molecules are commonly called integrins.
4. Integrins are two-chain molecules which are present on other cells as well.
5. In activated T cells, these molecules appear quite late (almost 48-76 hours) after antigen-antibody interaction.

F. γ, δ TCR

1. As mentioned earlier γ, δ TCR bearing cells appear quite early in the ontogeny of T cells.
2. γ, δ TCR is distinct from α, β TCR but similar to α β TCR, it is expressed in association with CD3 complex. Therefore, CD3 is also a marker for both α β as well as γ, δ TCR bearing cells.
3. γ δ T cells are very low in number (5% of total T cells) and in contrast to α β T-cells these cells are present in tissues.
4. γ δ T cells lack CD4 and CD8 coreceptors.
5. One of the well-characterized ligand for γ δ TCR is **heat shock proteins** present on microbes or parasites.
6. Various T cells associated molecules have been shown in Figure 13.2.

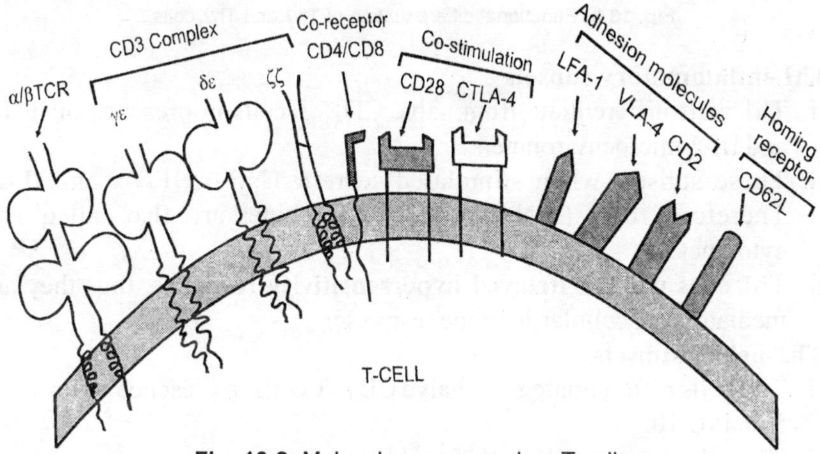

Fig. 13.2: Molecules expressed on T cells

V. T CELL SUBSETS

Functionally and phenotypically T cells (TCR α, β) contain two distinct subsets, CD_4^+ T-helper (Th) cells and CD_8^+ T-cytotoxic (Tc) cells.

A. T-helper (Th) Cells

1. T-helper cells are recognized on the basis of their CD4 membrane glycoprotein coreceptor.
2. These cells are MHC II restricted.
3. Functionally CD_4^+ Th cells can be subdivided into two, Th1 (inflammatory) and Th2 (helper) subsets (Fig. 13.3).

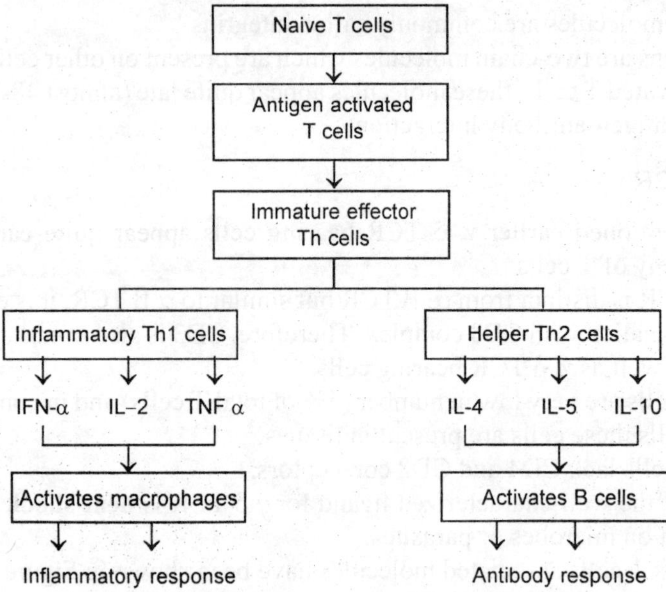

Fig. 13.3: Functional differentiation of Th1 and Th2 cells

a. Th1-inflammatory subsets

i. Th1 cells differentiate from naive CD_4^+ T cells in presence of IFN-γ and IL-2 microenvironment.

ii. These subsets when stimulated secrete TNF-α, IFN-γ and IL-2. Therefore, these (IFN-γ and IL-2) cytokines are also called Th1 cytokines.

iii. Th1 cells mediate **delayed hypersensitivity response**, thus they are mediators of cellular immune response.

b. Th2-helper subsets

i. Th2 cells differentiate from naive CD_4^+ T cells in presence of IL-4, IL-5 and IL-10.

ii. Th2 subsets secrete Il-4, IL-5 and IL-10 cytokines.

iii. These cells are **T-helper cells** and mediate humoral immune response.

B. T-cytotoxic (Tc) Cells

i. Cytotoxic T lymphocytes (CTL) are phenotypically recognised by its surface glycoprotein coreceptor molecules CD8.

ii. These cells interact with endogenous antigens bound to MHC I molecules. Therefore, these cells are also known as MHC I restricted T cells.

iii. CTLs are the efficient cytotoxic T-subsets which cause lysis of virus infected cells and tumor cells.

iv. It has been shown that CD_8^+ T cells can be activated by IL-2 even in absence of costimulatory molecules B 7.1 and B 7.2. It also requires IL-2 produced by CD_4^+ T cells for proliferation and survival.

v. **Cytotoxic T cell clones** (CTL clones) can be generated against tumor cells or virus infected cells by incubating naive T cells with tumor/viral antigens in presence of IL-2.

VI. T CELL ACTIVATION AND SIGNAL TRANSDUCTION

Activation of lymphocytes by means of antigen, mitogen or cytokines is the principal requirement for generation of effector cells of the immune response. Though there could be somc minor differences in activation pathways of B cells, Th cells and Tc cells, in general these cells follow almost similar signal transduction pathway. Signal transduction leads to generation of effector cells such as plasma cells to secrete antibody, Th cells for T-B co-operation or DTH response and CTL for cytotoxicity. Some of the common features of activation and signal transduction pathway are as below:

1. Unlike B cells which require antigen-BCR interaction for activation, T cells require an additional signal through B7.1 or B7.2—CD28 interaction for activation, called costimulatory signal.
2. Receptor-ligand interaction leads to changes in cytoskeleton, generation of second messenger inside the cells, release of transcription factor (that binds to regulatory elements of the gene) and production effector molecules (i.e. Ab, cytokines, etc.).
3. Protein kinases and phosphatases act as on and off switch in signal transduction pathway.
4. Signal transduction leads to amplification of signals by enzymatic cascade reaction.
5. Most of the signal transduction pathways involve activation of GTP binding (G-protein) protein.
6. Normally, signal transduction pathway remains switched off in the resting cells.

In this chapter we will learn about signal transduction in CD_4^+ T cells which is similar to B cell activation and Tc activation. This process is initiated with antigen recognition by TCR and activation of coreceptor CD4 and signal transductions unit CD3 molecule complex (Fig. 13.4).

A. Initiation

1. Antigen TCR interaction leads to accumulation of cholesterol anc splingolipids that forms lipid rafts through reorganization of cellular cytoskeleton. Lipid rafts mobilize intracellular signaling molecules.
2. Within seconds two proteins tyrosine kinases, **Lck** and **Fyn** which belongs to a family of protein kinase **Src** and is associated with the cytoplasmic tail of coreceptors, phosphorylate ITAM motifs of CD3 zeta (ζ) chains.
3. Phosphorylated CD3 ζ forms a docking site for another **Syk** family of tyrosine kinase known as Zap 70.

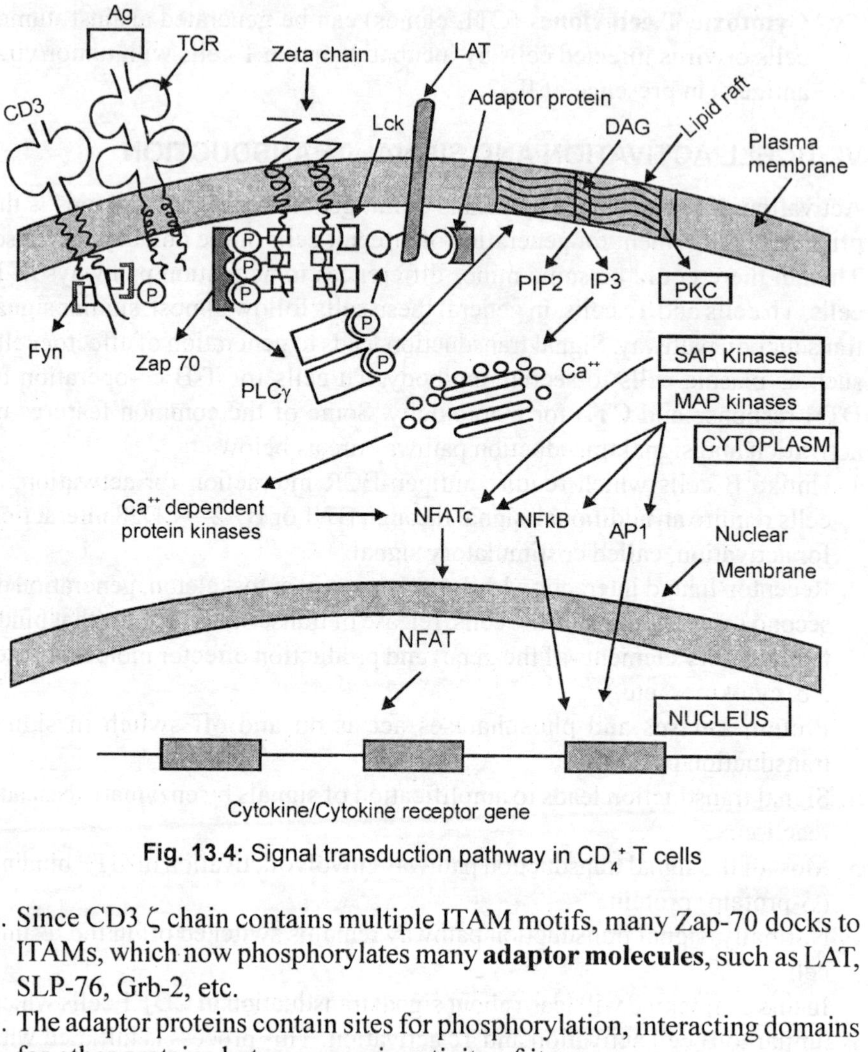

Fig. 13.4: Signal transduction pathway in CD_4^+ T cells

4. Since CD3 ζ chain contains multiple ITAM motifs, many Zap-70 docks to ITAMs, which now phosphorylates many **adaptor molecules**, such as LAT, SLP-76, Grb-2, etc.
5. The adaptor proteins contain sites for phosphorylation, interacting domains for other proteins but no enzymic activity of its own.

B. Generation of Second Messengers

Phosphorylation of adapter proteins by Zap-70 leads to generation of several second messengers and initiates several signalling pathways.

a. Phospholipase Cγ (PLC) Pathway

1. PLC activation takes place by membrane associated adaptor proteins which hydrolyses PLC-γ giving rise to two second messengers inositol triphosphate (IP3) and diacyl gycerol (DAG) through a series of chemical reactions.
2. IP3 causes opening of Ca^{++} channel on the cell membrane causing influx of Ca^{++} in the cell as well as release of intracellular Ca^{++} from Ca^{++} pool of the endoplasmic reticulum.

3. Ca^{++} activates Ca^{++} dependent protein kinases which induce movement of transcription factor NFAT (nuclear factor in activated T cells) from cytoplasm to nucleus.
4. DAG activates protein kinase C that phosphorylates multiple cellular targets.

b. Protein Kinase C (PKC) Pathway

1. PKC is a potent protein kinase which activates multiple cellular proteins.
2. The important function of PKC is the activation of transcription factor NF-κB (nuclear factor kappa B).
3. Generally NF-κB remains inactive in association with inhibitory kappa B (IκB).
4. PKC phosphorylates IκB which separate out from NF-κB and gets degraded by proteasomes.
5. Now NF-κB is released, gets phosphorylated and translocates to the nucleus.

c. Ras, MAP Kinase and SAP Kinase Pathway

1. Ras is a smaller GTP binding protein whose activation leads to activation of MAP kinase (Mitogen-activated protein kinase).
2. MAP kinase activates another transcription factor called AP-1.
3. AP-1 is a dimer of two proteins, Fos and Jun which remain bound to each other by **'Leucine zipper'.**
4. Phosphorylation of Fos and Jun by JNK (c-jun terminal kinase) activates AP1 that associates with other transcription factor such as NFAT.
5. Stress activated protein (SAP) kinase pathway is another signal transduction pathway which is activated by stress protein such as HSP, TNFα, etc. It activates SAP kinases that in turn phosphorylates nuclear factors that translocates to nucleus and binds to regulatory elements of cytokine genes.

d. Secretion of Cytokines

1. Activated transcription factors enter the nucleus and binds to the regulatory elements of cytokine genes and transcription of the gene is initiated.
2. In signal transduction, receptor-ligand interaction takes place repeatedly but only for milliseconds, however, it leads to tremendous amplification of second messengers.
3. The earliest protein produced during cell activation is transcription factors (within 30 minutes), followed by cytokines (1-24 hours) and finally adhesion molecules (3-7 days).
4. Cytokines thus produced activates different effector cells in autocrine or paracrine manner and the cells progress through cell cycle to generate large number of effector and memory cells.
5. Expression of adhesion molecules help in movement and homing of the effector cells to the site of action where antigen has been encountered and the effector functions take place in order to combat the infection.

VII SUPERANTIGEN AND T CELL ACTIVATION

Superantigens are bacterial or viral proteins which do not require processing but binds directly to the superantigen binding domain of α-chain of MHC II molecule and Vβ chain of TCR. Binding of these antigens leads to activation of a large number of T cells and their proliferation. Therefore, these are called superantigen.

Like other antigens, superantigens are also of two types: exogenous superantigens and endogenous superantigens.

A. Exogenous Superantigens

These are soluble bacterial proteins such as exotoxins secreted by gram-positive bacteria, staphylococcal enterotoxin (SE), toxic-shock syndrome toxins (TSST) and exfoliative-dermatitis toxin. These toxins bind to specific Vβ chain of T cell receptors of mice and humans (e.g. SEB is specific for Vβ8 in mice and Vβ3 in human while TST-1 specifically binds to Vβ 15,16 in mice and Vβ2 in human) and crosslink MHC II molecules (Fig. 13.5).

B. Endogenous Superantigens

Endogenous superantigens are cell membrane viral proteins produced by mammary tumor virus (MTV). Four such MTV products have been identified so far which have different Vβ specificity. These antigens can stimulate large number of T cells bearing particular Vβ specificity through binding to outside of TCR antigens binding cleft. Superantigens act as polyclonal activators. Such activated T cells undergo proliferation and secrete cytokines. Secretion of larger quantity of cytokines is toxic to the cells.

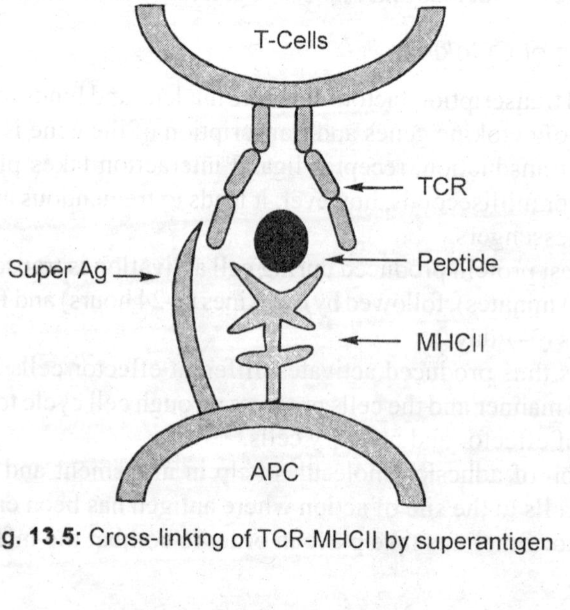

Fig. 13.5: Cross-linking of TCR-MHCII by superantigen

Like other antigens, interaction of superantigens during developmental stage of lymphocytes leads to negative selection of T cell repertoire.

VIII. EFFECTOR T CELLS AND THEIR FUNCTIONS

We have learned that antigenic stimulation leads to differentiation of T cells into effector cells and memory cells. Two effector subsets of T cells are recognized by expression of coreceptors CD4 and CD8. These two subsets are functionally different.

A. CD_4^+ Helper/Inducer T Cells

The principal effector functions of these cells are as follows:

a. **Cytokine secretion:** CD_4^+ effector T cells are potent cytokine producing cells. These cells are subdivided into two subsets (Th1 and Th2), on the basis of cytokine production. Th1 subsets produce IFN-γ, IL-2, INFα while Th2 subsets secrete IL-4, IL-5, IL-10, TGF-β and lymphotoxin. Cytokines produced by CD_4^+ T cells exert also its effect on other cells bearing cytokine receptor. Therefore, CD_4^+ T cells are also called **inducer** cells. These cells induce production of other biological molecules such as antibody, cytokine, cytotoxic receptor, MHC, etc. through its cytokine secretion.

b. **B cell co-operation in Ab synthesis:** CD_4^+ effector T cell subsets interact with T-dependent exogenous antigen and help B cells to produce antibody against the antigen, thus CD4 T cells are called helper T cells. In addition, activated CD_4^+ T cells produce IL-6 which is B cell growth and activating factor required for production of antibody by B cells. IL-4 and IL-5 produced by CD_4^+ T cells brings about immunoglobulin isotype switch from IgG to IgE, IFN-γ and CD40-CD154 interaction leads to isotype switching from IgM to IgG.

c. **Delayed type of hypersensitivity (DTH) response:** DTH response is another important effector function mediated by CD_4^+ T cell subset. DTH reaction is characterized by erythema and induration of the skin and granuloma formation as is observed in case with leprosy patients. This reaction is mediated by cytokines produced by Th1 cells. DTH reaction takes place after antigen exposure in two phases:

   i. In **sensitization phase,** the antigen is taken up by APCs (macrophage and dendritic cells) which moves to the regional lymph node where CD_4^+ T cells are sensitized. During sensitization phase mainly Th1 subsets of CD_4^+ T cells are activated. This phase lasts for 1-2 weeks.

   ii. **Effector phase** of DTH response begins with the re-exposure of the antigen which induces activation of Th1 typc of CD_4^+ T cells that elaborates cytokine response. This phase lasts for 48-72 hours during which large quantity of Th1 cytokines, cytokine receptor, chemokines and chemokine receptors are expressed. It leads to migration of macrophages to the site of antigen exposure and there is increased phagocytic response. The principal cytokines secreted during DTH response are IFN-γ, TNF-β and GMCSF. This reaction subsides when the parasites or microorganisms are destroyed. In experimental condition, DTH response can be induced by local application of chemicals such as Dinitrochlorobenzene (DNCB) and Dinitrofluorobenzene (DNFB) as well as by subcutaneous injection of PPD (purified protein derivative of Mtb).

B. CD_8^+ Suppressor/Cytotoxic T Cells

CD_8^+ T cell subsets are called suppressor cytotoxic subsets due to its cytotoxic activity on tumor cells, virus infected cells and cells infected with intracellular parasites. Due to their cytotoxic functions these cells are commonly called cytotoxic lymphocytes (CTL). There are mainly two mechanisms through which CD_8^+ effector T cells (CTL) mediates cytotoxic reactions.

a. **Cytotoxicity by perforins granzymes.** In this pathway, endogenous Ag activated CTLs express IL-2R and become responsive to IL-2 secreted by CD_4^+ T cells or CD_8^+ cells itself. Activated CD_8^+ T cells reorganize perfornis which polymerizes to form tube-like hole in the infected (target) cells through which vital substances leaks out from the cells resulting in cell death. CD_8^+ CTLs also synthesize **granzymes,** a serine protease that enters the cell through pores formed by perforins and leads to proteolytic cleavage of vital proteins resulting in cell death.

b. **Induction of Apoptosis:** Apoptosis is another mechanism which is induced by CTLs in the target cells. In this pathway, activated T cells express Fas ligand (CD178) on its surface which binds to Fas (CD95) molecules (that are constitutively expressed) on the target cells. Fas and Fas L interaction triggers a cascade of biochemical process in the target cells leading to activation of proteolytic enzyme called, **caspases.** The target cells undergo apoptotic death within hours of caspase activation.

Certain parasites, virus and tumor antigens activate both CD_4^+ T cell subset as well as CD_8^+ T cells by forming complex with MHCI and MHCII respectively. Such phenomenon is called **cross-priming.** In such situation, activated CD_4^+ T cells produce cytokines (IFNγ and IL-2) resulting in activation of CD_8^+ T-subsets to differentiate into CTL effector cells. In addition antibody response against such antigen also takes place. Thus infection is contained by both cellular and humoral immune response. After containment of infection activated CD_4^+ and CD_8^+ effector cells undergo apoptosis and the immune reaction stops.

STUDY QUESTIONS

1. Describe stages of T cell development with suitable diagrams. What is the role of cytokine in functional differentiation of Th cells?

2. Describe in brief T cell activation and signal transduction pathway.
3. Write short notes on
   a. DTH response
   b. CTLs
   c. Th1/Th2 cells.

MCQs

Select one most appropriate answer given for each questions below:

1. **Which of the following gene arrangement takes place first in the early stage of T cell development.**
   A. CD3
   B. CD4
   C. TCR β chain
   D. TCR α chain
2. **C Kit is expressed first in:**
   A. Pre-T cells
   B. Pro-T cells
   C. Double negative cells
   D. Double positive cells
3. **Positive selection of T cells during development depends on:**
   A. Expression of CD_4^+/CD_8^+ coreceptors
   B. Successful rearrangement TCR αβ chains
   C. Affinity for self MHC
   D. Affinity for foreign MHC and self antigens
4. **Negative selection of T cells depends on:**
   A. High affinity for self-antigens
   B. High affinity for self-MHC
   C. Low affinity for self-antigens
   D. Low affinity for self-MHC
5. **Th1 cells are characterized by**
   A. Production of perforins and granzymes
   B. Secretion of IFN-γ and IL-2
   C. Secretion of IL-4 and IL-5
   D. TCR γδ antigen receptors
6. **TCR cannot transmit intracellular signal because:**
   A. It has a short cytoplasmic tail
   B. It lacks signal peptide
   C. It is α–β– heterodimer
   D. It is expressed on the cell surface

7. **TCR diversity is generated by:**
   A. Hypermutation upon binding to antigens
   B. α, β chain gene rearrangement during development.
   C. Cytokine induced activation
   D. Signal transduction
8. **CTLA–4:**
   A. Antagonist of CD28.
   B. Agonist of CD28.
   C. Its binding activates T cells
   D. Its binding leads to apoptosis of T cells
9. **γ α T cells:**
   A. Express coreceptor CD4
   B. Express coreceptor CD8
   C. Express CD3 molecules
   D. None of the above
10. **Which of the following is the transcription factor involved in IL–2 gene expression:**
    A. NFATc
    B. NF κB
    C. Gro 1
    D. PLC
11. **Superantigens:**
    A. Can activate B cells to produce large quantity of antibody
    B. Interacts with vβ region of TCR
    C. Can activate few T cells but more B cells
    D. Have numerous B cell and T cell epitopes

1 C **2** A **3** C **4** A **5** B **6** A **7** B **8** A
9 C **10** A **11** B

14

Cytokines and Cytokine Receptors

I. INTRODUCTION

Cytokines are the low molecular weight glycoproteins secreted by white blood cells during effector phase of immune response. They act on target cells through binding to their receptors and regulate immune response. Cytokine deficiency or over production has been implicated in various diseases. Some cytokines are produced by monocytes, therefore, they are named as '**monokines**' while those produced by lymphocytes are named as **'lymphokines'.** Some cytokines are named as **interleukins** referring to its production by leukocytes and their action on another leukocytes. However, it has been seen that all these molecules share some of the common properties, therefore, they are commonly called as **cytokines**. Cytokines, similar to hormones, act on different target cells and unlike hormones these molecules are transiently produced in picomolar concentration during immune response, trigger effector function on the target cells and are exhausted in the process. Thus these are not present in the cells in the preformed state. Although different cytokines have their specific amino acid sequence but they share many biological functions and even receptors. These molecules are also called **biological response modifiers.**

II. GENERAL PROPERTIES

The general properties of cytokines include pleiotropy, redundancy, synergy, antagonism and induction of cascade reaction.

A. Pleiotropy

Pleiotropism refers to unique property of cytokines, i.e. same cytokine may act on different cells in different ways. For example, IL-1 acts on monocytes and macrophages, T cells and hypothalamus. Similarly, IL-2 acts as T cell growth and stimulatory factor and it also causes activation of NK cells. TNF-α is another versatile cytokine that acts as comitogen for T and B cells, activates macrophage, regulates haematopoiesis, induces fever by acting on hypothalamus and leads to production of other cytokines.

B. Redundancy

Many different cytokines may have similar function on the same target cells. This could be due to similar receptor molecules on the target cells. For example, IL-2, IL-4 and IL-6 induce proliferation of B cells. Similarly, TNF-α, IL-1 and IL-6 activate macrophage.

C. Synergism

Two cytokines may have additive effect on the same target cell function such unique effect of cytokines is called synergy, e.g. IFN-γ and IL-2 activate NK cells to become **Lymphokine-activated killer (LAK) cells.** Similarly, interferon-γ and TNF-α enhance expression of MHC II molecules on antigen presenting cells.

D. Antagonism

Antagonism refers to opposing effect of two cytokines on the same target cell. The example of antagonistic effect is IFN-γ and IL-10. IFN-γ activates macrophage while IL-10 downregulates macrophage function. Similarly, IFN-γ activates Th0 cells to differentiate into Th1 cells that promotes cellular immune response while IL-4 down-regulates Th1 function and promotes Th2 cell differentiation which mediates humoral immunity.

E. Cascade Induction

Cytokines induce a cascade of immune response through cytokine network. For example, TNF-α activates macrophage to produce IL-1 which in turn further activates macrophage to secrete IL-6. It becomes obvious when cells are activated with LPS. LPS induces production of TNF-α which induces IL-1 production and finally IL-1 induces IL-6 production. Thus LPS stimulated cells secrete all the three cytokines one by one in the form of a cascade reaction.

III. GROUP OF CYTOKINES

Broadly cytokines can be divided into four groups depending upon their functions.

1. Cytokines which mediate innate immunity (Table 14.1).
2. Cytokines that mediate lymphocyte growth differentiation and activation (Table 14.2).
3. Cytokines that mediate cellular activation and response (Table 14.3).
4. Cytokines that induce haematopoiesis (Table 14.4).

Table 14.1: Mediators of innate immunity

Cytokine	*Molecular size*	*Source*	*Target*	*Functions*
1. Type 1 interferon	18-20 kDa	Phagocyte and fibroblast	Phagocyte and fibroblast and other leukocytes	Activation of macrophage, antiviral and antiproliferative
2. Tumour necrosis factor (TNF-α)	17 kDa	Macrophage, T cells	Macrophage, T cells, liver, hypothalamus	Inflammation, fever, acute phase protein induction by liver cells
3. Interleukin 1 (IL-1)	14 kDa α, β isotype	Mononuclear phagocytes	T cell, B cell, endothelial cells, liver, hypothalamus	Bone marrow suppression, inflammation, induction of fever and acute phase proteins
4. Interleukin-6 (IL-6)	26 kDa	Mononuclear cells, T cells, endothelial cells	T cell, B cells, liver	Costimulator, acute phase reaction, growth of B-cells hybridoma
5. Interleukin-8 (IL-8)	8-10 kDa	Mononuclear cells, endothelial cells, fibrablast	Leukocytes	Activation and chemotaxis
6. Interleukin-12 (IL-12)	35-40 kDa	Macrophage, dendritic cells	T cells, NK cells	Interferon-γ production, Th 1 cell differentiation, cytolysis, activation of cell-mediated immune response.

Table 14.2: Mediators of lymphocyte activation, growth and differentiation

Cytokine	*Molecular size*	*Source*	*Target*	*Functions*
1. Interleukin-2 (IL-2)	14-17 kDa	T cells	T cells, B cells, NK cells	T cell growth and activation, antibody synthesis by B-cells MHC II expression.
2. Interleukin-4 (IL-4)	20 kDa	CD_4^+ Th 2 cells	T cells, mononuclear phagocyte, B cells	Th 2 cell activation, IgE class-switching, Expression of FcγR
3. Transforming growth factor-β (TGF-β)	14 kDa	T cells, macrophage	T cells, macrophage	Growth inhibition, regulation of growth of various cells, IgA class switching in mice

Table 14.3: Mediators cellular activation

Cytokine	*Molecular size*	*Source*	*Target*	*Functions*
1. Interferon-γ (IFN-γ)	21-24 kDa	T cell, NK cell	All leukocytes	Activation, Ig class switching, differentiation of Th 1 cells
2. Lymphotoxin (TNF-β)	24 kDs	T cells	Neutrophils, endothelial cells	Activation
3. Interleukin-5 (IL-5)	20 kDa	T cells	Eosinophils, B cells	Activation, IgA secretion, differentiation of Th 2 cells
4. Migration inhibition factor (MIF)	Unknown	T cells,	Macrophages	Migration inhibition
5. Interleukin-10 (IL-10)	20 kDa	T-cells, B cells, monocytes, keratinocytes	Macrophages, B cells	Inhibits cellular activation
6. Interleukin-12 (IL-12)	35-40 kDa	Macrophage	NK cells, T cells	Activation, growth and differentiation

Table 14.4: Mediators of haematopoiesis

Cytokine	*Molecular size*	*Source*	*Target*	*Functions*
1. c-kit Ligand	24 kDa	Bone marrow stromal cells	Pluripotent stem cells	Growth and differentiation
2. Interleukin-3 (IL-3)	20-26 kDa	T cells	Immature progenitors	— do —
3. Granulocyte-macrophage colony stimulating factor (GM-CSF)	22 kDa	T cells, macrophage, Endothelial cells	Progenitor cells	Differentiation of granulocytes, macrophage, dendritic cells, activation
4. Macrophage colony stimulating factor (M-CSF)	40 kDa	Macrophage, endothelial cells, fibroblast	Committed progenitor cells	Differentiation to mononuclear phagocytes
5. Granulocyte colony stimulating factor (G-CSF)	19 kDa	— do —	— do —	Differentiation to granulocytes
6. Interleukin-7 (IL-7)	25 kDa	Bone marrow stromal cells, Thymic stromal cells, fibroblast	Immature progenitors	$CD4^-$ $CD8^-$ T cells, B cells growth and differentiation

IV. CYTOKINE RECEPTORS

Cytokine acts on target cells through binding to their receptors. Many different types of cells express receptors for various cytokines. Therefore, one cytokine may act on different targets. Secondly, cytokine receptors are either heterodimers, consisting of one cytokine specific chain and another signal transduction unit or it may be heterotrimer (consisting of one cytokine specific chain and two signal transduction unit). Signal transduction chain of a group of cytokines is shared by several members of the family. Therefore, such cytokines have redundancy in their function. Cytokine receptors have been divided into four families based on their common structural features and nucleotide sequence (Fig. 14.1) as given below:

1. Immunoglobulin superfamily
2. Class I cytokine receptor family
3. Class II cytokine receptor family
4. TNF receptor family

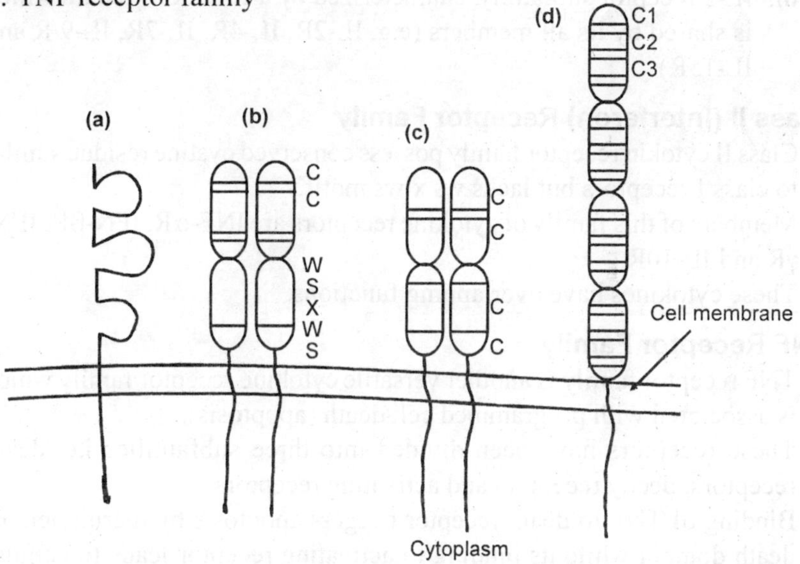

Fig. 14.1: Cytokine receptor family (a) Immunoglobulin superfamily (IL-1, M-CSF, c-Kit); (b) Class I cytokine receptor family (IL-2, IL-3, IL-4, IL-13, GM-CSF, G-CSF, CNTF, etc.); (c) Class II cytokine receptor family (IFN-α, β, γ, IL-10) and (d) TNF receptor family (TNF-α, β, CD-40, Fas, FasL, NGF)

A. Immunoglobulin Receptor Family

Members of this receptor family are IL-1, cKit and M-CSF. As the name indicates this receptor family is characterized by their immunoglobulin domains.

B. Class I Cytokine Receptor Family

1. This receptor family have largest number of members consisting of cytokines responsible for haematopoiesis such as GM-CSF, IL-3, IL-6, etc.

2. The main features of these receptors are: (a) **conserved cystine residue** and (b) conserved sequenc of **tryptophan-serine-x-tryptophan-serine.**
3. These receptors contain one cytokine specific α chain which is unique for each cytokine and two (β, γ) signal transduction chains-shared by other members of the family.
4. Signal transduction chains have a long cytoplasmic tail.
5. Class I cytokine receptor family has been further divided into three subfamilies:
   i. GM-CSF receptor subfamily, characterized by a common β-chain (e.g. GM-CSFR, IL-3R, IL-5R). Members of this family show redundancy.
   ii. IL-6 receptor subfamily, characterized by a common gp 130 signal transduction unit (e.g. IL-6, IL-11, CNTF (ciliary neurotrophic factor) and oncostatin M.
   iii. IL-2 receptor subfamily, characterized by a common γ-chain which is shared by its all members (e.g. IL-2R, IL-4R, IL-7R, IL-9/R and IL-15R).

C. Class II (Interferon) Receptor Family

1. Class II cytokin receptor family possess conserved cystine residue similar to class I receptors but lacks ws x ws motif.
2. Members of this family of cytokine receptors are INF-αR, IFN-βR, IFN-γR and IL-10R.
3. These cytokines have overlapping functions.

D. TNF Receptor Family

1. TNF receptor family is another versatile cytokine receptor family which is associated with programmed cell death (apoptosis).
2. These receptors have been divided into three subfamilies i.e. death receptors, decoy receptors and activating receptors.
3. Binding of TNF to death receptor triggers apoptosis by recruitment of death domain while its binding to activating receptor leads to cellular prolifration.
4. Mutation/polymorphism in TNF receptor has been implicated in certain autoimmune disorders.

V. POLARIZATION OF CYTOKINE RESPONSE

Cytokines induces polarization of T cell response either towards **cellular immune response** or towards **humoral immune response**. Naive CD_4^+ T cells are basically nonpolarized Th0 cells which secrete a large number of cytokines such as IL-α, IFN–γ, IL–4, IL–5, IL–6, IL–10 etc. However, under the influence of cytokines microenvironment Th_0 cells differentiate into IFN–γ and IL–2 producing Th_1 cells or IL–4, IL–5 and IL–10 secreting Th_2 cells

(Fig. 14.2). Experiments with gene knock out mice have suggested that of Th_0 cells, require IFN-γ and IL-2 for generation of Th_1 cells, contrarily presence of excessive amount of IL–4 leads to generation of Th2 cells. Although Th1 and Th2 cells are subsets of CD_4^+ T–cells producing IFNγ, IL–2 and IL–4, IL–5, IL–6 respectively, similar differentiation has been suggested for CD_8^+ T–cell subsets as well.

Th1 subsets have been shown to mediate inflammatory immune response such as delayed type of hypersensitivity (DTH), antibody dependent cell mediated cytotoxicity (ADCC), and macrophage activation which are features of CMI response. TH_2 clones however, have been implicated in humoral immunity such as Ig production, mucosal immunity and mast cell activation as well as allergic reactions. In the absence of polarizing signals, Th0 cells slow functionally less prominent cytokine response. It has been further suggested that a cross-regulation of Th–like cytokine response exists. Th_1 cytokines (IFN–γ, IL–2) down regulate functions of Th_2 cells, similarly Th_2 cytokines (IL–4, IL–5) down regulate functions of Th1 cells.

The important role of Th–like cytokine response has been implicated in disease susceptibility and resistance to infection. Imbalance in Th_1/Th_2 response leads to disease conditions. For example, we have observed a shift from Th_1 to Th_2 cytokine associated with advanced stage of **oral cancer**. Others have shown Th_1 associated response in **tuberculoid leprosy** which is characterized by a strong CMI response and granuloma formation. However, otherside of the spectrum of leprosy, i.e. **Lepromatous leprosy** which is characterized by down regulation of CMI response has been found to be associated with Th_2–like cytokine response. Similar dichotomy of cytokine response has also been suggested in AIDS patients.

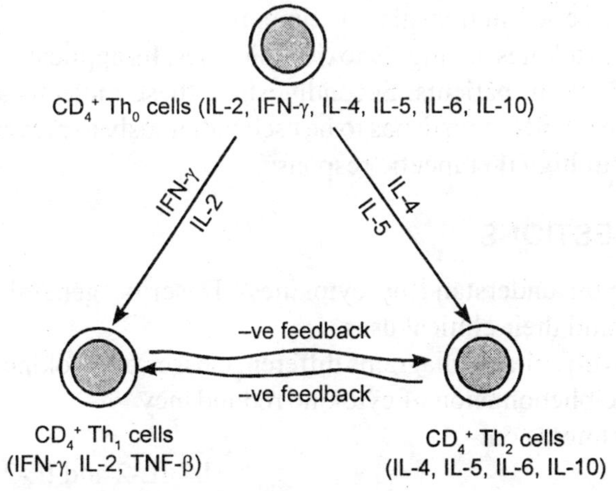

Fig. 14.2: Differentiation of Th_0 cells and cross-regulation of Th_1/Th_2 response

VI. CYTOKINE ANTAGONISTS

1. The proteins which bind to cytokine receptor and inhibit their activity are called cytokine antagonist.
2. Naturally occurring IL-1 antagonist has been identified which binds to the receptor IL-1Rα, (blocks IL-1 binding to its receptor) but has no activity.
3. The cytokine antagonists are produced by enzymatic cleavage of extracellular domain of cytokine receptors.
4. Some viruses produce proteins that mimic cytokine and binds to its receptor, e.g. adenovirus produce IL-10 like molecules.
5. Antagonists for IL-1, IL-2, IL-4, IL-6, IL-7, IFN-α, γ and TNF-β has also been identified.

VII. CYTOKINE AND CYTOKINE RECEPTOR THERAPY

Due to its role in immunoregulation cytokine and cytokine receptor has potential role in treatment of various diseases. Some of its therapeutic roles are as follows:

1. IL-2 and IFN-γ generates lymphokine-activated killer (LAK) cells which can lyse NK cell resistant tumour cells. Thus IL-2 and IFN-γ have been implicated in cancer therapy
2. Antibody against IL-2R (anti-TAC) has been used in transplant recipients.
3. Patients with X-linked agammaglobulinemia have been successfully treated with G-CSF that has been found to reverse *neutropenia.*
4. Patients suffering from **severe combined immunodeficiency (SCID)** respond to IL-2 treatment. This disease is characterized by deficiency of the enzyme, **'adenosine deaminase (ADA)'.**
5. CD25 antibody against IL-2R α-chain has been used in patients with leukemia/lymphoma.
6. Toxin conjugated anti CD25 antibody has been used in prevention of allograft rejection in transplant recipients.

However, cytokines are highly toxic molecules. Its application shows many toxic side effects in patients. Secondly, since these molecules have a short half-life in the bloodstream, it has to be used judiciously to prevent undesirable side effects but high therapeutic response.

STUDY QUESTIONS

1. What do you understand by cytokines? Describe general properties of cytokines and their clinical uses.
2. Describe with suitable diagrams different classes of cytokine receptors and explain the phenomenon of cytokine redundancy.
3. Write short notes on:
   a. TNF α
   b. IL-2 and IL-2 receptor
   c. IFN-γ
   d. GM-CSF
   e. Cytokine antagonists.

MCQs

Select one most appropriate answer given for each questions below:

1. **Th 2 cells produce:**
   A. IL-2
   B. IFN-γ
   C. IL-4
   D. IL-10
2. **IFN-γ influences the differentiation of:**
   A. Th 2 cells
   B. Th 1 cells
   C. Granulocytes
   D. Monocytes
3. **Cytokines are:**
   A. Glycoproteins
   B. Lipoproteins
   C. Polysaccharides
   D. Mucopolysaccharides
4. **Which of the following cytokine mediates inflammatory response:**
   A. IFN-γ and IL-2
   B. IL-1 and TNF-α
   C. GM-CSF and G-CSF
   D. IL-1 and IL-10
5. **Th 1 response takes place by:**
   A. TGF β and lymphotoxin
   B. IFN-γ and IL-2
   C. IL-2 and IL-4
   D. IL-4 and IL-5
6. **Which of the following cytokines shows redundancy:**
   A. IFN-γ and IL-10
   B. IL-3 and IL-5
   C. IL-2 and IL-4
   D. TNF-α and TGF-β
7. **The ligands for class II cytokine receptors are:**
   A. IL-2 and IL-4
   B. TNF-α and TNF-β
   C. IL-1 and c-Kit
   D. IFN-γ and IFN-α
8. **Which of the following statements is *False:***
   A. Patient suffering from severe combined immunodeficiency respond to IL-2 therapy
   B. IFN-γ causes immunoglobulin class switching
   C. IL-4 induces antiviral state in the host
   D. GM-CSF induces haematopoiesis

9. **Which of the following cytokines plays a role in terminating in flammatory responses?**
   A. IL–2
   B. IL–4
   C. TGF–β
   D. IFN–α

10. **IL–1, IL–6 and TNF–α are known to:**
    A. Cause increased vascular permeability
    B. Acute phase response
    C. Inflammation
    D. All the above

1 C 2 B 3 A 4 B 5 B 6 D 7 D 8 C
9 C 10 D

15

Chemokines, Chemokine Receptors, Cell Migration and Homing

I. INTRODUCTION

Immune cells are called surveillance cells because they continuously migrate through bloodstream and lymphatics to central and peripheral lymphoid organs. Continuous circulation provides them an opportunity to encounter foreign antigens that enter the host either from the external environment or are produced in the body by intracellular pathogens, virus or malignant cells. Some of the antigen specific cells upon exposure to antigen get activated and secrete *chemokines* that facilitate macrophage and other committed cells to mediate *inflammatory response*. Chemokines regulate and facilitate leukocyte migration to the site of action in order to contain the infection. Lymphocyte recirculation further allows maximum number of antigen specific committed cells to reach at the site of action. Chemokines such as *intracellular adhesion* molecules expressed by endothelial cells of the infected area creates stickly environment for leukocytes which gets adhered to the surface and accumulates at the site where chemokines further help these cells in the process of *extravasation* through which they enter the site of infection.

II. DEFINITION

Chemokines are a family of closely related low molecular weight chemo-attractants which are produced during inflammatory reaction and functions as potent attractors for effector cells of inflammatory process such as neutrophils, monocytes and lymphocytes. These molecules serve to increase the functional interactions between cells of the immune system.

III. CHEMOKINE SUPERFAMILY

Chemokine superfamily consists of four structurally related subfamilies of molecules known as **selectins**, **mucins**, **integrins** and **ICAMS (intracellular adhesion molecules).** These molecules share a common structural motif of four highly conserved NH_2-terminal cysteine amino acid residues; CXC, CC or CX3C represent cysteine (where X represents a non conserved amino acid residue and C represents cysteine. Chemokines function in concert with other inflammatory molecules. Some of the important chemokines have been shown in Table 15.1.

A. Selectin Family

1. Selectin family of glycoproteins consists of mainly three proteins (a) E-selectin, expressed on endothelial cells (b) P-selectin expressed on platelets and (c) L-selectins which are expressed on leukocytes (Fig. 15.1).
2. These molecules have a number of extracellular domains.

Table 15.1: Some of the important chemokines and their features

Receptors on cells	*Expression*	*Ligands on endothelium*	*Step involving interaction*	*Main function*
CLA or ESL-1	Effector T cells	E-selectin	Tethering/rolling	Homing to skin and migration into inflamed tissue
L-selectin	All leukocytes	GlyCAM-1, CD 34, MAdCAM-1	Tethering/rolling	Lymphocyte recirculation via HEVs to peripheral lymph nodes and migration into inflamed tertiary sites
LFA-1 (αLβ2)	Leukocyte subsets	ICAM-1, 2, 3	Adhesion/arrest	General role in lymphocyte extravasation via HEVs and leukocyte migration into inflamed tissue
LPAM-1 (α4β7)	Effector T cells, monoctyes	MAdCAM-1, VCAM-1	Rolling/adhesion	Homing of T cells to gut via mucosal HEV; migration into inflamed tissue
Mac-1 (αMβ2)	Monocytes	VCAM-1	—	Monocyte migration into inflamed tissue
PSGL-1	Neutrophils	E- and P-selectin	Tethering/rolling	Neutrophil migration into inflamed tissue
VLA-4 (α4β1)	Neutrophils, T cells, monoctyes	VCAM-1 MAdCAM-1 fibronectin	Rolling/adhesion	General role in leukocyte migration into inflamed tissue
VLA-6 (α6β1)	T cells	Laminin	—	Homing of progenitor T cells to thymus; possible role in T cell homing to non-mucosal sites

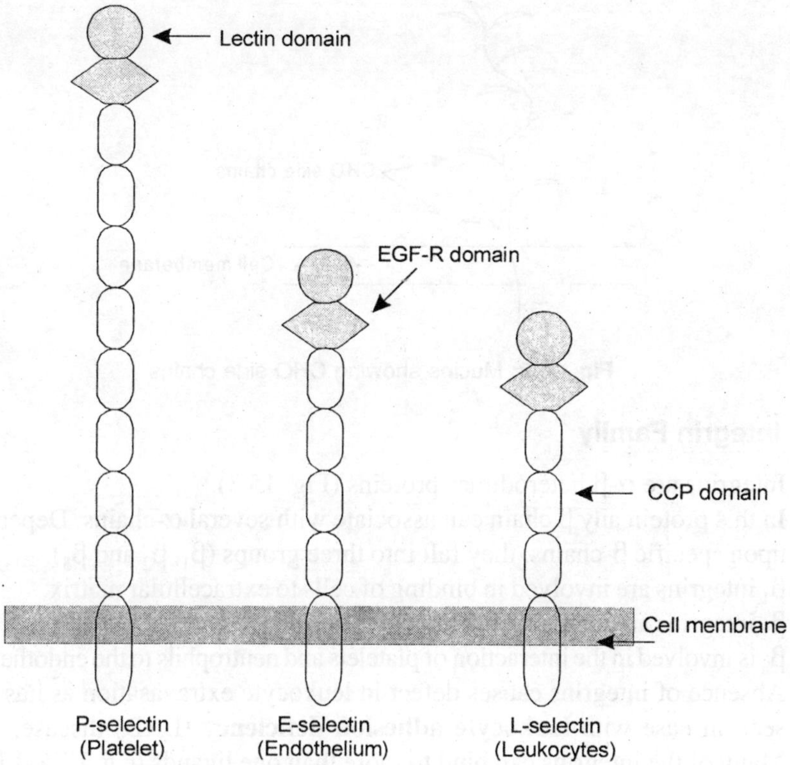

Fig. 15.1: Selectin family of molecules

3. The N-terminal of these molecules consists of carbohydrates (lectin) domain and also another domain related to the epidermal growth factor (EGF) receptor.
4. When selectins bind to their ligands on the circulating cells, the circulating cells slow down within the venule.

B. Mucin Family

1. Mucin family consists of serine and threonine rich glycoproteins.
2. These proteins are heavily glycosylated, therefore, they interact with lectin domain of selectins.
3. CD34 and Gly CAM-1 are expressed on lymph node endothelial cells.
4. Another member of this family P-selectin glycoprotein ligand (PSGL-1) are expressed on neutrophils and Th1 cells.
5. Mucins have characteristic thread like branched carbohydrate side chains (Fig. 15.2).

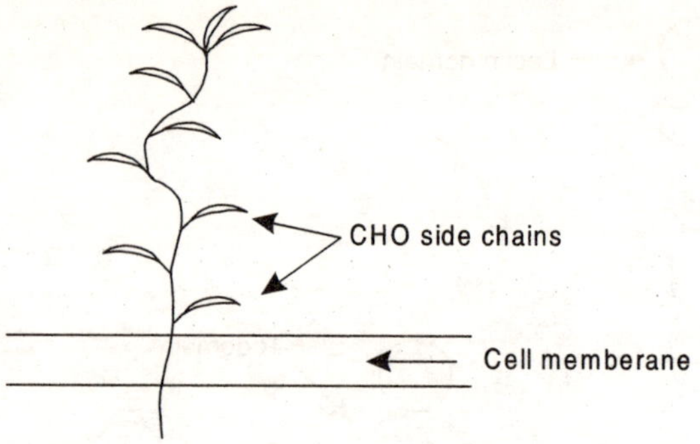

Fig. 15.2: Mucins showing CHO side chains

C. Integrin Family

1. Integrins are α-β heterodimer proteins (Fig. 15.3).
2. In this protein any β chain can associate with several α-chains. Depending upon specific β-chains, they fall into three groups (β_1, β_2 and β_3).
3. β_1 integrins are involved in binding of cells to extracellular matrix.
4. β_2 integrins are involved in leukocyte adhesion to endothelium
5. β_3 is involved in the interaction of platelets and neutrophils to the endothelium.
6. Absence of integrins causes defect in leukocyte extravasation as has been seen in case with **leukocyte adhesion deficiency** (LAD) disease.
7. Many of the integrins can bind to more than one ligands (e.g. LFA-1 binds to ICAM-1 and ICAM-2).
8. Integrins are expressed on large variety of cells such as T cells, macrophages, neutrophils, phagocytes, fibroblast, kidney and thyroid cells.

Fig. 15.3: General structure of integrin

D. ICAM Family

1. ICAMs (Intracellular adhesion molecules) belong to immunoglobulin superfamily. They are characterized by immunoglobulin like domains (Fig. 15.4).
2. These molecules (ICAM-1, ICAM-2, VCAM-1 and MAdCAM-1) are expressed on vascular endothelium.

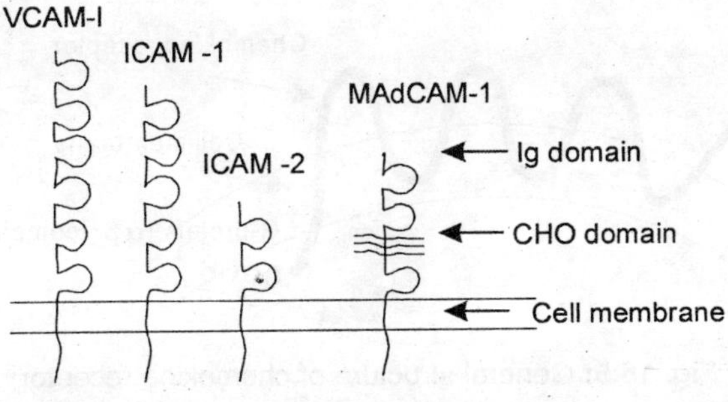

Fig. 15.4: Structures of intracellular adhesion molecules

3. MAdCAM-1 has both mucin-like and immunoglobulin-like domains. It is expressed in mucosal endothelium-where it directs lymphocyte entry into the mucosa.
4. MAdCAM-1 can bind to both selectins as well as integrins by virtue of its Ig like domain and CHO domains.

IV. FUNCTIONS OF CHEMOKINES

Chemokines play an important role in immune response such as cell-cell interaction, homing and development of certain organs. The chief functions of chemokines are as below.

1. Chemokines regulate various functions of leukocytes such as adhesion, chemotaxis, activation and cell trafficking.
2. Chemokines regulate development of brain and heart.
3. They also regulate development of blood vessels and wound healing.
4. Chemokines are induced by infections and upregulated by TNF-α.
5. They act through G-protein-linked receptors containing seven transmembrane helices.

V. CHEMOKINE RECEPTORS

1. All the chemokines act through their '**serpentine receptors**' which consist of seven transmembrane helices and are linked to G-proteins that cause cell activation (Fig. 15.5).
2. Most chemokines act via more than one receptor, similarly, one chemokine receptor respond to several chemokines (Table 15.2).
3. The chemokine receptors are selectively expressed on particular population of leukocytes, and it determines that which cell will respond to signal coming from which tissue.
4. Cells move from one place to another receiving successive signals for the next destination.
5. Chemokine receptor profile on a cell depends upon its type and stage of differentiation. For example, Thl cell expresses CCR5 and CXCR3 while TH2 cells preferentially express CCR3 receptor.

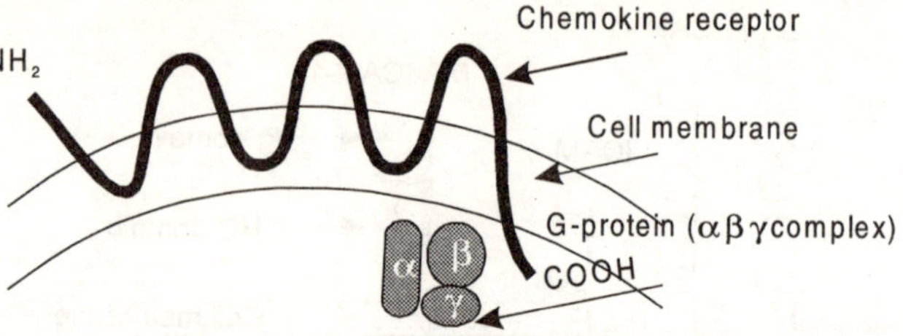

Fig. 15.5: General structure of chemokine receptor

Table 15.2: Chemokine receptors, chemokines and their distribution

Receptor	*Ligands*	*T cell*	*B cell*	*NK cell*	*Mono-cyte*	*Granulo-cyte*	*Other*
CCR1	RANTES, MCP-2, MCP-3, MIP-1α, MIP-3, MIP-5	+		+	+	E,B,N	Imm.DC
CCR2	MCP-1, MCP-2, MCP-3, MCP-4, MCP-5	*	+	*	+		
CCR3	RANTES, MCP-2, MCP-3 MCP-4, MIP-5, Eotaxins 1-3	TH2			+	E, B	
CCR4	MDC, TARC, MIP-1α	TH2		+			Imm.DC
CCR5	RANTES, MCP-2, MIP-1α, MIP-1β	TH1	+		+		Imm.DC
CCR6	MIP-3α,	+	+				Imm.DC
CCR7	SLC, MIP-3β,	Naïve	+				Mat.DC
CCR8	TARC, MIP-1β, 1309	TH2*				E, B	
CCR9	TECK	+			*		DC
CCR10	RANTES, MCPs-1-5, MIP-1β, Eotaxin-1, skinkine	M			+		
CCR11	MCD-1, MCP-2, MCP_4						
CXC1	IL-8, GCP-2	*				N	Endoth.
CXCR2	IL-8, Gro-α, Gro-β, Gro-γ, NAP-2	*	+			N	Endoth.
CXCR3	IP-10, Mig, I-TAC	TH1					Endoth.
CXCR4	SDF-1	Naïve	+		+		Many cells
CXCR5	BCA-1	+/-	+				
XCR1	Lymphotactin	+					
CX3CR1	Fractalkine	*		+	+		

VI. GENERAL FEATURES OF CELL MIGRATION AND HOMING

The leukocyte migration and homing take place in several steps which are followed by activation of endothelial cells by cytokines, expression of adhesion molecules, adherence of leukocytes at the site of action and extravasation of leukocytes into the tissue by crossing endothelial lining. Three-step model has been suggested for movement of leukocytes, i.e. rolling, triggering and latching that finally leads to transendothelial migration (Fig. 15.6).

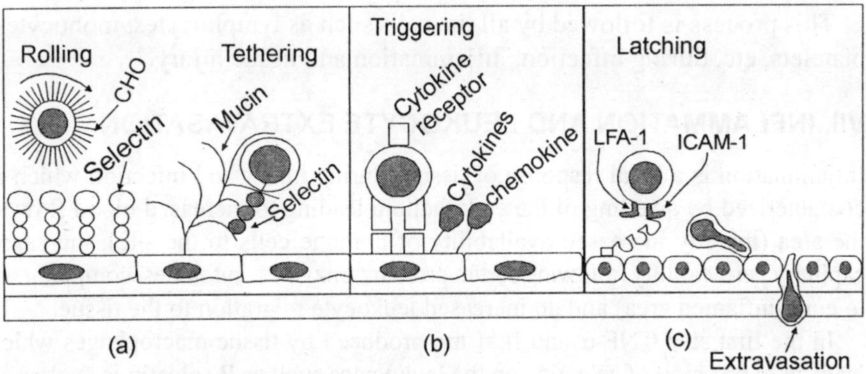

Fig. 15.6: Steps in leukocyte migration and homing

A. Rolling (Tethering)

Rolling or tethering is mediated by interaction of selectins expressed on the leukocytes and mucins (carbohydrate imoieties) present on the endothelial cells. This is a weak interaction which slow down the leukocytes that flow in the blood- stream. In this interaction, leukocytes tumble over the endothelial surface binding with mucins on the endothelial cell surface (Fig. 15.6a). In the endothelium, leukocytes respond to new sets of chemotactic stimuli present on subsequent cells.

B. Triggering

As the leukocytes roll over the endothelial cells, it get activated by various cytokines produced by the endothelial cells. These cytokines are basically chemoattractant (chemokines) such as interleukin-8, macrophage inflammatory proteins (MIP-1β), platelet activating factor and complement products (C3a, C5a, and C357).

C. Latching

Activation by cytokines causes conformational change in the integrin molecules which increases its affinity to intracellular adhesion molecules expressed on the endothelial cells. CAMs stabilize the adhesion of leukocytes to the endothelial cells (Fig. 15.6b). At this stge leukocytes are firmly attached to the endothelial cell surface.

The conformational changes in integrins in leukocytes as well as in endothelial cells alter its shape and leukocytes extravasate through the endothelial lining into the tissues (Fig. 15.6c). However, the detailed mechanisms of such transendothelial migration of leukocytes is not well understood. At this stage several proteolytic enzymes are released which digest proteins and collagen present between the interstitial space of the endothelium allowing leukocytes to migrate into the tissues.

This process is followed by all the cells such as lymphocytes, monocytes, platelets, etc. during infection, inflammation and tissue injury.

VII. INFLAMMATION AND LEUKOCYTE EXTRAVASATION

Inflammation is a local response of tissues during injury and infection which is characterized by swelling of the endothelium leading to increased blood flow to the area (thereby increased availability of immune cells to the site), increased capillary permeability (allowing serum proteins, enzymes, cytokines, complements to enter inflamed area) and an increased leukocyte migration to the tissue.

In the first step TNF-α and IL-1 are produced by tissue macrophages which leads to expression of selectins on the leukocytes such as P-selectin is expressed on platelets, leads to platelet aggregation, L-selectin expressed on leukocytes causes rolling and tethering movements of leukocytes and E-selectin expressed on endothelial cells slows down the movement of these cells so that more and more leukocytes and platelets remain at the site.

In the second step, neutrophils are triggered by chemokines to express chemokine receptors and endothelial cells to express more adhesion molecules so as to provide firm attachment to the leukocytes (lymphocytes, monocytes and neutrophils) on the endothelial surface. It is facilitated mainly by ICAMs and VCAMs on endothelial cells as well as their respective ligands expressed on the leukocytes. These molecules are induced by different cytokines and also they are expressed on specific tissues. For example, ICAM-1 is expressed at higher level in brain while VCAM 1 is expressed at higher level in skin endothelium. In addition, these molecules have differential affinities to their ligands on different cells. Such as ICAM-1 and ICAM-2 interact with LFA-1 (α_1, β_2 integrins) which is expressed on leukocytes where as VCAM-1 interacts with VLA-4 ($\alpha_4\beta_1$ integrins) which is upregulated in lymphocytes. Such interactions are probably aimed to selective attraction of various leukocyte subsets at the site of infection/ inflammation due to different causative agent.

Finally, there is extravasation of selective cells at the site of action through the transendothelial migration.

VIII. LYMPHOCYTE MIGRATION AND HOMING

1. Although migration and extravasation of lymphocytes to tissues are similar to that of the neutrophil extravasation, lymphocyte extravasate to selective tissues through the process named as 'lymphocyte traficking' or homing.
2. Lymphocytes migrate to lymph nodes, Peyer's patches and mucosal tissues through **high endothelial venule (HEV).**
3. High number of lymphocytes (almost 25%) which enters the lymph nodes are diverted to HEV.
4. Lymphocytes trafficking is mediated by selected adhesion molecules, chemokines and receptors (homing receptor) which directs lymphocytes to extravasate into selective lymphoid tissues.

5. HEVs are normally present in the lympoid tissues but they may be created at other sites as well during infection and inflammation.
6. HEV endothelial cells have peculiar shape and they express distinct-set of heavily glycosylated and sulphated adhesion molecules called addressins.
7. Naive and effector and memory lymphocytes follow different and distinct migration pathway.
8. Naive T lymphocytes express CCR7 and CXCR4 which allow them to respond to chemokines expressed on the endothelial cells.
9. The selectins expressed on naive T cells allow them to attach to the carbohydrate ligand on **HEV of mucosal tissues and peripheral lymph nodes**.
10. Naive B cells also express CCR7 and CXCR4 and migrate to the secondary lymphoid tissues.
11. Memory lymphocytes migrate and home to the tissues where they had first encountered the antigen.
12. Effector cells home to the region of infection by recognizing inflamed endothelium at the site of action. They are guided by the chemoattractant molecules produced during inflammation and immune response.

There are various other cells such as mast cells, basophils, and platelets as well as other molecules such as complement, complement receptor and decay accelerating factors (DAF), etc. that also participate in the inflammatory response. Ultimately, outcome of the inflammatory and immune response is to contain the infection. It has been seen that if the infection persists then leukocyte accumulation also continues leading to the development of chronic inflammatory response. The pattern of cytokine and chemokine expression varies between acute and chronic inflammation so are the leukocyte participation. Acute infection and inflammation is taken care of by neutrophils, platelets, mast cells, etc. while in case of chronic infection and inflammation the major effector cells are macrophage and lymphocytes.

STUDY QUESTIONS

1. Describe the general strategy of leukocyte migration and homing with suitable diagram.
2. What do you understand by chemokine superfamily? Describe briefly different members of chemokine superfamily.
3. Write short notes on
   a. Integrins
   b. HEV
   c. Chemokine receptor
   d. Selectins (3) Recirculation and homing of naive lymphocytes.

MCQs

Select one most appropriate answer given for each questions below:

1. **Which of the following molecule is a α/β heterodimer?**
   - A. Selectins
   - B. Mucins
   - C. Integrins
   - D. ICAMs
2. **Platelets express?**
   - A. E-selectin
   - B. P-selectin
   - C. L-selectin
   - D. P and L selectin
3. **Which of the following statements is false:**
   - A. Naive lymphocytes circulate indiscriminately to the peripheral lymphoid organs
   - B. Macrophages extravasate to the inflamed tissues.
   - C. Memory cells recirculate to the site of infection recognizing chemokines
   - D. Effector cells home to the region of infection
4. **Which of the following receptors is expressed by Th1 cells:**
   - A. CCR-5
   - B. CCR-4
   - C. CCR-3
   - D. CCR-2
5. **Which of the following receptors-ligand binds with high affinity?**
   - A. L-selectin-CD34
   - B. P-selectin-CD34
   - C. LFA-1-ICAM-1
   - D. Mucin-selectin

---

1 C 2 B 3 C 4 A 5 C

16

Autoimmunity

I. INTRODUCTION

The main aim of immune system is to discriminate between 'self' and 'non-self', preserve the 'self' and destroy non-self (foreign). Preservation of self depends upon tolerance to self-antigens which immune cells are educated during developmental stage. Similarly, destruction of non-self depends upon recognition of foreign antigens by cells of the immune system. However, self and non-self discrimination is not absolute. Therefore, some autoreactive cells persist in the host. In case with defect of self-tolerance or activation of auto-reactive cells due to genetic or environmental factors, such cells start destroying normal healthy tissues of the body by the immune mechanisms, called *auto-immunity*. Autoimmunity may precipitate in the form of localized disease such as myocardial infarction, myasthenia gravis, Hashimoto's thyroiditis; systemic diseases such as rheumatoid arthritis (RA), systemic lupus erythmatosus (SLE) and insulin dependent diabetes mellitus (IDDM), etc. Autoimmune disorders affect almost 5-7% of the population, often with severe disability. The affected persons are unfortunately in the prime of their life and due to unknown reasons, most victims are women than men.

II. FEATURES OF AUTOIMMUNE DISEASES

If we recall the mechanism of immune response, it becomes clear that the major player of immunity are antigen, antigen presenting cells (MHC molecules), effector cells (T and B lymphocytes) and their products, i.e. antibody, cytokines, receptors, etc. Thus autoimmunity could arise due to defects in one of them or many of them.

Human autoimmune diseases can be classified in several ways:

1. Most of the autoimmune diseases are strongly influenced by genes encoded within the MHC.
2. MHC II based diseases are:
   i. **Organ specific** which is characterized by their **autoantibody pattern** directed against the related organs (e.g. pancreatic β-cells in IDDM);
   ii. **Multisystem autoimmune diseases** which are characterized by a variety of antibody raised against nuclear or cytoplasmic components (anti- DNA Ab in SLE).

3. Some of the manifestations of systemic autoimmunity are due to **auto-antibodies** and others are due to deposition of **antigen antibody complexes** (e.g. glomerulonephritis).
4. MHC II based autoimmune diseases may be caused by both cellular or humoral immune effector cells. For example, IDDM appears mainly due to **CD_4^+ T cells**, whereas in Graves' disease and myasthenia gravis the effector molecules are anti-TSH receptor antibody and **anti-acetyl choline receptor antibody** respectively.
5. MCH I associated autoimmune disorders are few in number (e.g. ankylosing spondylitis, Reiter's syndrome and psoriasis vulgaris) and it may be characterized by the presence of auto-reactive CD_8^+ T cells and show **male predominance**.
6. Most of the autoimmune diseases are **polygenic,** i.e. showing involvement of multiple genes as well as **multifactorial.**
7. Most of the autoimmune diseases have two peaks of incidence; first at the time of puberty (may be related to hormonal changes and development of secondary sexual features), and second at 4th and 5th decade of life.
8. Almost all autoimmune diseases have a tendency for **spontaneous aggravation and remission** which could be due to fluctuating balance between positive and negative immunoregulatory forces.

III. IMMUNOLOGICAL TOLERANCE

During developmental stage both B and T lymphocytes are made unresponsive or tolerant to self-antigen. There are two sites where tolerance develop: (i), central lymphoid organs (thymus and bone marrow), (ii) Peripheral lymphoid organs (lymph nodes and spleen).

A. Central Tolerance

1. T and B lymphocytes originate from stem cells in the thymus and bone marrow respectively.
2. These cells inherit antigen receptor genes that reorganize to give rise to functional receptors through gene-rearrangement.
3. Lymphocytes with high affinity for self-antigens are deleted by negative selection and cells with low or intermediate affinity undergo positive selection and differentiate into mature T and B lymphocytes.
4. As a result some of these cells are capable of recognizing self-antigens but during maturation in central lymphoid organs, they are made tolerant to self-antigens.

B. Peripheral Tolerance

1. Self reactive lympocytes that escape above selection process are subsequently inactivated at peripheral lymphoid organs (lymph node and spleen) under the conditions which favour tolerance but not activation.

2. In peripheral lymphoid organs, lymphocytes are made tolerant by three mechanisms: (a) clonal deletion (b) clonal anergy (c) clonal ignorance.
   a. *Clonal deletion:* If lymphocyte self-antigen specific receptors bind to self-antigen with high affinity, they undergo apoptotic death, this process is called clonal deletion.
   b. *Clonal anergy*: In some situation, self-antigen reactive clones are presented self-antigens by inappropriate APC in absence of co-stimulatory molecules, they develop anergy. Such clones become unresponsive to self-antigen.
   c. *Clonal ignorance:* In this case, a particular lymphocyte clone may have functional antigen receptors for self-antigens but they simply do not respond to self-antigens. Such form of unresponsiveness is called clonal ignorance. The reasons could be either low affinity receptors, low abundance of self-antigens or inhibitory action of Th2 cells. Some of the factors associated with self-tolerance has been shown in Table 16.1.

Table 16.1: Self tolerance in T and B lymphocytes

Features	*T cells*	*B cells*
1. Principal site of tolerance	• Thymic cortex • Peripheral lymphoid organs	• Bone marrow • Peripheral lymphoid organ
2. Stage of maturatin	• Double positive stage (CD_4^+ CD_8^+) • Low TCR-CD3	Membrane IgM^+ IgD^-
3. Stimulus	• High affinity receptor • Bone marrow derived APC • Absence of co-stimulatory molecules	• High affinity receptor • Multivalent Ag • Lack of T cell help
4. Mechanisms for tolerance development	• Clonal deletion • Clonal anergy • Clonal ignorance	• Clonal deletion • Clonal anergy • Blockade of signal transduction
5. Duration of tolerance	Long lasting	Short lived
6. Antigen concentration	Relatively low	Relatively high

IV. ETIOPATHOGENESIS OF AUTOIMMUNE DISEASES

1. The mechanism of autoimmunity has been shown in Figure 16.1. The first requirement for autoimmunity is the presence of autoreactive T cells in the system. These cells could arise through defects in either central tolerance or peripheral tolerance mechanisms.
2. Although autoreactive T cells are deleted in the thymus through negative selection, some autoreactive T cells may be included in the repertoire which may undergo positive selection by critical residues encoded by MHC susceptible allele. The exact mechanisms of breakdown in central and peripheral tolerance is not clearly understood.
3. The third possibility could be that the target antigens must be available for presentation by susceptible MHC allele to autoreactive T cells to complete

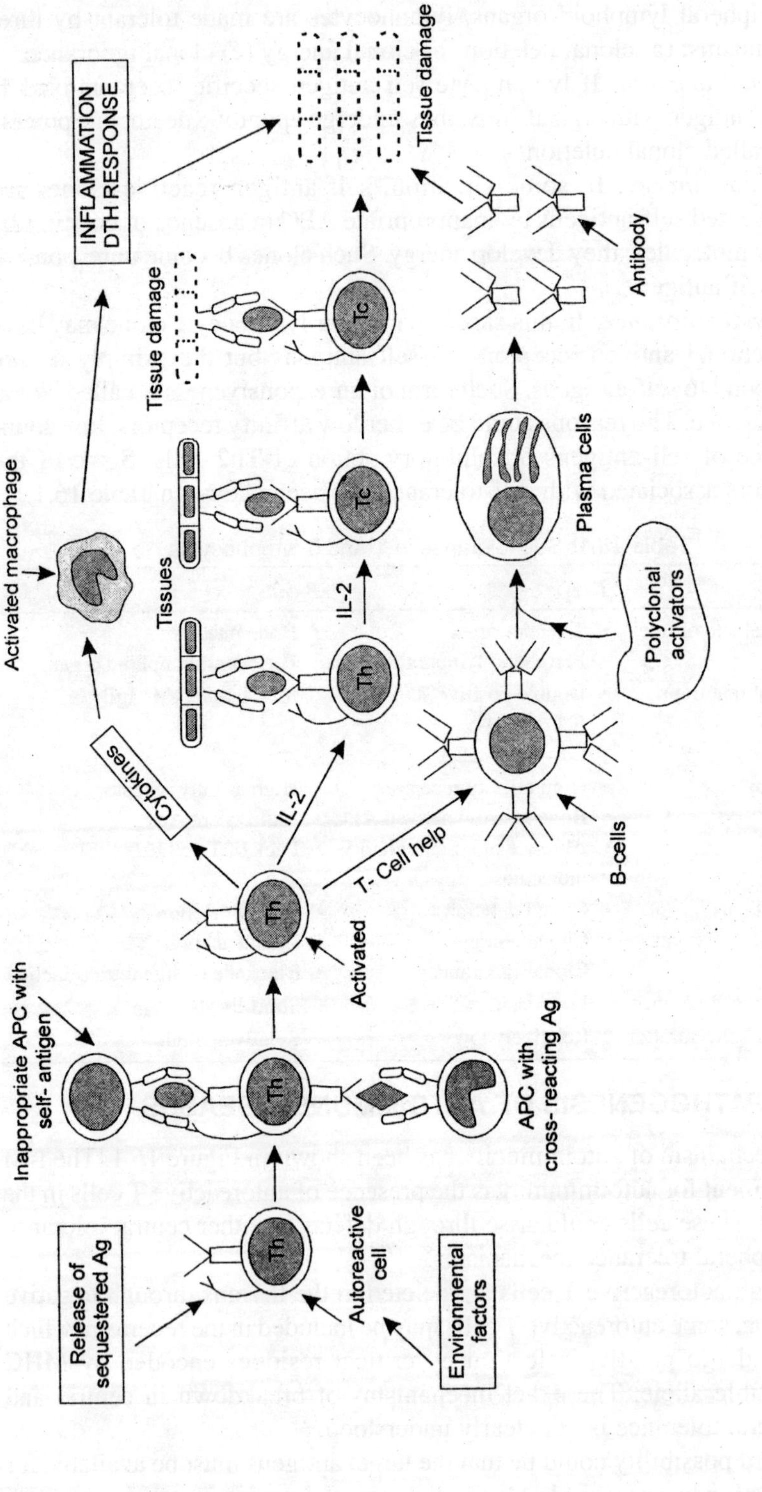

Fig. 16.1: Showing different mechanisms of induction of autoimmunity

the ternary complex consisting of antigen-MHC-TCR. The autoreactive antigens could be derived from anatomically sequestered antigens which are normally not exposed to the immune system.

4. In many diseases, it has been seen that autoimmunity is preceeded by tissue injury or environmental insults. Tissue injury or environmental insult by viral or bacterial infection could act through several mechanisms. It may allow:
   i. The release of sequestered antigens for immunologically priviledged site.
   ii. Local inflammation may cause ectopic expression of MHC II molecules and presentation of autoantigens.
   iii. Inflammatory reaction may lead to activation of Th1 cells and expansion of Th1 cytokine expression.
   iv. Bacterial and viral antigens may have molecular mimicry with self-antigens. For example, HLA-B27 molecule has been implicated in Reiter's syndrome and ankylosing spondylitis. The amino acid sequence of HLA-B27 has been shown to share a region of five or six amino acids with *Klebsiella pneumoniae* nitrogenase protein.

   Most of the critical events involved in breakdown of self-tolerance have been shown in Table 16.2.

Table 16.2: Genetic, immunological and environmental factors involved in autoimmunity

1. Presence of **MHC susceptibility alleles** which
i. Binds and selects target self-antigens
ii. Selects for autoreactive T cells repertoire
2. Existence of **autoreactive T cells** based on
i. Germline V, D, J and constant region elements.
ii. Somatic rearrangement of TCR.
iii. Positve and negative selection in thymus
iv. Escape from tolerance induction in periphery
3. Exposure of **self-antigens** to immune cells based on.
i. Release of sequestered antigen
ii. Ectopic expression of MHC II molecules on non-lymphoid cells.
4. Costimulatory signals and Th1 cytokines.
5. Immunoglobulin genes
6. Triggering by **environmental agents** due to:
i. Th1 cytokine production during infection.
ii. Molecular mimicry
iii. Epitope spreading

5. Ectopic production of MHC I and MHC II molecules in tissue other than professional antigen presenting cells may lead to in appropriate Th cell activation which may lead to autoimmune consequences. For example, individuals with IDDM express high level of MHC I and MHC II molecules on pancreatic β cells. Similarly, thyroid acinar cells in patients with Graves' disease express MHC II molecules on their surface whereas it is not expressed by normal thyroid acinar cells. Such inappropriate expression of

MHC II molecules, (which are otherwise only expressed by professional APCs), may serve to sensitize Th1 cells against self-antigen.
6. Polyclonal activation of B lymphocytes caused by bacterial and viral infection may lead to activation of large number of B cell clones that express IgM in the absence of Th cells. Such activation of β-cells may lead to its differentiation into plasma cells secreting autoantibody. For example, in case with **infectious mononucleosis** caused by EBV infection, a variety of autoantibodies are produced including those that react with T and B cells, rheumatoid factor and antinuclear antibody.
7. Imbalance between Th1/Th2 has been shown to be related to autoimmunity in experimental animals. It seems that Th2 cells keep a check on Th1 related autoimmune phenomena. However, when there is reduced number of Th2 cells, Th1 activity increases resulting in autoimmunity. Most of the human autoimmune disorders have been found to be associated with a higher Th1 cytokine production.
8. Autoimmunity has also been found to be associated with defect in genes regulating apoptosis such as BCL_2, Fas and Bax, etc.
9. Genetic predisposition is another factor where immunoglobulin and MHC genes have been implicated in the development of autoimmune diseases. Almost 25-70% of patients with autoimmunity have family history of the disease. A positive association of MHC haplotype has been reported in patients with rheumatoid arthritis (HLA-DR4) and IDDM (HLA-DR2, DR3, DR4, etc). Reports are coming on polymorphism of other cytokine and cytokine receptor gene's association with autoimmune diseases.

V. TREATMENT OF AUTOIMMUNE DISEASES

Although autoimmune diseases are self-contained but in some cases these are highly debilitating and cripling. Such patient needs immediate therapy to reduce the symptom if not to cure the diseases. Since autoimmunity is associated with aggravated immune response, most of the treatment modality is aimed to cause immunosuppression in the patients. However, immunosuppressive agents may increase the risk of infection and cancer as it causes generalized immunosuppressions in the patients. Some of the modality of treatment is as follows:

1. The common immunosuppressive drugs are corticosteroids, azathioprine and cyclophosphamide. These drugs inhibit cellular proliferation and are most active on proliferating cells including rapidly proliferating bone marrow cells. These drugs therefore, cause generalized immunosuppression in the patients.
2. The more selective drugs are **cyclosporine A** and **FK 506**. These drugs bind to immunophillin and FK-binding protein which inhibits IL-2 gene transcription by interfering the IL-2 transcription factor.
3. Removal of thymus (thymectomy) has produced positive results in patients with Graves' disease where thymic hyperplasia has been found to be associated with the disease.

4. Many autoimmune diseases have been found to have deposition of antigen-antibody complexes, such as myasthenia gravis, SLE and RA. In such cases, plasmapheresis has been found to provide significant remission.
5. In experimental encephalomyelitis, it has been found that the animal undergoes remission by injecting low dose of **myelin basic protein** which is a causative antigen in such disease. Thus oral dose of autoantigens have been found to give more therapeutic benefit in case with human autoimmune diseases.
6. More specific treatment has been imparted by using autoreactive T cells with cross-linking of cell surface receptor by paraformaldehyde or gluteraldehyde treatment. When cross-linked T cells were used in animals with EAE, a long-term remission of the symptoms was observed. It seems that cross-linking of membrane proteins elicits regulatory T cells specific Vβ-TCR determinants of the autoreactive T cell clones.
7. Recently, it has been shown that alteration of few amino-acid residue of autoantigen leads to blockade of TCR, causing inhibition of immune response. Such **peptide blockade** has been found to be effective in cases where amino acid sequence of the antigen is known.
8. Monoclonal antibodies against various effector cell molecules have also been found to be effective in patients with autoimmune diseases. Commonly used Mo Abs are anti tac (against IL-2R α chain), anti-CD4, anti-MHC II antigen as well as anti-TCR Vβ isotype antibody. Recently soluble TNF has been found to provide long-term remission in patients with rheumatoid arthritis.

STUDY QUESTIONS

1. What do you understand by immunological tolerance? Describe the mechanisms of immunological tolerance in human.
2. What is autoimmunity? Describe in brief factors influencing autoimmune diseases.
3. Writ short notes on (a) Peptide blockade (b) Central tolerance (c) Peripheral tolerance (d) Uses of monoclonal antibodies in the treatment of autoimmune diseases.

MCQs

Select one most appropriate answer given for each questions below:

1. **Autoimmune disease are mainly caused by activated:**
   A. CD_4^+ T cells **B.** CD_8^+ T cells
   C. Monocytes **D.** NK - cells
2. **Cyclosporin A is used in the treatment of autoimmune disease because:**
   A. It inhibits TCR expression **B.** It induces T cell anergy
   C. It inhibits IL-2 production **D.** It inhibits IL-2 R expression

3. **Which of the following statements is true?**
   A. Large amount of peptide injection can cause peptide blockade in patients with autoimmune disease:
   B. Autoimmunity is associated with TH1 cell activity
   C. Autoimmunity is associated with Th2 cell activity
   D. Anti Tac antibody down regulates IL-2 secretion
4. **Which of the following reactions is not associated with autoimmunity?**
   A. DTH response
   B. Inflammation
   C. Immediate hypersensitivity
   D. CD_8^+ T cells mediated cytotoxicity
5. **Antinuclear antibodies are generally present in:**
   A. Systemic autoimmunity
   B. Organ specific autoimmune disease
   C. Localized autoimmune diseases
   D. Rheumatoid arthritis
6. **Rheumatoid found in synovial fluid of patients with RA is most frequently:**
   A. IgM reacting with K chains of IgG
   B. IgM reacting H chains of IgG
   C. IgE reacting with bacterial antigen
   D. Antibody to collagen
7. **Autoimmune diseases due to antibody may occur:**
   A. Due to Ag-Ab complex
   B. Due to super antigen exposure
   C. Defective antigen presentation
   D. Enhanced T cell help
8. **Possible mechanisms for the development of autoimmune diseases are following *except*:**
   A. Infection with bacteria bearing cross-reactive antigen
   B. Leakage of sequestered self-antigen
   C. Loss of CD_4^+ T cell activity
   D. Alteration of self-antigens
9. **SLE develops due to:**
   A. Mutation in ds–DNA
   B. Multiple organ involvement
   C. Immune complex deposition
   D. T cell hyperactivity.

1 A 2 C 3 B 4 C 5 A 6 B 7 A 8 C
9 D

17

Transplantation Immunity

I. INTRODUCTION

Transplantation is a surgical procedure by which cells, tissues or organs of one individual (donor) are grafted on the other (recipient) in order to cure the disease or replace the diseased organ and tissues to resume normal function. In such cases, acceptance of grafted cells, organ and tissue depends upon the **immunological compatibility (also called histocompatibility)** of donor and the recipient. Rejection of the graft is an immunological phenomenon in which grafts are destroyed by the recipient's effector cells and molecules of the immune system if donor and recipient are immunologically incompatible. The graft rejection process could be acute (days and weeks) or chronic (months and years). Grafts from genetically identical twins are always accepted while that from genetically different individuals are always rejected. Additionally, graft can activate almost all immuno-regulatory mechanisms in the recipients, thus understanding the immunology of transplantation becomes more important.

II. HISTORY

1. In 1908, Alexis Carrel interchanged both kidneys in a group of nine cats. Some of the recipients maintained urinary output until 25 days and eventually all cats died.
2. In 1935, a Russian surgeon performed first human kidney transplantation but it failed due to mismatching of blood group.
3. In 1947, Sir Peter Medawar performed skin transplantation on pilots with burn injury during second world war. He observed that skin from different region of the same individual was always accepted but skin from other individual was rejected. Through a series of animal experiment later, he proved that rejection of graft is an **immunological phenomenon**.
4. The first successful human kidney transplantation was performed in Boston in 1954 between the identical twins which opened up the possibility of clinical transplantation in human beings.
5. Today with the development of understanding of transplantation immunology and discovery of immunomodulatory agents, clinical transplantation

of kidney, pancreas, heart, lung, liver, bone marrow and cornea has become a reality.

III. TYPES OF GRAFT

Transplantation immunologists have developed following scientific term to define grafts.

A. Autografts

The graft taken from an individual and transplanted on the same individual is called autograft. Autografts always survive.

B. Isograft

A graft transplanted between genetically identical individual is called isograft. Since such individuals are syngeneic, such grafts are also called **syngeneic** grafts. Survival of isograft is always longer and mostly it is accepted.

C. Allograft

A graft transplanted between genetically different members of the same species is called allograft such grafts contain **alloantigens** which induce immune reaction in the host leading to rejection of the graft.

D. Xenograft

Tissues transferred from one species to another is called xenograft such grafts present xenoantigen which is more violent, always leads to rejection of the graft.

Transplantation of the tissue on their natural anatomical site is called **orthotropic transplantation** while those at different anatomical location is called **heterotropic transplantation**.

IV. KINETICS OF GRAFT REJECTION

1. The graft rejection process shows specificity and memory.
2. In general, skin graft is rejected faster than the other organ grafts such as kidney or heart.
3. If skin is grafted on genetically different animal (allograft), the rejection process shows two types of rejection response.
   i. For the first time allograft is rejected in 12-14 days, called **first set rejection.**
   ii. If the skin is obtained from same donor and grafted on the same recipient, second rejection is quite quick and takes place within 5-6 days. This rejection response is called **second set rejection.**

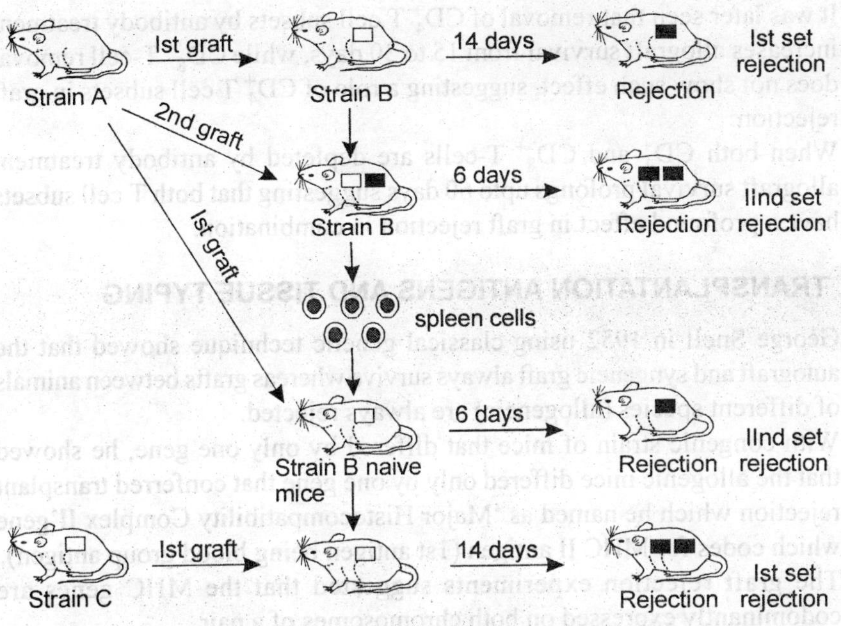

Fig. 17.1: Kinetics of graft rejection showing first set and second set of rejection as well as memory and specificity

iii. If immune cells are taken from syngeneic recipient mice which has previously rejected the graft, and injected in the naive recipient, it directly induce **second set rejection** of the graft when skin is grafted from the same donor, showing a **memory response** imparted by immune cells.

iv. If the same recipient receives graft from another strain of mice, it shows **first set rejection response** (Fig. 17.1).

V. GRAFT REJECTION IS MEDIATED BY CELL MEDIATED IMMUNITY

1. In 1950, Mitchison through adoptive transfer experiments showed that the graft rejection is a T cell mediated phenomenon. In his experiment, he showed that when T lymphocytes are transferred from grafted animal to the naive animal of same strain followed by grafting from the same donor, second set of rejection of graft rejection takes place. Thus the graft rejection process is accelerated by adoptively transferred sensitized T cells.
2. He also showed that when cell free serum is injected in the naive recipient, the animal shows 1st set of rejection response, suggesting that graft rejection process is T cell mediated phenomena but not the antibody mediated response.
3. Athymic nude mice does not reject graft from allogenic donors.

4. It was later seen that removal of CD_4^+ T cell subsets by antibody treatment increases allograft survival from 15 to 30 days, while CD_8^+ T-cell removal does not show such effect, suggesting a role of CD_4^+ T cell subsets in graft rejection.
5. When both CD_4^+ and CD_8^+ T-cells are depleted by antibody treatment allograft survival prolongs upto 60 days suggesting that both T cell subsets have a profound effect in graft rejection in combination.

VI. TRANSPLANTATION ANTIGENS AND TISSUE TYPING

1. George Snell in 1952 using classical genetic technique showed that the autograft and syngeneic graft always survive whereas grafts between animals of different species (allogeneic) are always rejected.
2. With congenic strain of mice that differed by only one gene, he showed that the allogenic mice differed only by one gene that conferred transplant rejection which he named as 'Major Histocompatibility Complex II' gene which codes for MHC II antigen (1st antigen being blood group antigen).
3. The graft rejection experiments suggested that the MHC genes are codominantly expressed on both chromosomes of a pair.
4. Codominant expression of MHC II genes also means that when A and B strains of mice produce (A × B) F_1 hybrid, it will recognize either of the parental graft A or B as self-antigen while (A × B) F_1 graft will be recognized by either of the parent as foreign. This forms the basis of host vs. graft reaction.
5. Additionally, ABO blood group antigens are also present on endothelial and epithelial tissues, therefore, unmatched blood group antigens may also lead to production of antibody in the recipients leading to complement mediated lysis.
6. Therefore, in clinical transplantation, in both blood group matching and HLA-typing (MHC matching) is done.
7. There are basically two methods for tissue typing. (A) Microcytotoxicity assay using monoclonal antibody (B) Mixed lymphocyte reaction using leukocytes.

   A. HLA-typing (Microcytotoxicity assay)

   i. In this assay, leukocytes from potential donors and recipient are distributed in a series of wells of microtitre plate.
   ii. Monoclonal Ab specific for various HLA-I and HLA-II antigen is added to the wells.
   iii. After incubation, complement is added to the wells and cytotoxicity is assessed using a chemical dye (Trypan blue) that stains dead cells.
   iv. If the leukocytes of the individual express particular HLA antigen to which Ab has been added, it undergoes complement mediated lysis.
   v. Percent matching can be scored between donors and recipient as shown in Figure 17.2.

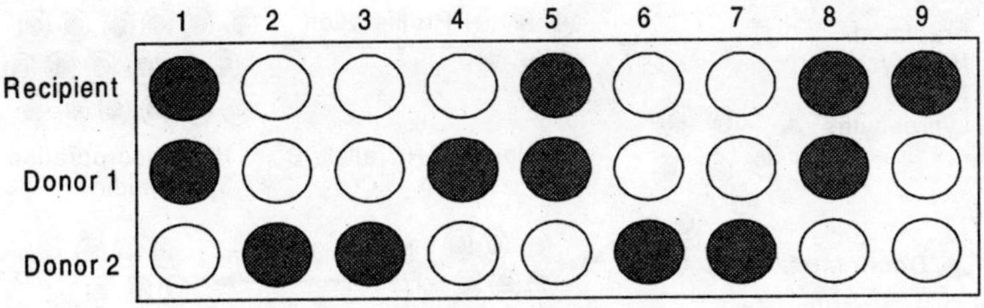

Fig. 17.2: Microcytotoxicity assay using monoclonal Ab. Donor 1 shows 3/4 (75%) matching of HLA antigen while donor 2 shows completely different HLA-type

B. HLA-typing (Mixed Lymphocyte Reaction)

i. When suitable donor is not available as per microcyte-toxicity assay, one way MLR reaction can be used to find out donors for successful transplantation.

ii. In this reaction, lymphocytes from potential donor is made unresponsive by mitomycin C treatment and plated in the wells of tissue culture plate.

iii. Recipient's lymphocyte is added to it and the plate is incubated.

iv. In case with HLA unmatching, recipients' lymphocyte recognizes donors MHC II molecule as foreign antigen and proliferates.

v. Proliferation is measured by radio-active **thymidine incorporation assay** in which proliferating lymphocytes incorporate exogenous radiolabelled thymidine in their DNA which is measured by radioactivity counting (Fig. 17.3).

vi. MLR is based on HLA-II matching and provides a better measurement of Th cell proliferation in response to recipients HLA molecule.

vii. It takes longer time (several days) where as microcytotoxicity assay could be performed in hours.

8. However, matching of major histocompatibility complex is not sufficient to predict survival of the graft. Graft can be rejected by differences in other **'minor histocompatibility antigens'** which are presented in context with self MHC molecules. Differences in the minor histocompatibility complex leads to chronic rejection of the graft which may take place in months and years.

VII. MECHANISM OF GRAFT REJECTION

Allograft rejection is principally a cell mediated immune response against alloantigens present in the graft which takes place in two stages.

A. Stage I: Sensitization of the Host Immune Cells

In sensitization phase, immune cells such as Th and Tc cells are sensitized against alloantigen causing activation and differentiation of these cells into effector cells. Such reaction takes place in following ways.

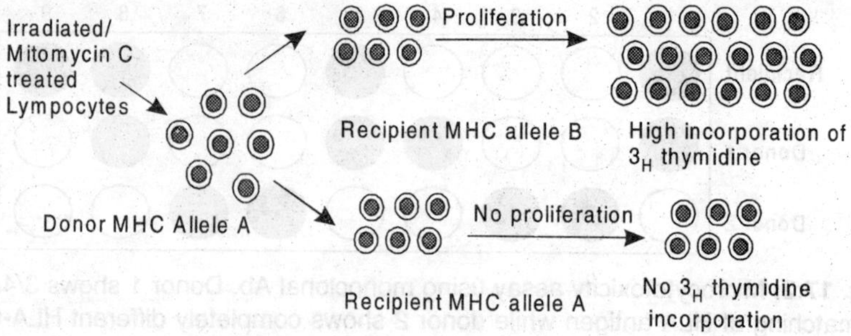

Fig. 17.3: Showing one way mixed lymphocyte reaction (MLR) in response to MHC II antigens

a. Direct antigen presentation

1. In the beginning of the transplantation, transplanted tissues have their own APCs such as **dendritic cells**, **Langerhans cells** and endothelial cells which express high level of allogenic MHC II and normal level of MHC I molecules on their surface occupied with allogenic peptides.
2. Such allogenic MHC molecules and associated peptides are directly presented by **host APC** to respective hosts' lymphocytes **without** further processing and such process is called **direct antigen presentation**.
3. In response to direct Ag presentation, host Th cells and Tc get activated and differentiate into respective effector cells.
4. In addition, some of the host's proteins are also processed by donors APCs it gets activated and migrates to the lymph node of the host. As mentioned, these are the donors dendritic, Langerhans and endothelial cells which are called **passenger leukocytes**.
5. Passenger leukocytes' also activate some of the hosts' autoreactive cells which can react to graft's cross-reacting antigens.
6. Donors alloantigen may have MHC molecules showing molecular mimicry with the host MHC molecules, such donor's MHC also activates autoreactive cells of the host.
7. In addition, donor's alloantigen may contain several T-cell epitopes that may induce proliferation of large number of T cell clones.

b. Indirect antigen presentation

1. In this case, allogenic peptides including allogenic MHC molecules may be processed thoroughly endocytotic pathway of antigen presentation like any other peptide antigen and it leads to activation and proliferation of host' effector cells.

B. Stage II. Effector Functions of the Host Immune Cells

Such activation and proliferation of host's CD_4^+ cells cause cytokine production such as IL-2, IFN-γ, IL-4 and IL-5 which may amplify and mediate a number of immune functions. The major effector immune response in graft rejection is DTH response, cell mediated cytotoxicity, complement mediated cytotoxicity, ADCC and inflammatory response as shown in Figure 17.4.

1. DTH response is mediated by effector CD_4^+ T cells subset. These cells produce cytokines that attract other cells such as monocytes and macrophages infiltrating the graft. These cells along with activated CD_4^+ can be seen histologically infiltrating the graft and they also express high amount of MHC I and MHC II molecules on their surface.
2. IL-4 and IL-5 cytokine produced by CD_4^+ T cell subsets also activate B cells to differentiate into plasma cells that secrete antibodies against alloantigen which leads to destruction of graft by complement mediated cytotoxicity and ADCC.

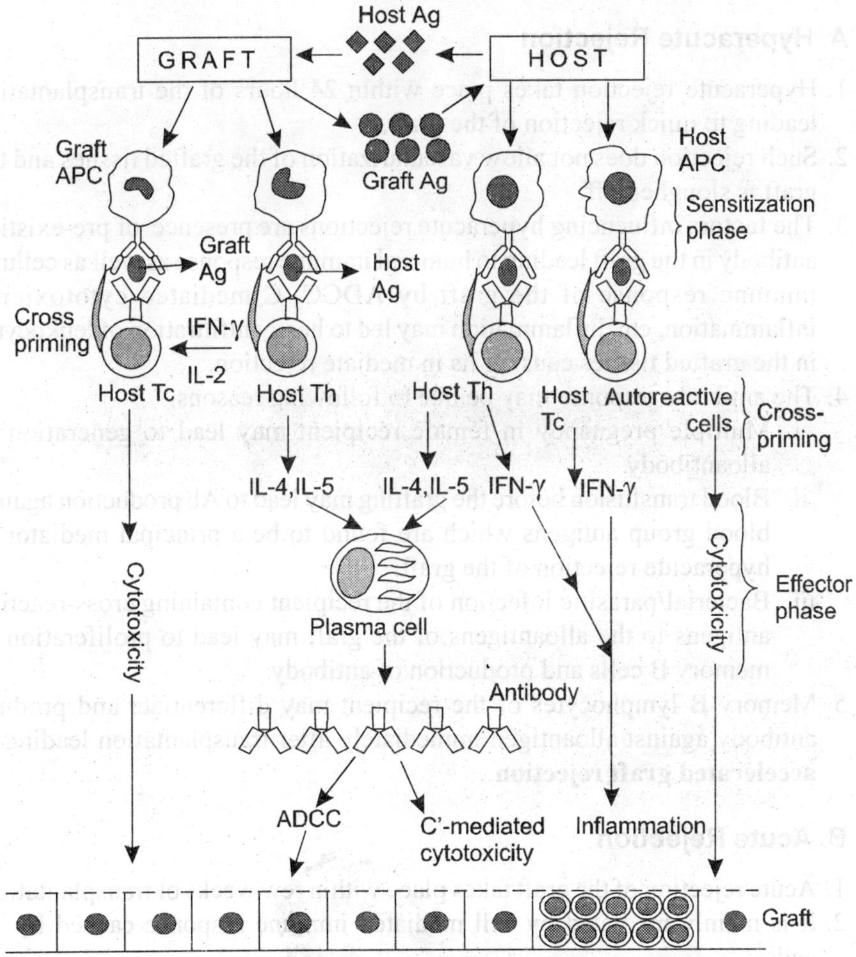

Fig. 17.4: Mechanism of transplantation rejection reaction

3. IL-2 and IFN-γ secreted by hosts' CD_4^+ T cell subsets lead to activation and proliferation of CTLs that mediate cytotoxic reaction to the graft.
4. IFN-γ secreted by CD_4^+ T cell subsets also leads to inflammatory response in the graft ultimately causing destruction of the transplanted tissues.

It has also been seen that passenger leukocytes can generate a state of tolerance of graft by mean of formation of chimeric leukocytes or by inducing negative selection (thymic deletion) of T cells possessing high affinity receptors for cross-reacting self-antigens. Therefore, these cells are also used to develop tolerance in the transplant recipients.

VIII. CLINICAL MANIFESTATIONS OF GRAFT REJECTION

Clinically, graft rejection phenomenon has been divided into hyperacute, acute and chronic rejection depending upon time course of survival of the graft.

A. Hyperacute Rejection

1. Hyperacute rejection takes place within 24 hours of the transplantation leading to quick rejection of the graft.
2. Such rejection does not allow vascularization of the grafted tissues and the graft is sloughed off.
3. The factors influencing hyperacute rejections are presence of pre-existing antibody in the graft leading to humoral immune response as well as cellular immune response of the graft by ADCC, C'mediated cytotoxicity, inflammation, etc. Inflammation may led to heavy infiltration of leukocytes in the grafted tissues causing its in mediate rejection.
4. The antibody response may be due to following reasons.
   i. Multiple pregnancy in female recipient may lead to generation of alloantibody.
   ii. Blood transfusion before the grafting may lead to Ab production against blood group antigens which are found to be a principal mediator of hyperacute rejection of the graft.
   iii. Bacterial/parasitic infection of the recipient containing cross-reacting antigens to the alloantigens of the graft may lead to proliferation of memory B cells and production of antibody.
5. Memory B lymphocytes of the recipient may differentiate and produce antibody against alloantigen immediately after transplantation leading to **accelerated graft rejection.**

B. Acute Rejection

1. Acute rejection of the graft takes place within few weeks of transplantation.
2. It is mainly mediated by cell mediated immune response caused by Th cells.
3. Such graft histology show infiltration of T cells and macrophages.

4. The mechanism are DTH response and Tc cell mediated cytotoxicity, as described earlier.
5. Acute rejection of the graft can be managed by immunosuppressive drugs or other modality of treatment.

C. Chronic Rejection

1. Chronic rejection of the graft takes place in months and sometimes in years.
2. It may take place even in MHC (HLA) matched tissue.
3. The major reasons for chronic rejection are the differences in **minor histocompatibility antigens** between donor and recipients.
4. Chronic rejections are commonly seen in case with kidney transplantation patients.
5. Such graft rejections are difficult to be managed by immunosuppressive drugs.

IX. TREATMENT

We have learned that the graft rejection is an immune response mediated phenomenon involving almost all immune cells in one form or the other. Therefore, treatment of allograft rejection involves causing immunosuppression of the host which could be by inducing **generalized immunosuppression**, **specific immunosuppression** or induction of tolerance in the recipients.

A. General Immunosuppressive Treatment

Generalized immunosuppression can be induced by using different drugs and immunomodulators as follows:

1. **Mitotic inhibitors:** These are the drugs which cause immunosuppression through suppressing bone marrow cell divisions. These drugs are basically mitotic inhibitors such as methotrexate, cyclophosphamide and azuthioprine. Methotrexate is a folic acid antagonist which blocks purine biosynthesis while cyclophosphamide is an alkylating agents that inserts into the DNA helix leading to disruption of the DNA chain. Azathioprine blocks synthesis of inosinic acid which is precursor of purines. Therefore, it prevents DNA synthesis in the cells and act on S-phase of the cell cycle.
2. **Corticosteroids:** Corticosteroids, such as dexamethasone and prednisone are potent anti-inflammatory agents. These compounds lead to immunosuppression. Therefore, these drugs are used to prevent acute rejection of the graft.
3. **Immunomodulators:** The common immunomodulators used in clinical transplantation patients are fungal metabolites, cyclosporin A and FK 506. Both these drugs are potent inhibitors of IL-2 gene expression and has been found to significantly prolong the graft survival. Another recently

discovered drug is rapamycin. These drugs reduce activation of Th cells, Tc cells, NK cells, macrophages and even antibody producing B cells and these are widely used.

4. **Irradiation:** It is known that T lymphocytes are highly sensitive to X-rays. Thus X-ray exposure to thymus, spleen and lympho nodes has been found to prolong graft survival if given before transplantation. Although bone marrow continue to supplement immune cells in such patients, these new cells becomes easily tolerant to the alloantigens present in the graft.

B. Specific Immunotherapy

Various strategies can be adapted for specific immunotherapy of the transplantation patients. These are most commonly imparted through monoclonal antibodies against various immune components such as CD3, CD4, CD 25 (IL2 receptor α-chain), IL-2, adhesion molecules, (ICAM-1, ICAM-2) costimulatory molecules like CD28 or CTLA, 4 etc. These antibodies prevent functions of the respective cells and molecules involved in the transplantation rejection reactions.

C. Immunotolerance

Tolerance to allograft can be developed using various strategies such as:

i. Low dose exposure of the alloantigens repeatedly to patients lead to development of tolerance.
ii. Adoptive transfer of low number of donor leukocytes leads to formation of chimeric leukocytes that do not react to transplants.
iii. Transplantation at immunologically privilege sites such as cornea, anterior chamber of the eye may lead to prolonged survival of the graft.
iv. Intrathymic injection of pancreatic cells has been found to develop tolerance to pancreatic cells in diabetic patients.
v. In one study, islet cells encapsulated in the semipermeable membrane was found to survive in the recipient for longer time.

Thus these strategies have made clinical transplantation possible even in unrelated recipients.

STUDY QUESTIONS

1. Describe molecular mechanisms involved in transplantation rejection with a suitable diagram.
2. Describe different types of graft. What strategy would you adopt to prolong survival of each type of graft?
3. Write short notes on:
   a. Passenger leukocytes
   b. Immunotolerance

c. Cyclosporin A
d. Cytokine effect in graft rejection
e. Mixed lymphocyte reaction.

MCQs

Select one most appropriate answer given for each questions below:

1. **Tissue typing involves:**
   A. Blood group matching
   B. Matching of HLA
   C. Matching of minor histocompatibility complex
   D. Matching of histological features of tissues
2. **Hyperacute rejection of graft takes place because of:**
   A. Passenger leukocytes
   B. IL-2 and IFN-γ
   C. Pre-existing antibody against the graft
   D. NK-cells
3. **Cyclosporin A inhibits**
   A. IL-2 receptor expression
   B. IFN-γ production
   C. IL-2 gene transcription
   D. DTH response
4. **Cyclophosphamide is:**
   A. An alkylating agent
   B. DNA binding protein
   C. Immunomodulator
   D. Transcription factor
5. **Chronic rejection of the graft takes place due to:**
   A. Mismatch in major histocompatibility complex between donor and recipients
   B. Passenger leukocytes
   C. Chronic infection
   D. Differences in minor histocompatibility complex
6. **Inclinical transplantation cytotoxic antibodies:**
   A. cause delayed rejection of the transplant
   B. Are responsible for hyperacute rejection
   C. When present in the donor, cause rejection
   D. May be directed against TCR
7. **GVH disease requires the followings *except*:**
   A. MHC differences between donor and recipient
   B. Immuno-competent donor cells
   C. Activated suppressor T cells
   D. Activated helper T cells

8. The most common serologic test used for the detection HLA antigen on lymphocyte is:

A. The complement fixation test
B. Micro-cytotoxicity test
C. Mixed lymphocyte reaction
D. Graft vs. host reaction

9. Transplant rejection may involve the followings *except*:

A. ADCC
B. CMI
C. Delayed hypersensitivity reaction
D. Immediate hypersensitivity reaction.

1 B **2** C **3** C **4** A **5** D **6** C **7** C **8** B
9 D

18

Tumour Immunity

I. INTRODUCTION

All the living beings maintain homeostasis, i.e. a balance between cell renewal and cell death. Normally, a cell have a fixed lifespan, regional and territorial identity and regulated cell division. If due to certain reason such regulation is lost (due to environmental and genetic factors), cells undergo uninterrupted growth and forms tumour. Some tumours revert back to normal phenotype once the stimulus (factors) ceased to exist, such tumours are called *benign tumours*. However, some tumours progress further called, malignant tumours which do not revert back to normal phenotype-rather they grow continuously and invade neighbouring or distant organs. Immune system plays an important role in preventing tumour growth through elimination of mutated or viral infected cells. Immune system consists of multiple antitumour cells such as macrophage, NK cells and cytotoxic T cells which remain in constant surveillance of the system and eliminate abnormal cells by their respective effector functions. Unfortunately, tumour cells arise, establish and grow even then in some individuals. In this chapter, various antitumour effector mechanisms, their evasion by tumour cells and different immunological strategies to combat cancer will be discussed.

II. MALIGNANT TRANSFORMATION AND FEATURES OF MALIGNANT CELLS

Malignant transformation of normal cells or carcinogenesis is a multistep process (Fig. 18.1). Environmental factors such as irradiation, carcinogens or virus initiates the process of carcinogenesis which leads to change in the genome (mutation). Initiation process is followed by promotion by tumour promoting factors (e.g. phorbol ester) causing rapid cell division of the initiated cells that form tumours. Malignant cells have following characteristics.

1. They become anchorage independent and loose their territorial identity.
2. They require low or negligible amount of growth factors, as opposed to normal cells which require growth factors in larger amounts for their growth.
3. Malignant cells become density independent. Even single cell can form appreciable tumour mass whereas normal cell require appreciable density of cells to grow.

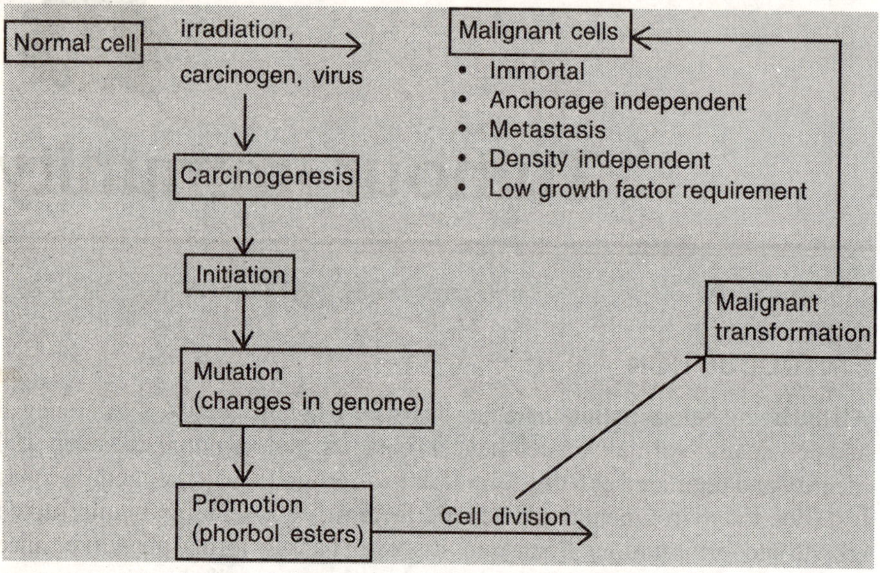

Fig. 18.1: Multistep process of carcinogenesis

4. Tumour cells are immortal and grow continuously.
5. They invade neighbouring or distant tissue, survive and thrive in new territory (metastasis).
6. Tumour cells initiate angiogenesis, i.e. they grow newer blood vessels for continuous supply of nutrients.

III. NOMENCLATURE AND CLASSIFICATION

Tumours may arise from almost any type of tissues. Thus they have been named after their nature of origin adding the suffix 'oma' to the tissue involved.

- *Carcinoma:* These are the tumours of the epithelial cells, however, almost all types of tumour are called carcinoma.
- *Sarcoma:* These are the tumours of mesenchymal cells such as fibroblast, muscle cells and fat. Tumours of fat cells are also known as lipoma.
- *Lymphoma:* The tumour cells originated from lymphoid tissues are called lymphoma. Lymphomas could be either T or B cell lymphomas.
- *Leukemia:* Tumour cells derived from hematopoietic cells are called leukemia. Leukemias are liquid tumours while all the above tumours are solid tumours.

IV. EVIDENCES OF ANTI-TUMOUR IMMUNE RESPONSE

Histological observations of tumours suggest that immune response against tumour takes place in the host:

1. Mostly solid tumours are infiltrated with effector cells or the immune system such as lymphocytes, monocytes, macrophages and NK cells. In addition,

the degree of lymphocytic infiltration in the tumour has been used as predictive marker for outcome of the treatment.

2. Tumour draining lymph nodes show hyperplasia and inflammation of the cells.
3. Histological sections of the tumour often show necrosis of the tissues.
4. Tumour infiltrating lymphocytes show higher expression of MHC molecules as well as ICAMs which are normally induced by cytokines, suggesting that cytokine secretion may have taken place.
5. High incidence of tumour development in patients with immunosuppression suggests a role of immunity in preventing tumour growth. Athymic nude mouse shows frequent tumour development. Similarly, Kaposi's sarcoma frequently arise in patients with acquired immunodeficiency syndrome (AIDS).

V. TUMOUR ANTIGENS

Immune response to tumours suggests that tumour express antigens to which T cells get sensitized and activated. In case of MHC I associated tumour antigen, there is proliferation and activation of cytotoxic T lymphocytes (CTL). Therefore, CTLs generated through various tumour antigen (peptides) can be used as a marker for tumour antigens. Tumour antigen specific CTLs can be generated against immunogenic tumours or viral induced tumours. On the other hand some tumour antigens may lead to activation of Th1 cells and antibody production by B cells. Such tumour antigens can be identified by using specific antibody. The third type of tumour antigen could be self-protein and may not induce antibody response in the host but its expression can be checked by antibody raised against such proteins.

Tumour antigens can be used for diagnosis of tumours, monitoring disease activity as well as an indicator of response to treatment. Tumour antigens can be broadly divided into three groups.

1. Mutated gene product
2. Viral gene product
3. Deregulated expression of normal gene.

A. Mutated Gene Product

This group of tumour antigens arise due to mutation of certain gene leading to development of tumours. Such antigens may be of two types.

A. Tumour specific antigens
B. Tumour associated antigens.

1. Tumour Specific Antigens

Tumour specific antigens are those antigens which are specific for one particular type of tumour but are not shared by any other tumours. The most common example of such antigens are **tumour specific transplantation antigens**

(TSTA) which are present on the physically or chemically induced tumours. When a naive animal of similar genetic background is sensitized with chemically induced tumour cells, and challenged with tumour cells later on, the sensitized animals rejects tumour cells by the mechanism similar to the transplantation rejection reaction. It has been shown that such sensitized animals generate memory cells against tumour antigen which participate in rejection of tumours whereas nonsensitized animals do not show such reaction.

It has been shown that such tumour transplant rejection is mediated by CTLs which are specific to that particular type of tumour cells which can proliferate in response to the same tumour antigens. Thus such CTLs could be used as marker cells to identify such tumour antigens in *in vitro* assay.

2. Tumour Associated Antigens (TAA)

Tumour associated antigens are those antigens which are expressed by more than one type of tumour cells. Such antigens may be mutated gene products or could be normal gene products, over expressed in certain type of tumours.

i. Mutated normal gene products has been seen in case with murine mastocytoma which is produced by point mutation in p19A gene.
ii. Another group of mutated gene products are oncogenes. Human chronic myeloid leukemia has been shown to arise due to bcr/abl gene rearrangement giving rise to protein p210. Similarly, p21 ras is another oncogene which shows point mutation in amino acid 12. P53 mutation has been found to be associated with many epithelial tumours and also in case with neuronal tumours.
    The oncogene products are also present in normal cells. Human breast cancer cell show elevated levels of oncogene encoded **Neu** protein which is a growth factor receptor. However, in normal breast tissue the level of Neu protein expression is significantly low.
iii. Other tumour antigens in this group are **silent gene products**. These genes are normally not expressed in normal cells but over expressed in tumour cells. One of the common example of such antigen is MAGE (melanoma associated gene product protein). These proteins are expressed in almost 50% of melanomas and 25% of breast cancer cells (MAGE-1). A number of MAGE proteins has been identified so far which has been found to be immunogenic and over expressed in many epithelial cancer such as glioma cell lines, non-small cell lung cancer and head and neck cancer (MAGE-1, 2). Since these antigens are shared by many tumours these can be potential candidate for development of tumour vaccine.

B. Viral Gene Products

Several murine and human tumours are induced by viruses. Such tumours express viral proteins on its surface, called viral antigens. Viral antigens are also shared antigen which may be present in more than one type of tumour cells. For example, Epstien-Bar Virus causes Burkitt's lymphoma as well nasopharyngeal carcinoma in human. Therefore, EBV antigens are expressed on both these tumours. Similarly, human papilloma virus (HPV) has been implicated in cervical cancer in woman. SV40 has been found to be associated in some murine tumours. These viruses code for T antigens (tumour antigen) which are shared by other viruses of the same group.

Antigens of virally induced tumours mostly elicit a strong humoral and cell mediated immunity. Therefore, these are potential candidate for development of antibodies for diagnostic and therapeutic uses as well as generation of CTL clones for above purpose. Following two strategies have been developed recently for characterization of such antigens.

i. **Isolation and purification of tumour associated antigen:** It has been successfully used to characterize melanoma associated antigens and generating CTL clones for these antigens (Fig. 18.2). In this strategy, tumour antigens (peptides) bound to MHC I molecules are eluted with low pH solution and purified by HPLC (High pressure liquid chromatography) and amino acid sequence is determined. In addition, these peptides are used to activate CTLs by coculturing with PBMC. CTLs specific for different peptides are labelled and stored in the form of CTL a library for diagnosis of tumour antigens of other tumours of unknown origin. Since these CTLs can lyse tumour cells, it can also be used for therapeutic purposes.

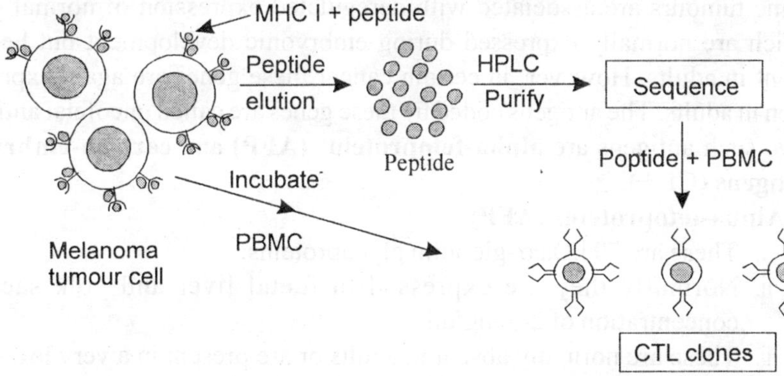

Fig. 18.2: Isolation of tumour antigens to generate CTL clones

ii. **Preparation of cDNA library from tumour cells:** In the second approach, a complementary, DNA library can be prepared using tumours cells as shown in Figure 18.3. These cDNA libraries are transfected transiently into monkey kidney cells (COS) with the SV40 large T-antigen gene. When these cells are transfected with plasmid containing tumour cell DNA and SV40 origin of replication, they provide about 10^4-10^5 copies of plasmid DNA resulting in high expression of tumour antigen

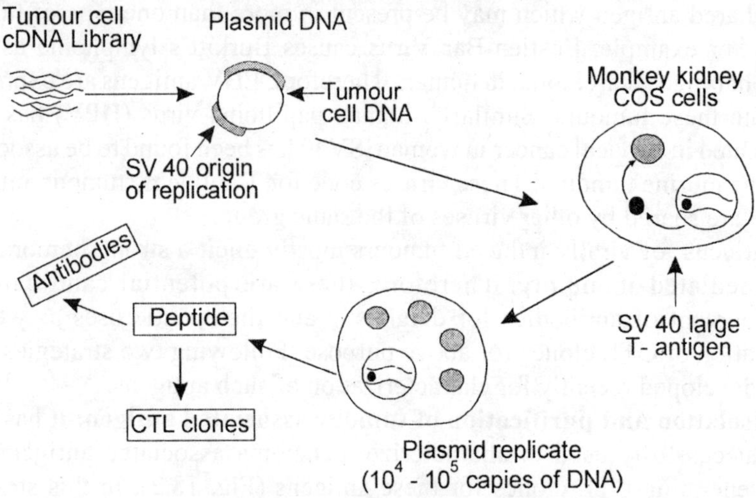

Fig. 18.3: Preparation of DNA library from tumour cells

DNA. Such tumour antigen may further be used to produce monoclonal antibodies or to generate CTL clones which may be used for diagnostic as well as therapeutic purposes. Sequence of such DNA has also indicated the type of mutations in such genes as well as their source of origin.

C. Deregulated Expression of Normal Gene (Oncofoetal Antigens)

Some tumours are associated with deregulated expression of normal genes which are normally expressed during embryonic development but become silent in adults. However, in certain cancer these genes are again expressed even in adults. The antigens coded by these genes are called oncofetal antigens. Two such antigens are **alpha-fetoproteins (AFP)** and **carcino-embryonic antigens (CEA)**.

a. Alpha-fetoproteins (AFP)

i. These are 70 kDa α-globulin glycoproteins.

ii. Normally they are expressed in foetal liver and yolk sac in a concentration of 2-3 mg/ml .

iii. These are normally absent in adults or are present in a very low level.

iv. These are over expressed in patients with liver cancer and germ cell tumours.

v. These are also expressed in case with cirrhosis of liver.

vi. AFP can be used to monitor disease status in patients after treatment and reappearance of serum AFP indicates recurrence of the tumour.

b. Carcino-embryonic antigen (CEA)

i. These are 180 kDa integral membrane glycoproteins.

ii. A member of immunoglobulin superfamily.

iii. These are present in foetal liver, gut and pancreas in third trimester.

iv. These are over expressed in almost 90% of patients with colon and colorectal carcinoma.

v. CEA is used for monitoring progression or recurrence of colorectal carcinoma.

VI. IMMUNOSERVEILLANCE AND ANTITUMOUR EFFECTOR MECHANISMS

Paul Ehrlich in early 1950s proposed the theory of **immunosurveillance** which suggests that the immunocompetent cells remain in constant surveilance and eliminate mutated and tumour cells as and when they arise. This theory was later expanded by Sir Mac Farlane Burnet and Lewis Thomas in 1960s. The proposed theory envisage the role of macrophages, NK cells, B and T lymphocytes in antitumour immunity. It is further strengthened by the observation that tumour frequently developes in immunosuppressed host, a thymic mice and elderly individuals with senescent immune functions. Thus it seems reasonable to presume that both innate immune component and adaptive immune component have potential role in antitumour response, which is evident from spontaneous regressions of tumours in ceratin hosts. The effector cells of antitumour response and their actions are as follows.

A. T Lymphocytes

T lymphocytes are the major antitumour cells of adaptive immune system. These cells express antigen receptors on their surface. The major evidence of cytotoxic T cells (CD_8^+) has been seen in case with virally induced tumours and immunogenic tumours such as melanoma where CTLs recognize tumour antigens in association with MHC I molecules and destroy it by various lytic mechanisms. It has been also seen that a major proportion of tumour infiltrating lymphocytes (TILs) are CD_8^+ Tc cells which can be expanded in tissue culture and functions as antitumour cells.

CD_4^+ T cells generally are not cytotoxic cells but activated CD_4^+ T cell produces a number of cytokines such as IFN-γ, IL-2 and TNF-α which potentiate cytotoxic response in various ways. IFN-γ leads to enhanced expression of MHC I molecules on tumours cells making them more susceptible to CTL mediated lysis, on the other hand it also activates NK cells. Similarly, IL-2 and TNF-α secreted by Th1 type of CD_4^+ T cell subsets activate CTLs, macrophage and NK cells. Thus both CD_4^+ T cell and CD_8^+ T cell subsets play a major role in immuno-surveillance and anti tumour immunity.

B.NK Cells

NK cells perform antitumour response in several ways.

i. These cells do not require prior sensitization and can kill tumour cells instantly through granzymes or perferin mediated lysis.

ii. NK cells also express low affinity Fc γ RIII (CD16) receptor for IgG molecules. Thus these cells can lyse antibody coated tumour cells by antibody dependent cell mediated cytotoxic mechanism (ADCC).
iii. IFN-γ and IL-2 activated NK cells are differentiated as LAK cells (Lymphokine activated killer cells), can lyse NK cell resistant tumour cells.

There has been a great deal of interest in recent days in the area of LAK cell research. It has been found that LAK cells generated in *in vitro* condition from tumour derived NK cells (TILs) cultured with IL-2 and/or INF-γ have increased capacity to lyse tumour cells when reinjected in the tumour bearing host. Therefore, LAK cells are potential candidate for adoptive immunotherapy of cancer.

C. Macrophages

Macrophages are also important cells of natural immunity against cancer. Activated macrophages produce TNF (tumour necrosis factor) which are directly cytotoxic to tumour cells through generation of superoxide. Additionally, it acts as an anti angiogenesis factor and prevent vascularization of tumour cells. TNF-α has also been found to induce apoptosis in the target cells through Fas-FasL interaction. Macrophages possess Fcγ receptors, thus they can efficiently kill antibody coated tumour cell by ADCC. In addition macrophages destroy tumour cells by phagocytic mechanisms as in case with infected cells.

D. Antibodies

Although natural antitumour antibody have limited role in antitumour immune response, antibodies may facilitate destruction of tumour cells through ADCC and complement mediated lysis. Human cancer patients produce antibody against tumour cell which can be used for diagnosis of limited tumour types since these antibodies in general are non-specific type.

VII. EVASION OF ANTITUMOUR IMMUNITY

Inspite of the above antitumour immune mechanisms, tumours continue to grow in the host suggesting that the tumours are able to evade host defence mechanisms. Current area of research, also aims to understand the various evasion mechanisms adopted by tumour cells in order to survive and thrive in the host. Some of the known immune mechanisms are as below.

A. Kinetics of Tumour Growth

It has been seen that in the beginning tumour grows slowly unless it attains certain size. Once appreciable number of tumour cells are attained (10^4, 10^5, cells), its growth becomes faster and doubling time of tumours are reduced as shown in Figure 18.4.

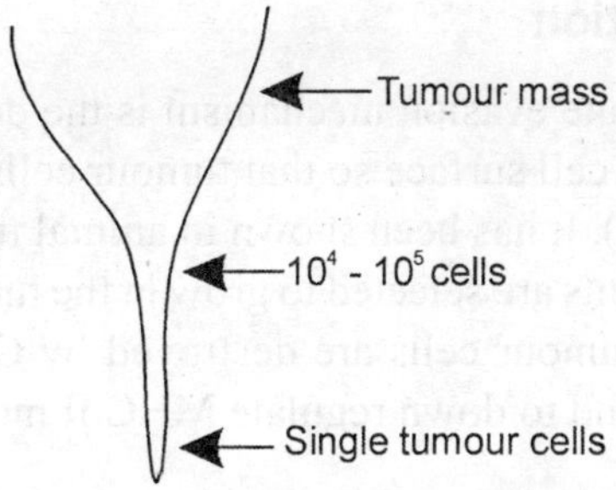

Fig. 18.4: Showing kineties of tumour growth

B. Immunosuppression by Antitumour Antibody

Hellström (1971) suggested that serum of tumour bearing host contains immune-inhibitory factors, which was later discovered as immune complexes (AgAb complex) and he named it as **blocking factor**. It was later shown that antibodies mask tumour antigens by binding to the tumour antigens. Blocking factors have been found to be associated with a number of tumours. It was further shown that Ag Ab complex activates Ts (Suppressor T-cells). In addition, antitumour antibody modulates tumour specific antigen by endocytosis (Fig. 18.5).

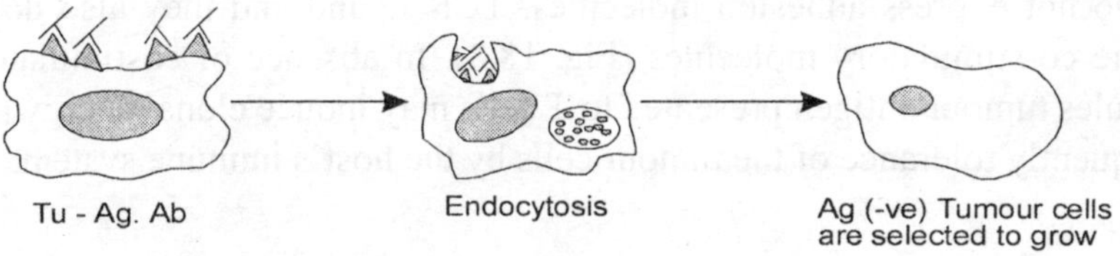

Fig. 18.5: Antigenic modulation by antitumour antibody

C. Immunosuppressive Factors

Tumour cells have been found to secrete many immunosuppressive factors. One of the most characterized immunosuppressive factor is TFG-β which has been found to be produced in larger quantity by several tumours. TGF-β inhibits lymphocyte and macrophage functions (Fig. 18.6).

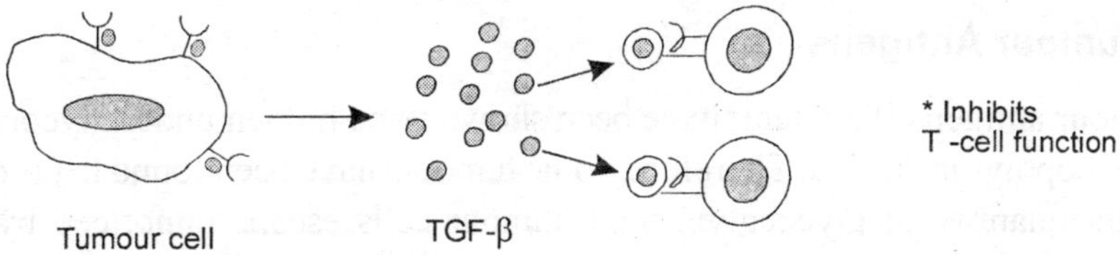

Fig. 18.6: Tumour induced immunosuppression

D. MHC I Down Regulation

One of the important immune evasion mechanism is the down regulation of MHC I molecules from the cell surface so that tumour cells escape attack by cytotoxic T cells (Fig. 18.7). It has been shown in animal tumours that MHC I negative tumour cell variants are selected to grow in the tumour bearing host while MHC I expressing tumour cells are destroyed by CTLs. In addition, tumour cells have been found to down regulate MHC II molecules as well.

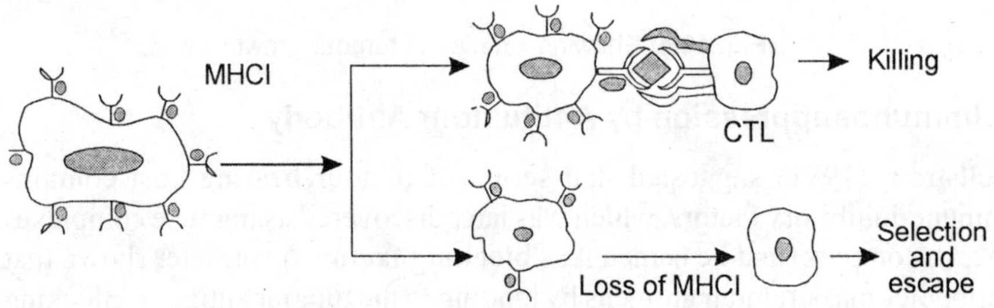

Fig. 18.7: MHC-I down regulation

E. Low Immunogenicity

Spontaneously developed tumour cells are mostly weakly antigenic. These cells do not express adhesion molecules, TCR ligand and they also down regulate co-stimulatory molecules (Fig. 18.8). In absence of costimulatory molecules tumour antigen presented to T cells may induce clonal anergy and consequently tolerance of the tumour cells by the host's immune system.

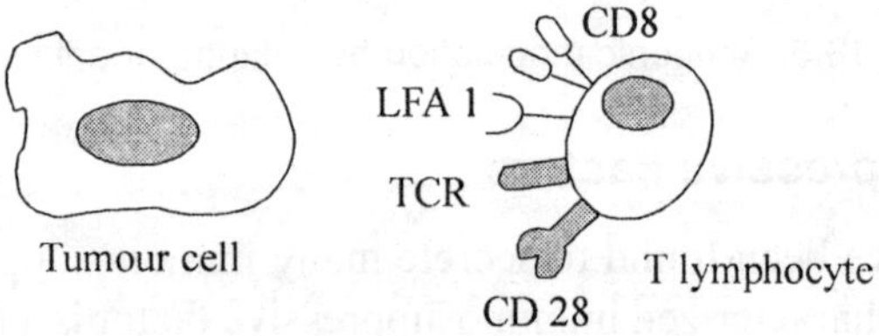

Fig. 18.8: Low immunogenic tumour cells do not express B7, adhesion molecules or TCR ligand (MHC)

A. Tumour Antigens

Tumour antigens if present, have been shown to be hidden under glycocalyx of mucopolysaccharide. Therefore, some tumours have been found to produce higher quantity of glycocalyx. Such tumour cells escape unnoticed by the immune cells.

G. Fas Ligand

Most of the tumour bearing host have been found to express Fas ligand which can induce apoptosis in constitutively Fas expressing tumour infiltrating lymphocytes.

H. Immunosuppressive Cytokines

Cancer patients show disregulated expression of Th2 cytokines such as IL-4, IL-5, IL-10 and TGF-β which down regulate protective cellular immune response of Th1 subsets and CTLs.

VIII. CANCER IMMUNOTHERAPY

The most common treatment of cancer is radiotherapy, chemotherapy and surgery. Both chemotherapy and radiotherapy are toxic to other cells of the body particularly rapidly dividing bone marrow cells and immune cells. Surgical treatment has its own limitations and it cannot be performed in cases where tumour metastasis have occurred in the multiple organs/tissue. Therefore, immunotherapy has the potential for tumour specific therapy and may be highly useful in patients with disseminated tumours and recurrent tumours. A number of immunotherapeutic modalities are currently under investigation as well as under clinical trials. Aim of cancer immunotherapy is to augment immune response against tumour in patients. Cancer immunotherapy may be divided into following five categories.

A. Nonspecific Immunotherapy

It is one of earliest tested approach in which immunization of tumour bearing host with heat killed M tb product BCG was found to augment immune response of the host and consequently it lead to regression of tumours. This early observation laid the foundation of cancer immunotherapy.

B. Active Immunization with Tumour Vaccine

Active immunization of tumour bearing individuals with modified autologous tumour cells have been found to be effective in generating protective antitumour immunity. Various modifications of tumour cells and tumour antigens are as follows:

i. **Irradiated tumour cell vaccine:** It has been found that introduction of irradiated autologous tumour cells provide augmentation of immune response and development of resistance against recurrent tumours.
ii. **Tumour specific antigen/pulsed dendritic cell vaccine:** Immunization of host with tumour specific antigen or tumour antigen pulsed dendritic cells have been reported to generate tumour specific CTLs in tumour bearing host.

iii. **TSTA + Costimulatory molecule tumour vaccine:** Tumour cells transfected with TSTA gene have been found to effectively express tumour antigens on the surface, however, it was found that such modified tumour cells leads to enhancement of tumour growth (Fig. 18.9A). It was also noted that such tumour antigens are recognized by T cells, however, since tumour cells do not express costimulatory molecules (B 7.1, B 7.2), in absence of costimulatory signal the T cells become anergic. In such situation, a rapid tumour growth takes place. When the tumour cells were transfected with both TSTA gene + B 7.1/B 7.2 genes, an effective antitumour response was observed (Fig. 18.9B). Thus transfection of TSTA and costimulatory molecule gene seems to be an effective mode of immunotherapy of tumours.

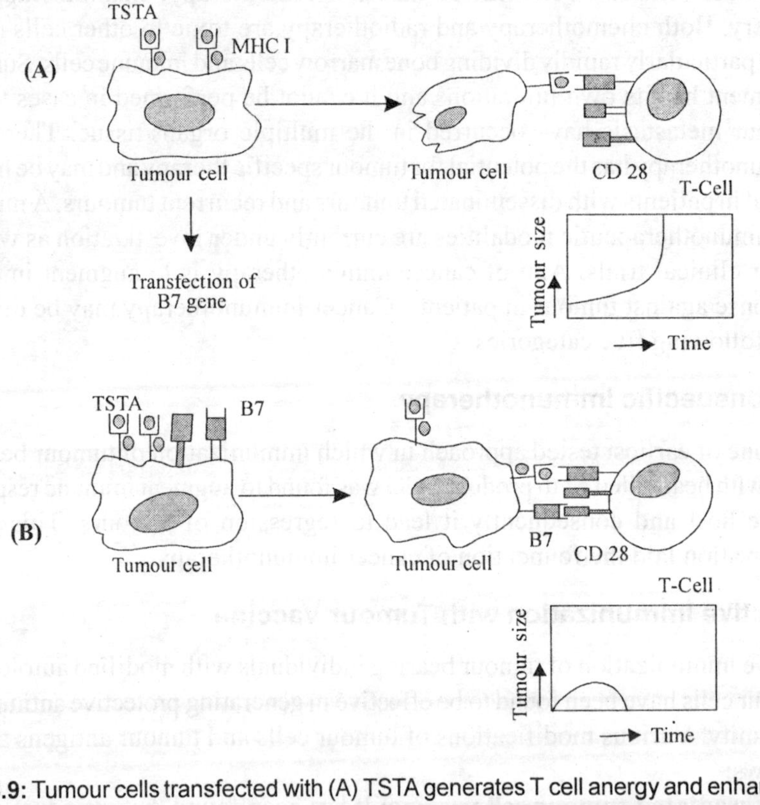

Fig. 18.9: Tumour cells transfected with (A) TSTA generates T cell anergy and enhancement of tumour growth (B) TSTA + B7 gene transfection induces antitumour immunity

iv. **Tumour vaccine transfected with GM-CSF gene:** Active immunotherapy with tumour cells transfected with GM-CSF gene seems to be another important mode of immunotherapy. In this situation, the tumour cells secrete GM-CSF which induces proliferation, differentiation and activation of dendritic cells in tumour bearing host. Since dendritic cells are an efficient antigen presenting cells, this strategy has been found to enhance tumour antigen presentation by dendritic cells to CTLs that lead to an increased tumour cytotoxicity (Fig. 18.10). Tumour cells transfected with other cytokine genes such as IFN-γ, IL-2, TNF, etc. have been tried in patients with melanoma and renal cell carcinoma.

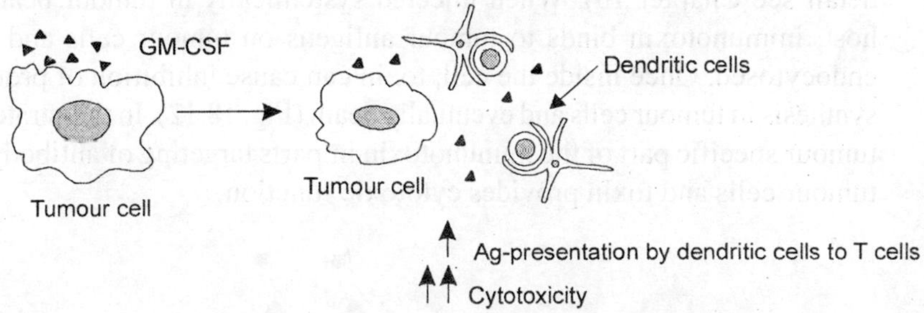

Fig. 18.10: Tumour cells transfected with GM-CSF enhance antigen presentation and cytotoxicity in tumour bearing host

C. Adoptive Immunotherapy with LAK Cells

Adoptive immunotherpay with lymphokine activated killer (LAK) cells has been widely studied in patients with different types of tumours such as melanoma, renal cell carcinoma, prostate cancer, etc. It was found that autologous peripheral blood mononuclear cells or TILs when incubated *in vitro* with IL-2 or IFN-γ generates LAK cells which show enhanced antitumour activity. It can lyse NK-resistant tumour cells as well. It was later found that such cells are cytokine activated NK cells. Various clinical trials have been done using LAK cells in cancer patients.

D. Passive Immunotherapy

Passive immunotherapy involves use of tumour specific antibody into the tumour bearing host. Such antibody binds to tumour antigens present on the tumour cell membrane and causes destruction of tumour cells by a variety of cytotoxic mechanisms (Fig. 18.11).

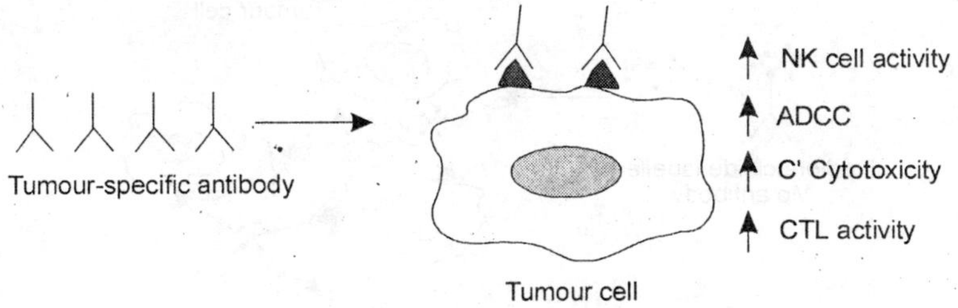

Fig. 18.11: Cytotoxic mechanism by antibodies

In such strategy, monoclonal antibodies have been used in a variety of way for successful immunotherapy of cancer. In this strategy, mostly **anti-idiotype antibodies** to tumour cells are used.

i. **Immunotoxin:** Immunotoxins have been found to be one of the important variation of monoclonal antibody based therapy for cancer. In this strategy, immunotoxins are prepared by conjugating toxin part of diphtheria or ricin toxin with tumour specific Fab portion of monoclonal antibody (for detail see Chapter 10). When injected systemically in tumour bearing host, immunotoxin binds to tumour antigens on tumour cells and are endocytosed. Once inside the cell, toxin can cause inhibition of protein synthesis in tumour cells and eventually death (Fig. 18.12). In this strategy, tumour specific part of the immunotoxin imparts targeting of antibody to tumour cells and toxin provides cytotoxic function.

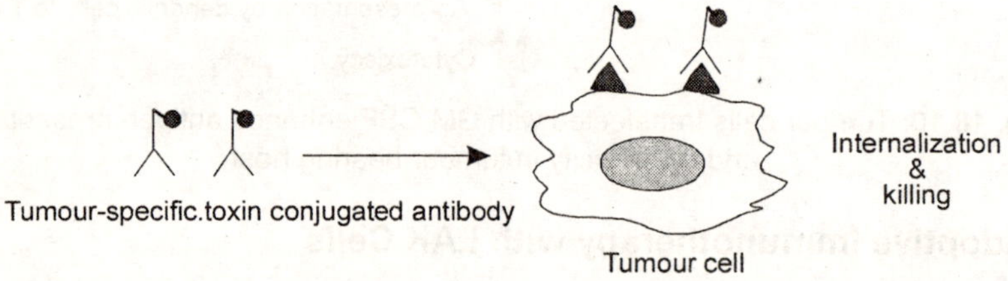

Fig. 18.12: Cytotoxicity by immunotoxins

ii. **Radionucleide conjugated antibody:** Monoclonal antibody against tumour antigen labelled with radionucleide has been useful in the treatment of deeply situated tumours as well as disseminated tumours. In this strategy, Ab is labelled with α-particles (which have higher energy) and systemically injected in the tumour bearing host. Ab is localized to the tumour area and provides localized irradiation to tumour cells. In addition to the targetted tumour cells, it can also kill the neighboring tumour cells (Fig. 18.13). In this strategy, dose of antigen as well as radiation is drastically reduced and therapeutic effect is more with less side effect.

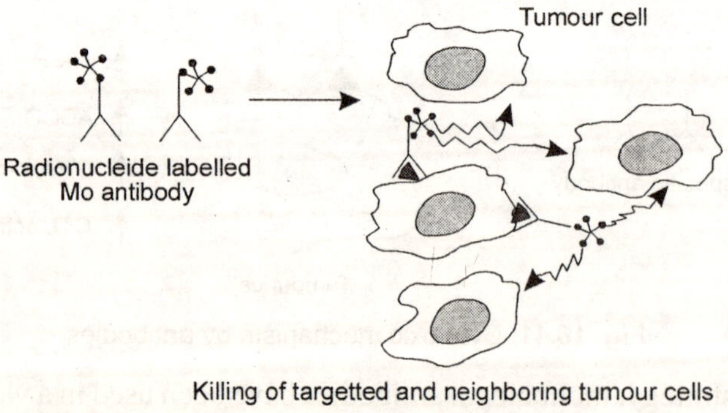

Fig. 18.13: Cytotoxicity by radionucleide labelled monoclonal antibody

E. Use of Immunomodulators and Cytokines

Nonspecific immunotherapy with immunomodulators and cytokines has been other area of immunotherapy of cancer. Immunotherpay with IL-2, TNF-α, IL-12, IL-6, GM-SCF has been tried in a variety of tumour system, with limited success in patients with melanoma, sarcoma, renal cell carcinoma and colon carcinoma. Major drawback of this therapy has been the high rate of toxicity of these cytokines. Such cytokine therapy has been reported to induce septic shock, hypotension, pulmonary oedema and abnormal liver functions.

STUDY QUESTIONS

1. Describe different antitumour immune component and their role in anti-tumour immunity.
2. How tumours evade immune system of the host? Describe strategies for development of tumour cell vaccine.
3. Writ short notes on
   a. Alpha-fetoprotein
   b. Carcinoma embryonic antigen
   c. LAK cell
   d. Immunotoxins

MCQs

Select one most appropriate answer for the questions given below.

1. Carcinoembryonic antigen is an example of:

A. Tumour specific transplantation antigens
B. Tumour specific antigens
C. Oncofoetal antigens
D. Tumour associated transplantation antigens

2. LAK cells are:

A. Cytokine activated NK-cells
B. Lymphokine activated CD_8^+ T cells
C. Cytokine activated CD_4^+T cells
D. Lymphokine activated macrophages

3. Immunotoxins are:

A. Antitoxin antibody
B. Toxin conjugated antibody
C. Toxins acting as antibody
D. Antibodies acting as toxins

4. α-fetoprotein is elevated in the serum of:

A. Pregnant women
B. Women with breast cancer
C. Women with cancer cervix
D. Patients with liver cancer

5. **Frequent development of primary lymphoreticular tumours are due to:**
   A. Hypergammaglobulinemia
   B. Acquired haemolytic anaemia
   C. Impairment of CM1
   D. Corticosteroid abuse
6. **Tumour antigens have been shown to cross-react immunologically in cases of:**
   A. Chemically induced tumours
   B. Viral induced tumours
   C. UV induced tumours
   D. Spontaneously arising tumours
7. **Tumour progression is favoured by:**
   A. CTL
   B. Suppressor T cells
   C. CD_4^+ inflammatory cells
   D. Interferons.

1 C 2 A 3 B 4 D 5 C 6 B 7 B

19

Infection Immunity and Strategies for Vaccine Development

I. INTRODUCTION

In the previous chapters, we have learned that immune system is armed with innate and adaptive components of host defence to combat infection, cancer and other diseases. However, infections remain to be a leading cause of death in the **Third World**. It is estimated that over 1 billion people are infected worldwide, resulting in more than 11 million deaths every year. Despite the alarming death rate, estimated expenditures for research on infectious diseases prevalent in the third world is less than 5% of total health research expenditure world wide. India is spending almost negligible amount on research on infectious diseases and cancer as compared to that on other areas. Unfortunately, not only the age old infectious diseases are present in the developing country but some of these are emerging and remerging in newer forms. Many of the infectious agents undergo mutations so frequently that before the drugs are developed and marketed the bugs develop resistance.

Immune system is organized to prevent the entry of infectious agents through mucosal barrier. Once the bug enters, it is prevented to colonize through various humoral components such as enzymes, complements and cellular components such as macrophage, NK cells and their products (interferons, TNF, etc). If colonized, the adaptive immune system takes over to destroy the infections through humoral and cell mediated effector mechanisms. Unfortunately, parasites have also developed various strategies to evade immune functions of the host so as to avoid attack by the immune system. Many pathogens shed their antigens or mimic to the self-components of the host. Such antigenic variations in pathogens may arise by the gradual accumulation of mutations, and adaptation to the hostile host environment.

In this chapter, we will learn about the various non-specific and specific host defence mechanisms against infections, different immune evasion mechanisms adopted by pathogens and finally the strategies for vaccine development against infections.

General Features of Infection Immunity

The role of immune system in protecting host from infections may become evident by the following observations in humans:

1. Neutrophil deficiency is commonly associated with bacterial and fungal infections, suggesting its role in prevention of such infections.
2. Antibody deficiency is associated with repeated and chronic infections with pyogenic bacteria.
3. Patients with complement deficiency suffer from bacterial infections, most frequently involving *N. gonorrhoeae* and *N. maningitidis*. It appears therefore, that activation of complement components is important for lysis of these bacteria.
4. Patients with primary deficiency of CMI usually suffer from chronic or recurrent fungal, parasitic and viral infections, suggesting a significant role of CMI in immune response against broad range of pathogens.
5. Combined (cellular and humoral) immune deficiency has been found to be associated with all types of infections involving viruses, bacteria, fungi and parasites.

It seems, therefore, that both innate and adaptive immune response protects the host from various infections and that the adaptive immune system plays a major role in containment of a wide range of infections.

Infective organisms that cause disease fall into five groups: viruses, bacteria, fungi, protozoa and helminths. Protozoa and worms are often grouped together and called parasites. These pathogenes can be found in different body compartments where they are combated with different host defense mechanisms as shown in Table 19.1.

Table 19.1: Distribution of pathogens in the host and respective host defence mechanisms

Site of infection	*Intracellular*		*Extracellular*	
	Cytoplasmic	*Vesicular*	*Interstitial*	*Epithelial*
Organisms	Viruses *Clamydia spp.* *Rickettsia* spp. *Listeria monocytogen* Protozoa	Myobacteria *Salmonella typhi* *Leishmania* spp *Listeria* spp. *Trypansoma* *Legionella* *Cryptococcus* *Yersinia pestis* Histoplasma	Viruses Bacteria Protozoa Fungi Worms	*N. gonorrhoea* Worms *Myocoplasma* *S. pneumoniae* *Vibrio cholerae* *E. coli* *Candida albicans* *Helicobacter pylori*
Protective immune mechanisms	CTL ADCC NK cells	T-cell dependent Macrophage activation	Antibodies complement phagocytosis neutralization	Inflammatory cells IgA antibody Mucosal immunity

Common features of some of the infective agents, host defence mechanism and their mode of evasion of immune system are given below.

II. VIRAL INFECTION

Viruses are obligate intracellular parasites which uses host's biochemical system to synthesize proteins and metabolize carbohydrates. Some viruses have RNA genome, e.g. Retroviruses, Reoviruses, Flavoviruses, etc; while others have DNA genome encoding about 200 genes, e.g. adenovirus, herpesviruses, Parvovirus, etc. Viroids are infectious agents of plant and Prions are essentially **infectious proteins** associated with neurodegenerative diseases in human and cattles.

A. Mode of Infection

i. Viruses bind to host cell through specific receptors such as CD4 (HIV), ICAM-1 (rhinovirus) and CD46 (measles, herpes), etc.
ii. Following entry it uncoats, releasing nucleic acids.
iii. It uses hosts machinery to synthesize viral proteins.
iv. Viral particles (varions) are assembled and released to infect neighbouring cells (Fig. 19.1).

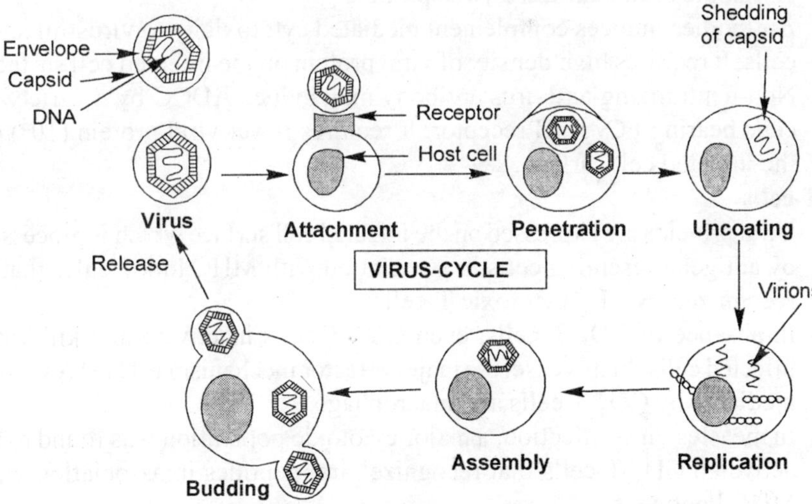

Fig 19.1: Infection and replication of viruses

B. Antiviral Immunity

Antiviral immune response consists of innate immunity mediated by interferons and NK cells and adaptive immunity mediated by T and B cells.

a. Interferons

i. Viral infections lead to production of IFN-α and β which activates antiviral mechanism in neighbouring cells that are said to be in antiviral state.

ii. IFN-α and β when bound to its receptors activates JAK-STAT pathway of signal transduction causing transcription of several genes.
iii. One of the gene encode 2'-5' adenylate cyclase synthetase which activates **ribonuclease I** causing lysis of RNA viruses.
iv. Other genes induce a specific **protein kinase** called ds-RNA dependent protein kinase which inhibits synthesis of viral proteins.
v. Cytokines activate NK cell enhancing its lytic activity to virus-infected cells.

b. NK cells and macrophages

i. Virus induces inhibition of MHC I expression in infected cells.
ii. Absence of MHC I releases NK-inhibitory receptor thus NK cells can effectively lyse virus-infected cells.
iii. NK cells can kill virus infected cells by antibody dependent cell mediated cytotoxicity (ADCC).
iv. Macrophages show antiviral activity by ADCC and DTH responses. It also produces TNF-α and IFN-γ which are key cytokine in its activation and antiviral action.

c. Antibodies (B cells)

i. Antibodies are generated against viral envelope proteins.
ii. Antibodies can neutralize virus particle.
iii. Antibodies induces complement mediated cytotoxicity to virus-infected cells. It requires high density of viral protein on the infected cell surface.
iv. Non-neutralizing antivirus antibody may induce ADCC by a variety of cells bearing FCγ R III receptor. It requires fewer viral protein (10^3) on the infected cell surface.

d. T cells

i. Viral proteins are expressed on the infected cell surface which is processed by antigen presenting cells in association with MHC I molecules that is recognized by CD_8^+ cytotoxic T-cells.
ii. In absence of CD_8^+ T cells, even CD_4^+ T-cells inactivate and kill virus infected cells. In this case, the major effector mechanism is DTH response mediated by CD_4^+ T cells and macrophage.
iii. In measles virus infection, a major cytotoxic population was found to be activated CD_8^+ T-cells that recognize viral peptides in association with MHC I antigens.

Figure 19.2 shows antiviral immune mechanism induced by different components of innate and adaptive immunity.

C. Evasion of Immune Defences

Viruses have evolved a number of mechanisms to evade hosts's immune defences.

1. **Antigenic drift:** Antigenic drift is one of the important immune evasion mechanisms adopted by viruses which is caused by **point mutation** in

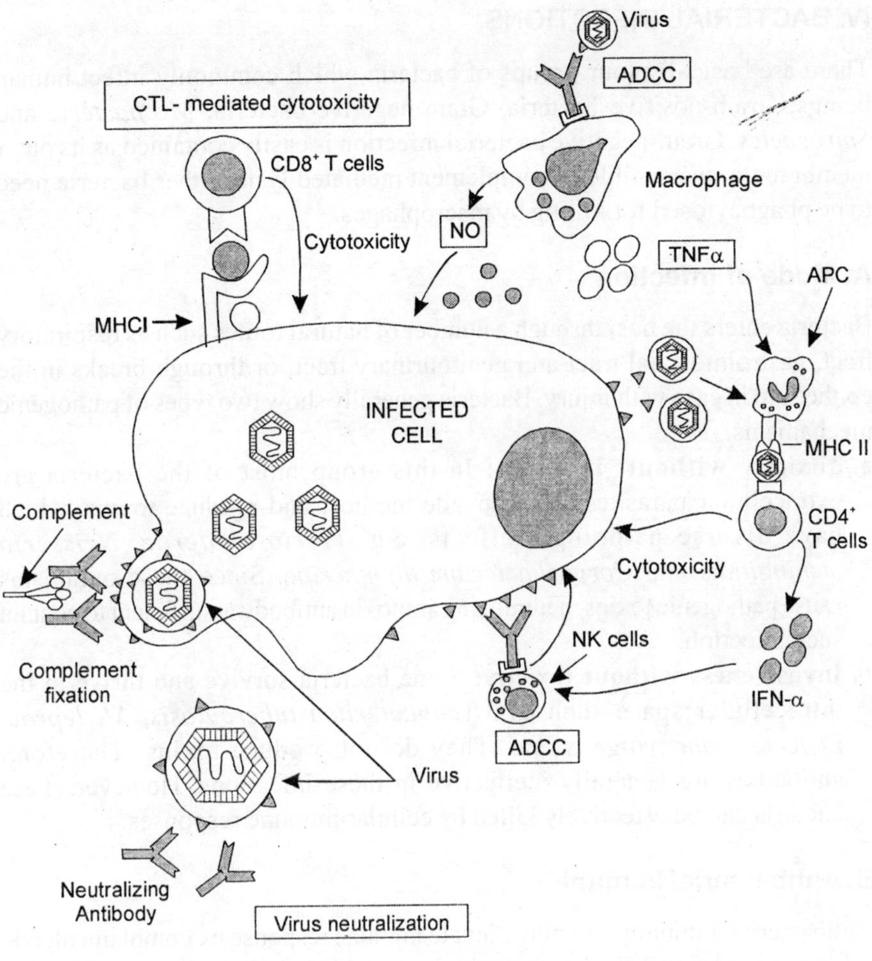

Fig. 19.2: Antiviral immune effector mechanism

the gene. For example, influenza virus evades neutralization by antibody through point mutation in gene encoding haemagglutinin and neuraminidase. It results in appearance of a new variant of the virus in every two years.

2. **Antigenic shift:** In this case, there is **no assortment of the segmental RNA genome** of the virus leading to a significant change in viral proteins on the infected cell surface. Thus the virus is poorly recognized by either antibody on T cells leading to severe infections.
3. **Down regulation of MHCI:** Viruses can block assembly of MHC I molecules leading to unstable MHCI and poor antigen presentation.
4. **Cytokine antagonist:** Some viruses produce molecules similar to cytokines which may bind to cytokine receptors and inhibit signal transduction as well as compete with the natural cytokines. For example, Poxviruse produces IFN-γ, TNF, IL-1 and IL-8 antagonists.

IV. BACTERIAL INFECTIONS

There are basically four groups of bacteria which commonly infect human beings; Gram-positive bacteria, Gram negative bacteria, *Myobacteria* and *Spirochetes*. Gram-negative bacterial infection is easily contained as its outer membrane is susceptible to complement mediated lysis. Other bacteria need to be phagocytosed for killing by macrophages.

A. Mode of Infection

Bacteria enters the host through a number of natural routes such as respiratory tract, gastrointestinal tract and genitourinary tract, or through breaks in the epithelium in case with injury. Bacteria generally show two types of pathogenic mechanisms.

a. **Toxicity without invasion:** In this group most of the bacteria are extracellular parasites. They invade the host and produce toxins, which have diverse pathologic effects, e.g. *Vibrio cholerae, Neisseria meningitidis* and *Corynebacterium diphtheriae*. Since these organisms cause pathogenic toxins, neutralizing antitoxin antibody is sufficient to contain such infection.

b. **Invasiveness without toxicity:** Some bacteria survive and thrive in the intracellular space such as *Mycobacterium tuberculosis, M. leprae, Listeria monocytogens,* etc. They do not produce toxins. Therefore, antibodies are generally ineffective in these infections. However, these bacteria can be effectively killed by cellular immune responses.

B. Antibacterial Immunity

Antibacterial immunity constitute innate immune response by complement, NK cells, macrophage, γ/δ T cells and adaptive immune response by B and T lymphocytes.

a. **Innate immunity by nonspecific mechanisms**

1. There are several nonspecific immune mechanisms by components of innate immunity such as temperature, pH, lysozyme, complement component, etc. (as described in chapter 7) that provide antibacterial immunity.
2. C3b component of complement acts as opsonins and enhances phagocytosis.
3. C3a, C4a and C5a act as anaphylatoxin and induce mast cell degranulation leading to bacterial lysis.
4. Many bacteria are destroyed by pH of stomach and vagina.

b. **NK cells and macrophages**

1. Bacterial infection can be contained by NK cells which can kill extracellular bacteria by ADCC or by direct cytotoxic mechanisms.
2. Activated NK cells produce IFN-γ, and IL-12 which can activate macrophages.

3. Macrophages engulf bacteria and kill them by a variety of cytotoxic mechanisms.
4. Bacterial lipopolysaccharide induces production of TNF-α, IL-1 and IL-6 which induce effector immune mechanisms in several ways (Fig. 19.3).

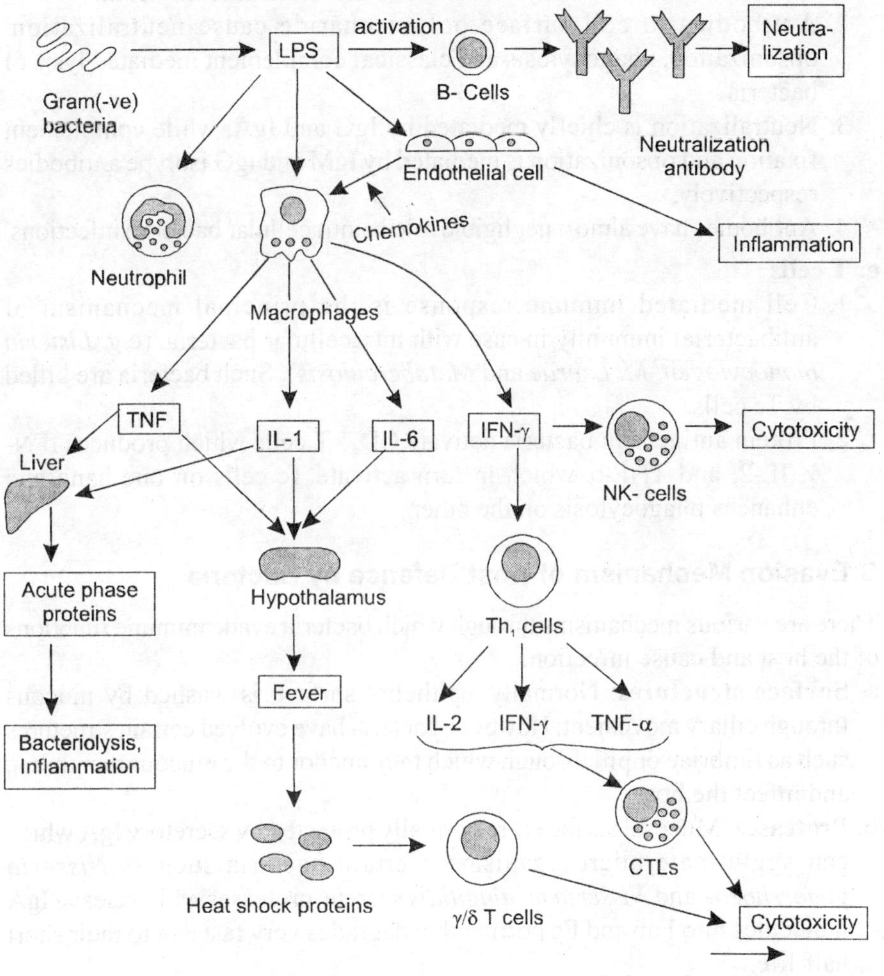

Fig. 19.3: Antibacterial immune mechanisms induced by LPS

c. γ/δ T cells

1. γ/δ T cells are important for local immunity in the gut epithelium.
2. These cells recognize antigens in homogenous way as there is no receptor gene rearrangement in these cells.
3. Although its mechanism of action is not known, it seems to recognize alterations in infected cells.
4. These cells respond to heat shock protein which is produced during infection.

5. Infected epithelial cells express MHC IB which is also recognized by γ/δ T cells.

d. Antibodies (B lymphocytes)

1. Antibody mediated immune response is one of the principal mechanism of antibacterial immunity in case with **extracellular bacteria.**
2. Antibodies to cell surface polysaccharide cause neutralization, opsonizaiton, phagocytosis and classical complement mediated lysis of bacteria.
3. Neutralization is chiefly mediated by IgG and IgA, while complement fixation and opsonization is mediated by IgM and IgG isotype antibodies respectively.
4. Antibodies have almost negligible role in intracellular bacterial infections.

e. T cells

1. Cell mediated immune response is the principal mechanism of antibacterial immunity in case with intracellular bacteria (e.g. *Listeria monocytogen, M. Leprae* and *M. tuberculosis*). Such bacteria are killed by Tc cells.
2. Protein antigens in bacteria activate CD_4^+ T cells which produces IFN-γ, IL-2, and TNF-α which in turn activate Tc cells on one hand and enhances phagocytosis on the other.

C. Evasion Mechanism of Host Defence by Bacteria

There are various mechanisms through which bacteria evade immune functions of the host and cause infection.

a. Surface structures: Normally, epithelial surface is washed by mucous through ciliary movement. However, bacteria have evolved certain structures such as fimbriae or pili through which they anchor to the mucous membrane and infect the host.

b. Proteases: Mucosal surfaces are normally protected by secretory IgA which can agglutinate microorganisms. Certain bacteria such as *Nisseria gonorrhoeae* and *Nisseria meningitidis* secrete proteases which cleave IgA molecules into Fab and Fc portion that degrades very fast due to their short half-life.

c. Inhibition of phagocytosis: Some bacteria possess surface structures which inhibit phagocytosis. For example, *Streptococcus pneumoniae* possesses polysaccharide capsules in their cell wall that prevents their phagocytosis.

d. Changing surface antigen: Some bacteria evade immune response by changing their cell surface antigens. For example a large variation of pilin amino acid sequence has been observed in *N. gonorrhoeae*. In this organism, mini cassette of silent gene causes variation in surface antigens.

e. Inhibition of complement activation: Bacterial proteins inhibit complement pathway thereby protect themselves from complement mediated lysis. The capsule of many gram-positive and gram-negative bacteria contain sialic acid residues that inhibits alternate pathway of complement activation.

V. IMMUNITY TO PARASITES

Parasite infections present a major medical problem as it infects millions of people around the globe. Parasitic protozoan, malaria for example, infects about 1-2 million people every year while intestinal worms infect about one-third of the worlds population. Protozoan parasites live in the gut, erythrocytes, macrophages and liver while worms infect gut, blood vessels and lymphatics. Many parasites pass through complicated life cycles, including migration through various parts of the host.

A. Mode of Infection

Parasites may gain entry to the cell by using mannose fucose receptors on macrophages; Leishmania uses complement receptor to gain entry to the macrophage while malaria parasite merozoites invade RBC through specialized organelles. Since these parasites are larger in size, they contain a great variety of antigens which are stage specific. For example, protein coat antigen of malaria sporozoite induces antibody production in the host, which do not react to the blood stage of parasite. Similarly, *T. spiralis* display different antigens in different stage of its life cycle. Some general characteristics of parasite infections are as follows:

1. Susceptibility to many parasitic infection is linked to MHC haplotype of the host.
2. Merozoites of *Plasmodium vivax* uses **Duffy antigens** present on erythrocytes for its entry. Therefore, some African population who lack such antigen are resistant to infection.
3. It is not in the interest of the parasites to kill the host unless it has ensured transmission to another host. Therefore, most parasitic infections are chronic.

B. Antiparasite Immunity

Protective immunity of parasitic infections constitute both nonspecific (innate) and adaptive immune mechanisms.

As discussed earlier, most effective mode of immune effector mechanisms for extracellular parasites is humoral immune response through antibody while cellular immune response is effective against intracellular parasites.

a. Innate Immunity

1. Macrophage, neutrophils, eosinophils and platelets form the first line of defence against parasitic infections such as trypanosomes and malaria. Trypanosomes and malaria parasites that infect blood cells are phagocytosed and killed by macrophages, through nitric oxide (NO), speroxide anion, ADCC etc. In addition, macrophage produces IL-1, IL-6, IL-12 and TNF-α which enhance immunity by activating other cells of the immune system.

2. IFN-γ, IFN-α and GM-CSF produced by immune cells activate neutrophils which can engulf larger parasites and kill them by both oxygen-dependent and oxygen independent mechanisms.
3. Eosinophils can kill worms (*S. mansoni* and *T. spiralis*) through increased release of its lytic granules. It has been seen that eosinophil inactivated (through antibody) mice develope more cysts when infected with *T. spiralis* as compared to control mice whose eosinophils are active.
4. Platelets can kill a large number of parasites through its Fc receptor by IgE mediated ADCC.
5. NK-cells activated by IL-12 released by macrophage kill parasite infected cells. These activated NK-cells secrete IFN-γ which amplifies antiparasite effector mechanisms by other cells of the immune system. Various non-specific and specific immune responses to parasite have been shown in Figure. 19.4.

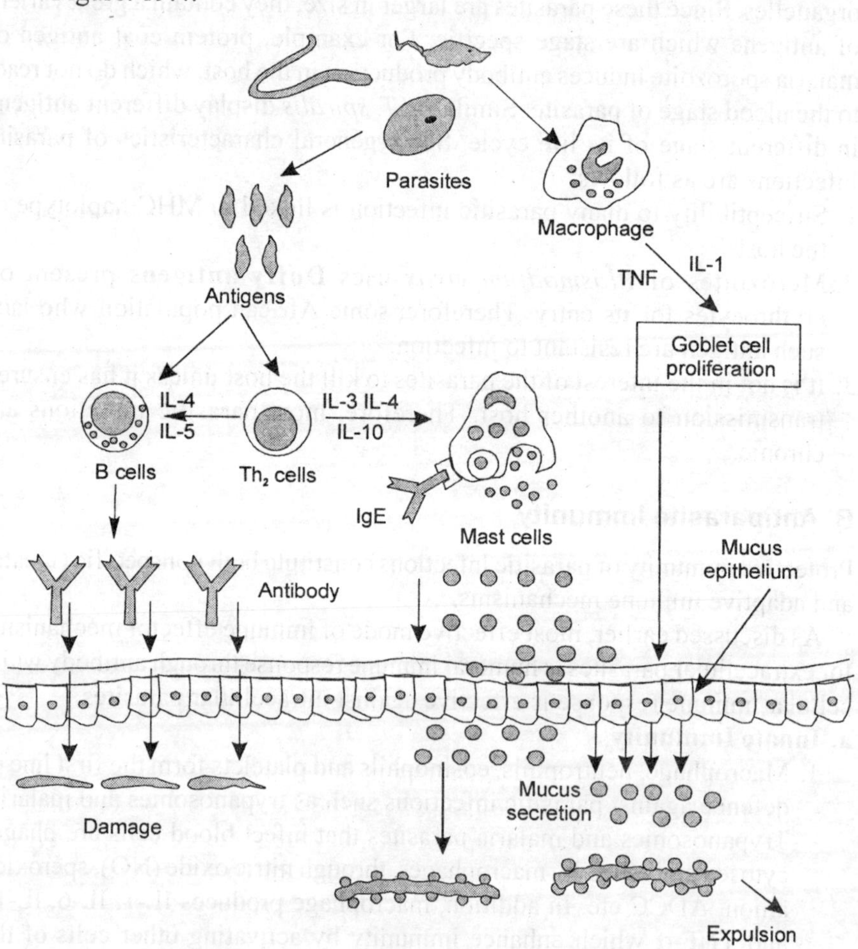

Fig. 19.4: Immune response against intestinal parasites

b. Antibody

1. Many parasitic infections induce nonspecific antibody production through its released antigens, toxins or metabolic products. Increased local IgM has been seen in case with malaria and trypanosoma infected patients and IgG in malaria and Leishmania infections.
2. Antibodies can act directly on the parasites causing complement mediated lysis.
3. Antibodies neutralizes parasites by interfering its attachment and entry to the host epithelium.
4. Antibody enhances Fc receptor mediated phagocytosis and ADCC by various cytotoxic cells such as macrophage, platelet, NK cells and Tc cells.

c. T cells

In case with parasitic infection, Th 2 cells play an important role in immune response. Parasite antigens activate Th 2 cells which produce IL-3, IL-4, IL-5 and IL-10. These cytokines enhance antibody synthesis by B-cells as well as activate mast cells. In addition, IgE produced by plasma cells binds to Fc receptors on mast cells causing its degranulation. Mast cell granules are cytotoxic to parasites. Nonspecific inflammatory molecules produced by activated macrophage induce goblet cell proliferation which secrete large quantity of mucus. Mucus coating to worms prevents its spread and causes their expulsion from the site of infection.

A definitive role of TH1 and TH2 cytokines in resistance and susceptibility to *Leishmania* infection has been shown by Mosmann. It has been shown in mice than IFN-γ produced by TH1 cells provides protection to infection while IL-4 produced by Th2 cells favours infection in susceptible strain of mice. Administration of anti-IL-4 antibody has been shown to reduce cutaneous lesions in infected mice through removal of IL-4 and concomitant increase in IFN-γ production by Th1 cells (Fig. 19.5).

C. Immune Evasion by Parasites

Parasites have evolved various mechanisms to evade its destruction by the immune components. In fact some of the parasites use immune machinery for its entry into the host and also for their survival. **Leishmania** for example, uses complement receptor for entry into the macrophages while *T. bruci* uses IFN-γ as its growth factors.

a. Complement inactivation

Different parasites inactivate complement mediated lysis in different way. For example, *Leishmania major* activates complement cascades which are shed off so that it avoids complements. *T. cruzi* bears a surface glycoprotein which has molecular mimicry with decay accelerating factor (DAF) that limits the complement reaction.

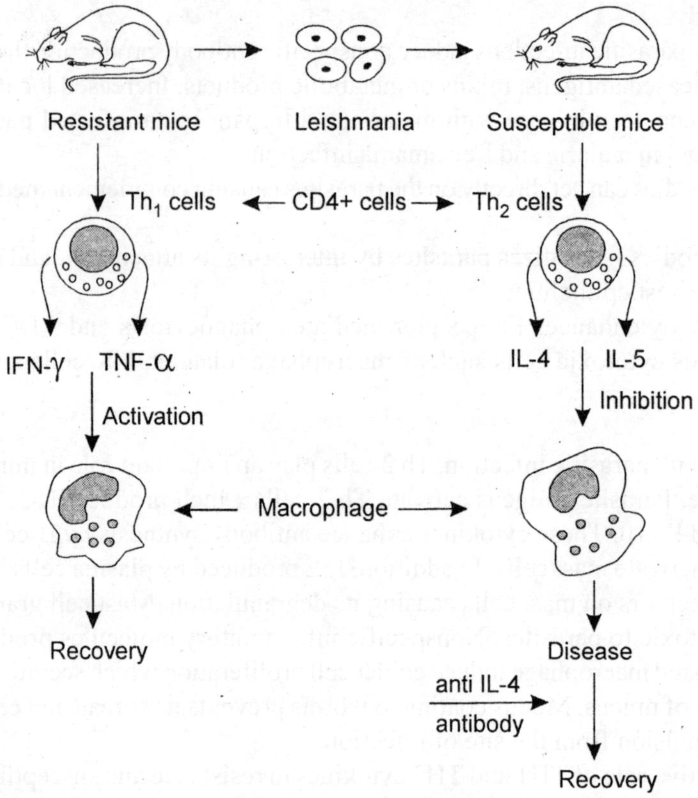

Fig. 19.5: Resistant mice produce Th1 cytokines that activates macrophage which destroy leishmania leading to recovery from infection, while susceptible mice produce Th2 cytokines that inhibits activation of macrophage leading to disese. The suscepitble mice recover after anti IL-4 antibody reatment

b. Inhibition of macrophage activity

Some macrophage inhabiting parasites protect themselves by cytotoxic activity of macrophage. For example, *T. gondii* penetrates macrophage through non-phagocytic pathway to avoid the triggering of oxygen burst by macrophages. *Leishmania* posses superoxide dismutase enzyme which quench oxygen radical. Additionally, it also secretes a protein Gp63 that inactivates the action of lysosomal enzymes.

c. Antigenic variation

Many parasites change their antigenic coat very frequently so that adaptive immune response is not effective against them. Malaria parasite and African trypanosomes undergo antigenic variation very frequently. It is interesting to note that some of the parasite acquire host protein on its surface to prevent attack by the immune cells. For example, *Schistosoma* cultured in human serum and RBC acquires surface molecules containing A, B and H blood group antigenic determinants. They can also acquire MHC molecules of the host. Some parasites have evolved protective coat around them (e.g. *Entamoeba).*

d. Immunosuppression

One of the important mechanisms by which parasites protect itself in hostile environment of the host is by causing suppression of both humoral and cellular immunity of the host. Some of the parasites can disrupt immune cells directly (e.g. larvae of *T. spiralis*) by releasing soluble lymphotoxic factors. Most of the parasite contain phosphoryl choline (PC) which has been shown to inhibit both T and B cell proliferation. It has been shown that parasites release large quantity of antigens which causes **immune distraction** as well as antigen specific shift of protective Th1 response to immunosuppressive Th2 response as seen in case with *P. falciparum* and *L. donovoni* infections respectively.

VI. STRATEGIES FOR VACCINE DEVELOPMENT

The clinical use of immunology in preventing infectious diseases dates back to late 18th century when Edward Jenner in 1878 developed cowpox vaccine and Louis Pasteur in 1881 performed the successful immunization experiment in sheep with attenuated anthrax bacillus. There has been significant reduction in incidence of several infections such as diphtheria, measles, mumps, rubella, etc. Smallpox has been eradicated so far and poliomyelitis is soon going to be wiped out. It has been possible due to successful vaccine development against these organisms. However, other infections such as HIV, tuberculosis and malaria are on the rise. WHO estimates that almost 20-35% mortality in children could be prevented by effective use of existing vaccines.

A. Qualities of Successful Vaccine

To be successful a vaccine should have following qualities:

- It should have similar protective effect in human beings as in experimental animals.
- Manufacturing cost should be low so that it is affordable to people of all socio-economic groups.
- It should be stable for sufficient time so that it could be delivered efficiently to at risk population.
- It should have acceptable side effects.

Recent developments in immunology, molecular biology and recombinant DNA technology have led to promising strategies of finding new vaccine candidates.

B. Active and Passive Immunization

Immunity can be achieved either by natural process through previous infection or artificially through injecting inactivated/killed organisms, called **active**

immunization. **Passive immunization** include artificial transfer of immunity through antibodies or vaccines. The characteristic of active and passive immunization has been shown in table 19.2.

Table 19.2: Features of passive and active immunization

Features	*Passive immunization*	*Active immunization*
1.	Natural through maternal antibody	—
2.	Induced by antibody (immunoglobulin)	Whole organism
3.	Antitoxin	Macromolecules, natural antigen, cloned antigen
4.	Neutralization	Multivalent complex
5.	No memory, short term protection	Memory cells, long term protection
6.	CMI or humoral immunity absent	CMI and humoral immunity present
7.	Toxic to host	Variable response
8.	Do not cause infection	May cause active infection

Although passive immunization is most commonly used for a number of infections such as botulism, tetanus, diphtheria, hepatitis, measles and rabies, it should be used with caution due to its side affects. Since antibodies are produced in another species such as horse, the recipients sometimes show serious serological reactions. Similarly, active immunization with pertussis attenuated bacterial vaccine have been found to induce seizure, encephalitis, brain damage and even death.

C. Designing Vaccines for Active Immunization

One of the important aspect of vaccine development is to provide sustained immune response to the specific microorganism for longer time. It is important therefore, to generate memory cells, which can be attained by active immunization of the host. Various strategies can be adopted for active immunization.

a. Whole Organism Vaccine

Two principal strategies may be adopted for active immunization of the host:

i. **Attenuated** vaccine by growing pathogenic microbes in abnormal conditions so that it loses pathogenicity but retain immunogenicity, is one strategy. The great advantage of attenuated vaccine is that they elicit all the innate and adaptive immune response. For example, BCG (Bacillus Calmette-Guerin) was developed by growing *Mycobacterium bovis* is increasing concentrating of bile so that it becomes non pathogenic but retains immunogenicity. One of the possibility of attenuated vaccine is that they revert back to a virulent form.

ii. **Inactivated** vaccine may be produced by inactivating the organisms by heat or chemical treatment. The salk polio vaccine is produced by inactivating the organism by formaldehyde treatment. However, killed/inactivated vaccine often require booster injection since it generates mainly humoral immunity, where as attenuated vaccine provides long lasting response in only one injection. The features of attenuated and inactivated vaccines has been shown in table 19.3.

Table 19.3: Comparison of attenuated and killed vaccine

Characteristics	*Attenuated*	*Inactivated*
1. Production	Grown in adverse condition and a virulent mutant is selected	Inactivation by heat or chemical treatment
2. Booster dose	Not required	Required
3. Stability	Less	More
4. Type of immunity	Humoral + CMI	Humoral
5. Reversal tendency	Possible	No
6. Contamination	Possible	No

b. Purified Antigen (Subunit) Vaccine

The risks involved with attenuated and inactivated vaccines have led to the development of subunit vaccines consisting of purified antigens from the organisms. Three different strategies have been adopted (i) using purified bacterial polysaccharide antigen as vaccine (ii) using inactivated bacterial toxin as vaccine and (iii) using recombinant peptide as vaccine.

i. Polysaccharide Vaccine

The current vaccine for **S. pneumoniae** and **H. influenzae** are composed of bacterial polysaccharides. Although polysaccharides are inefficient inducers of B cell memory, these vaccine provide long term memory by producing high affinity antibody of IgM class. These are unable to activate CD_4^+ T cells.

It has been seen that if polysaccharide is covalently linked to a protein carrier such as tetanus toxoid, it enhances opsonization and phagocytosis of the organism, activates CD_4^+ T cells, generates memory cells and causes IgM to IgG class switching.

ii. Toxoid Vaccine

Some bacterial pathogens such as Diphtheria and tetanus produce exotoxins. These exotoxin produces many of the disease symptoms. These toxins are highly immunogenic. Therefore the exotoxins are purified and then heat inactivated or treated with paraformaldehyde to form '**toxoid**'. Such toxoid induces potent neutralizing antibody when injected in the host. Immunization of children with toxoid preparations have drastically reduced diphtheria and tetanus infections in children.

iii. Recombinant Antigen Vaccine

This strategy involves cloning and expression of genes coding immunogenic proteins of bacteria, viruses, and parasites so that large quantity of proteins are obtained for immunization of the host. First vaccine developed by this strategy was by cloning gene for the major surface antigen of hepatitis B virus (HBsAg). The recombinant hepatitis B antigen has been found to induce production of protective antibody in the host.

c. Recombinant Vector Vaccine

A recent approach has been to introduce gene that encodes major antigens of various pathogens, into attenuated viruses or bacteria and to infect the host with recombinant vector. The vector replicates in the host expressing the genes of the pathogens against which immune response including CTL response is generated. Commonly used vector vaccine is vaccinia virus vaccine which was widely employed to replicate in the host. The strategy has been shown in Figure 19.6. The genetically engineered vaccinia express high level of the inserted gene.

One of the practical problem with vaccinia virus vector vaccine is that the virus may infect various host cells. Although the virus is non-pathogenic, it may produce antigen which generate may CTLs that can kill infected host cells.

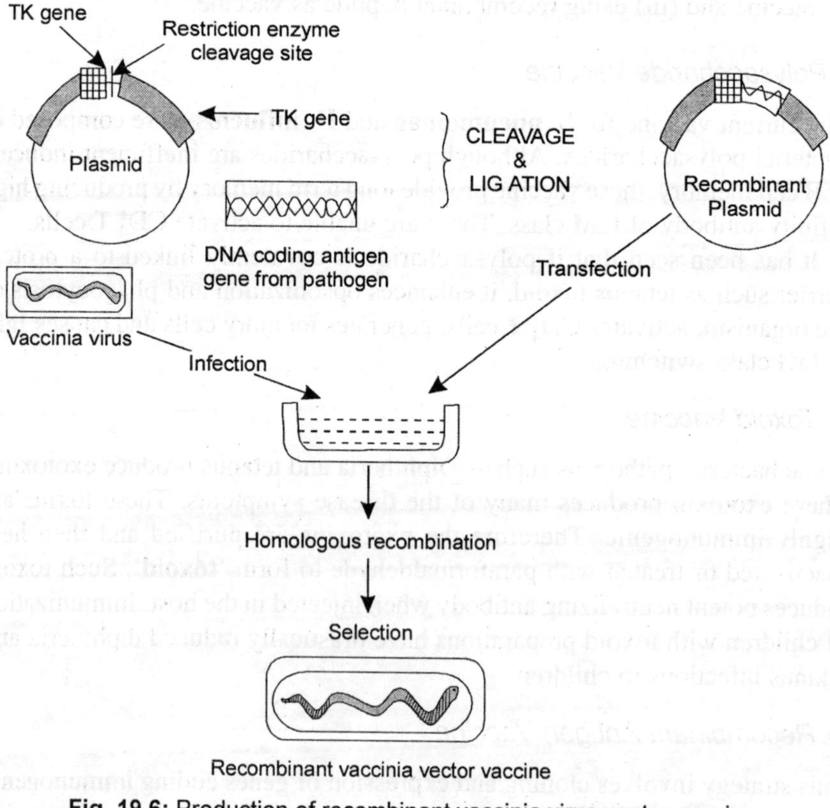

Fig. 19.6: Production of recombinant vaccinia virus vector vaccine

d. DNA Vaccines

In one of latest approach plasmid DNA encoding antigenic protein is directly injected in muscle. Muscle cell integrate DNA very quickly and encode protein antigen which elicit both humoral and cell mediated immune response of the host. Since the muscle cells express low level of MHCI but not the constimulatory molecules, local dendritic cells present these antigens, providing an efficient immune response.

DNA vaccines have several advantage over the conventional vaccines.

1. The antigenic protein is directly expressed in the host, no denaturation or modification take place.
2. It induces both humoral and cellular immune response.
3. Refrigeration is not required, so it can be easily delivered and there is no problem of handling and storage.
4. Plasmid vectors can be tailor made.
5. Cost is drastically reduced.
6. A gene gun can be applied so that complexity of delivery may be reduced and needles and syringes will not be required.

e. Multivalent Subunit Vaccines

One of the limitations of subunit purified antigen is the development of short lived humoral immune response. Therefore currently, a new strategy of multivalent vaccine have been developed using both B cell and T cell epitopes of the antigen combined together on the solid matrix—Antigen complex (SMAA). In this strategy monoclonal antibodies are attached to the solid matrix and then it is saturated with the desired antigens. By using multiple monoclonal antibodies, a number of B cell and T cell epitopes of antigen can be combined on the matrix (Fig. 19.7A). Such multivalent subunit vaccine have been shown to induce a vigorous humoral and cell mediated immune response. Their particulate nature has been found to induce phagocytosis as well.

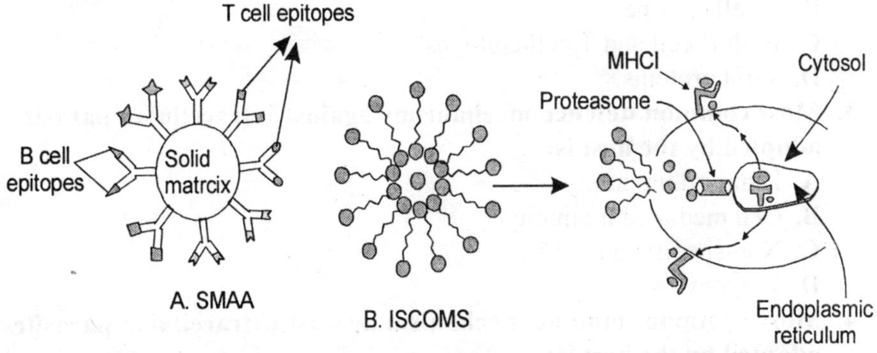

Fig. 19.7: Multivalent subunit vaccine (A) SMAA and (B) ISCOMS showing cytosolic antigen processing and presentation with MHC I complex.

Another strategy has been to introduce multivalent subunit vaccine in liposomes (Lipid vesicle) or ISCOMS (immunostimulating complexes). ISCOMS are lipid carriers prepared by mixing protein with detergent and a glycoside called quil. A ISCOMS fuses with plasma membrane of the antigen presenting cells, where it is processed by cytosolic pathway and thus induce cell mediated immune response as shown in Figure 19.7B.

STUDY QUESTIONS

1. Describe various strategies adopted by viruses to evade host immunity. What are the immune effector mechanisms against viral infections?
2. Describe mode of infections and immune evasion mechanism adopted by bacteria. How bacterial infections are contained by the host.
3. What are the immune effector mechanisms against parasites? How parasites evade immune response of the host.
4. What are the current strategies of vaccine development? What strategy would you adopt to generate both humoral and cell mediated immune response by vaccines.
5. Write short notes on (A) DNA Vaccine (B) ISCOMS (C) Th1 cytokines in infection immunity.

MCQs

Select one most appropriate answer for the question given below.

1. **The mice showing protection against Leishmania infection produce**
   A. Th1 cytokines
   B. Th2 cytokines
   C. Complement proteins
   D. Heat shock proteins
2. **Multivalent subunit vaccine contains**
   A. B cell epitope
   B. T cell epitope
   C. Both B cell and T cell epitopes
   D. Viral proteins
3. **Most common defence mechanisms against intracellular parasites adopted by the host is:**
   A. Humoral immunity
   B. Cell mediated immunity
   C. Non-specific immunity
   D. Phagocytosis
4. **Most common immune mechanism against extracellular parasites adopted by the host is:**
   A. Phagocytosis
   B. Cell mediated immunity

C. Humoral immunity
D. Non specific immunity

5. **Toxoids are:**
A. Bacterial toxins
B. Antitoxin antibody
C. Peptide fragments of toxin
D. In activated toxin

6. **First vaccine produced by recombinant DNA technology in human use is made against:**
A. Polio
B. Measles
C. Hepatitis
D. Diphtheria

7. **Antigenic variation is a mechanism of immune evasion that results in:**
A. Interference with attachment to host cells
B. Induction of immune suppression
C. Alteration in surface antigen leading to escape from immune attack
D. Mutations in surface antigen

8. **The pneumococcal polysaccharide vaccine should be administered to all except one of the following:**
A. Individuals with chronic cardio respiratory diseases
B. The elderly subjects of >60 years of age
C. Children < 2 years of age
D. Persons with chronic renal failure

9. **Which of the followings is the least likely side effect of immuno-prophylaxis:**
A. Measles vaccine
B. Pneumococcal vaccine
C. BCG
D. Mumps and rubella vaccine

10. **The usual sequence of events in the development of an effective immune response to a viral infection.**
A. IFN secretion → Ab synthesis → CM1 → NK cell
B. Ab synthesis → IFN secretion → NK Cell → ADCL
C. NK cell → ADCC → IFN secretion → Ab synthesis
D. IFN secretion → CM1 → Ab synthesis

---

1 A 2 C 3 B 4 C 5 D 6 C 7 C 8 C
9 B 10 D

PART III:

MOLECULAR BIOLOGY AND RECOMBINANT DNA TECHNOLOGY

•

DNA Replication and Repair *270*

•

Transcription *309*

•

Translation *336*

•

Regulation of Gene Expression *364*

•

Recombinant DNA Technology *380*

20

DNA Replication and Repair

DNA REPLICATION

I. INTRODUCTION

DNA is the genetic material. The transfer of genetic information from the parents to progeny organisms requires a faithful replication of DNA. In simple terms, replication involves the separation of the parental strands of the DNA double helix each of which then serves as a template for the addition of each of the four deoxyribonucleotides by complementary base pairing. However, replication of the simplest DNA molecule is a complex, multistep process requiring the interplay of several proteins. The complexity of the process is due to following reasons:

1. The two strands of the DNA are antiparallel. New nucleotides must be added in such a way as to maintain the reverse polarities.
2. A supply of energy is needed to unwind the helix.
3. Single stranded DNA tends to form inter- and intra-strand base pairs when it finds complementary sequence and the separated strands of double-helical DNA lead to quickly re-anneal.
4. Several enzymes are required for replication.
5. Errors do occur during replication and mechanisms do exist to repair them.
6. Both circularity and enormous lengths of DNA molecules impose physical constraints on replication.

Inspite of the complexity of the process, once replication has started it continues until the entire genome has been duplicated.

II. DNA REPLICATION: AN OVER VIEW

1. Replication of DNA is *semi-conservative.* The DNA double helix unwinds and each parental single strand is a template for the synthesis of a new daughter strand. Thus, as replication proceeds, two new double-helices

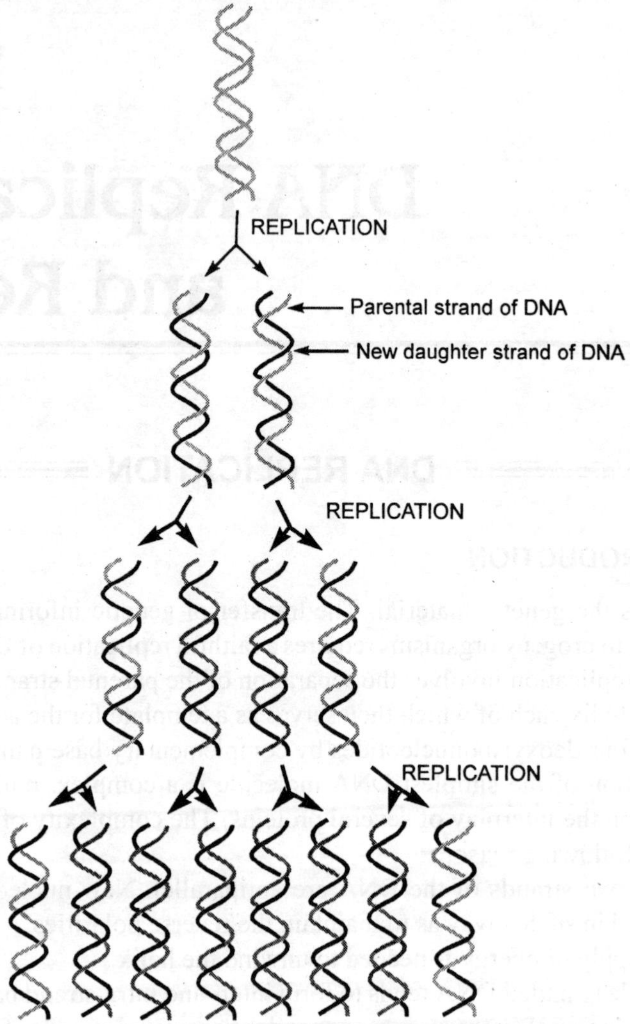

Fig. 20.1: The semi-conservative model of DNA replication

are formed each of which contains one original parent strand and one newly formed daughter strand. This process is shown schematically in Figure 20.1.

2. The DNA of both prokaryotes and eukaryotes is highly coiled and folded into a compact structure so as to fit conveniently in a cell. The DNA molecule is unwound and prepared for synthesis by the action of *DNA gyrase* (topoisomerase), *helicase* and *single-stranded DNA binding proteins* (SSB). Template preparation involves at least three steps:
   i. Unwinding of the double helix
   ii. Breakage of the H-bonds holding the two strands together
   iii. Exposure of single-stranded templates for new synthesis and mechanisms to prevent them from re-annealing.

3. Initiation of replication occurs at specific base sequences, which comprise the *"origin of replication"* (ori). In prokaryotes, replication is a single event involving a unique *ori* on the bacterial chromosome. In eukaryotes, DNA replication requires many start sites on the large, linear chromosomes in a timely fashion as shown in Figure 20.2. There, the initiation of replication must be carefully co-ordinated to ensure that all sequences are replicated exactly once.

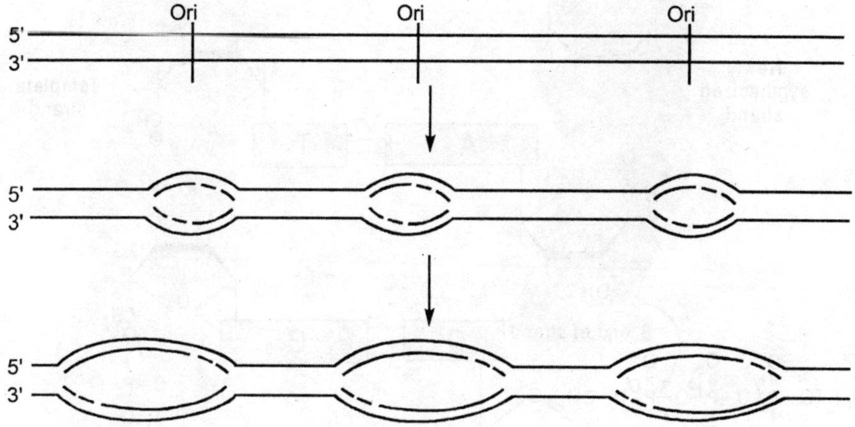

Fig. 20.2: Large eukaryotic DNA has multiple origins of replication

4. DNA synthesis requires the four deoxyribonucleoside triphosphates-dGTP, dCTP, dATP and dTTP.
5. A free 3'OH group is required for replication, but no such group exists when the strands separate. So, an *RNA primer* comprising of 1-60 bases (depending upon the organism) is synthesized by a *primase*. The free 3' OH group of the primer annealed to the DNA template is used to start replication.
6. The synthesis of DNA is catalyzed by the enzyme *DNA Polymerase III* which is "template dependent" in that it will *"read"* the sequence of bases on the template strand and then *"synthesize"* the complementary strand. The template strand is *always* read in the 3' to 5' direction so that the new complementary strand *must be synthesized* in the 5' to 3' direction. The polymerization can only occur by the addition of the precursor dNTPs (dATP, dGTP, dCTP, dTTP) to the available 3'-OH group of the primer. The phosphodiester bond is formed when the hydroxyl group at the 3' end of the primer attacks the α-phosphoryl group of the incoming deoxyribonucleoside triphosphate. The overall chemical reaction catalyzed by the DNA polymerase is:

$$\text{Poly (nucleotide)}_n\text{-3'-OH} + \text{dNTP} \longrightarrow \text{Poly (nucleotide)}_{n+1}\text{-3'-OH} + \text{PP}$$

Where PP is the pyrophosphate which arises from the release of the β- and γ-phosphates of the nucleotide substrate. The overall chemical synthesis of DNA is illustrated in Figure 20.3.

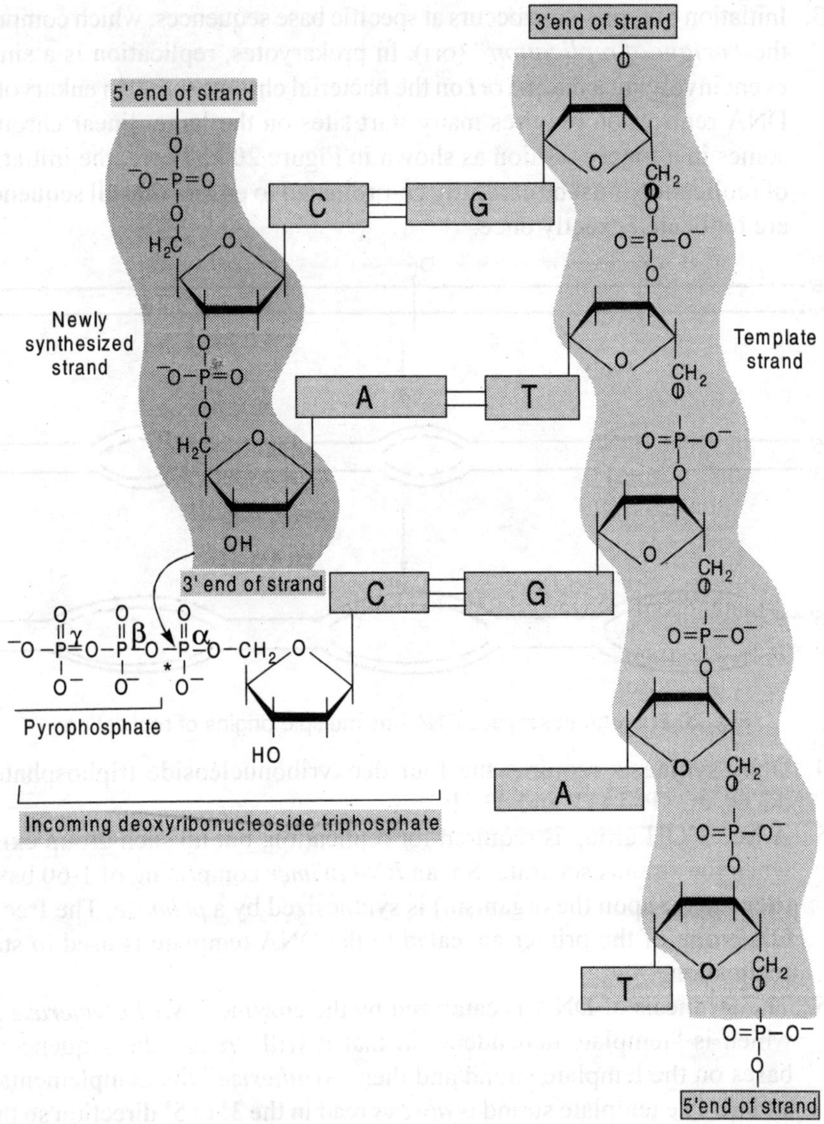

Fig. 20.3: Chemistry of DNA-synthesis

7. In the cell, both the strands of DNA duplex are replicated together, at the same time. The point at which replication occurs consists of the two newly separated template strands and the duplex region of the unreplicated DNA. This junction is known as the *replication fork.* The replication fork always moves in the direction of the duplex DNA. Replication may be *unidirectional* or *bidirectional* depending on whether one or two replication forks are formed at the origin.
8. Since the two strands of the duplex DNA are antiparallel and since DNA can be synthesized only in one direction by elongating a 3' end, only one

of the separated template strands can be replicated continuously in the direction of the movement of replication fork. The newly synthesized DNA complementary to this strand is known as the *leading strand*. The other strand is synthesized *discontinuously* as short fragments called the *Okazaki fragments* and form the *lagging strand* that is synthesized in a direction opposite to the direction of movement of the replication fork. Figure 20.4 illustrates the direction of movement of the replication fork and the newly synthesized leading and the lagging strands of DNA.

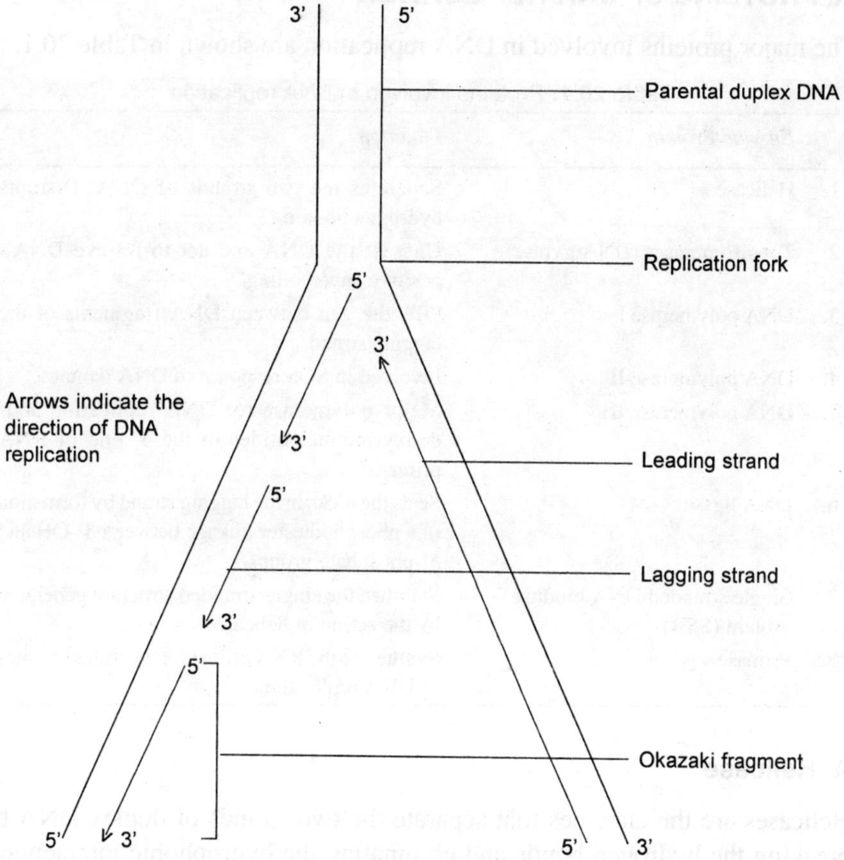

Fig. 20.4: A growing replication fork showing the direction of growth of the leading and lagging strands

9. The final product does not have RNA stretches in it. The RNA primers used for the initiation of DNA replication are removed by the enzyme *RNase H.* RNase H can cleave only the bonds between two ribonucleotides and so the ribonucleotide directly linked to the DNA end remains. This is removed by the 5' to 3' exonuclease action of DNA polymerase 1.

10. Removal of RNA primers leaves gaps in the newly synthesized DNA that are filled by DNA polymerase I. DNA Polymerase does not have the ability to form a bond between the 3'-OH and 5' phosphate of the repaired strand. The "nick" in the DNA backbone is sealed by the action of the enzyme *DNA ligase* that obtains energy for its action by hydrolyzing ATP or NAD. The DNA synthesis is complete only after all RNA primers are removed, the gaps are filled and the nicks are sealed.

III. PROTEINS OF DNA REPLICATION

The major proteins involved in DNA replication are shown in Table 20.1.

Table 20.1: Proteins involved in DNA replication

	Enzyme/Protein	*Function*
1.	Helicase	Separates the two strands of DNA. Disrupts hydrogen bonding.
2.	Topoisomerases (DNA gyrase)	Untwist the DNA and act to relieve DNA's positive supercoiling.
3.	DNA polymerase I	Fills the gap between DNA fragments of the lagging strand.
4.	DNA polymerase II	Involved in SOS response to DNA damage.
5.	DNA polymerase III	Major polymerase for DNA replication adds deoxyribonucleotides to the 3' end of RNA primer.
6.	DNA ligase	Seals the nicks in the lagging strand by formation of a phosphodiester linkage between 3'-OH and 5'-phosphate groups.
7.	Single stranded DNA binding protein (SSB)	Stabilize the single-stranded structure generated by the action of helicases.
8.	Primase	Synthesize the RNA primers at the initiation sites of DNA replication.

A. Helicase

Helicases are the enzymes that separate the two strands of duplex DNA by breaking the hydrogen bonds and eliminating the hydrophobic interactions. There are a number (about 10) of proteins in *E.coli* that have helicase activity. The one involved in DNA replication is the product of DnaB gene. The activity of this protein is shown in Figure 20.5. The energy required for the movement of helicase enzyme comes from the hydrolysis of ATP. Hexamers of helicase bind and encircle one of the two single strands at the replication fork in the form of a ring. Helicases act processively and unwind multiple base pairs of DNA. They have *polarity* and move along the single stranded DNA in a defined 5' → 3' and 3' → 5' direction depending upon the strand of DNA bound by the enzyme.

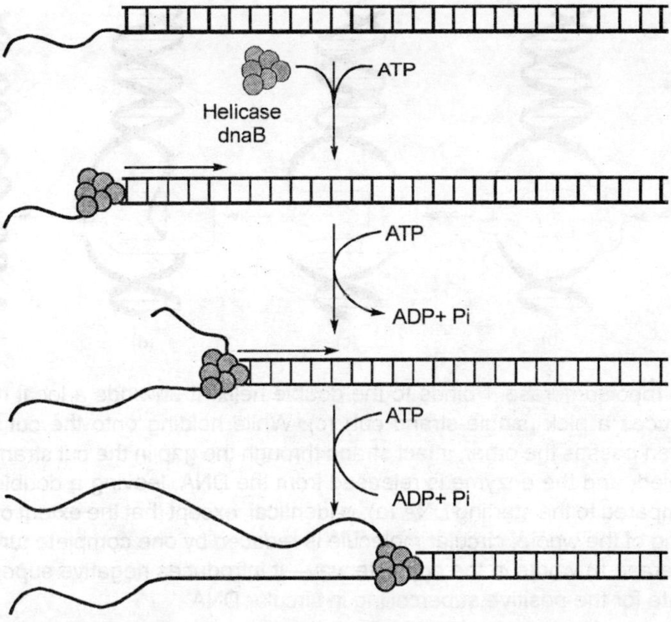

Fig. 20.5: DNA helicases separate the two strands of the double helix. When ATP is added to a DNA helicase bound to ssDNA, the helicase moves with a defined polarity on the ssDNA. In this figure the helicase has a 5' → 3' polarity. This means that the helicase would be bound to the lagging strand at replication fork.

B. Topoisomerases

These enzymes are responsible for the unwinding of DNA that is tightly twisted at 10 bp. Bidirectional replication of circular *E. coli* chromosomes unwinds about 100,000 bps/min twisted at 10 bp/turn, this is 10,000 turns of DNA per min or over 160 per second. If there were nothing to relieve this stress, DNA would be a tangled mass in seconds. Relief of torsional stress is essential for DNA replication to occur. During replication, the unwinding of DNA may cause the formation of tangling structures, such as *supercoils* or *catenanes*. The major role of topoisomerases is to prevent DNA tangling. Topoisomerases are enzymes with a "swivel" mechanism that can relieve torsional stress.

There are two types of topoisomerases: type I and type II.

Type I topoisomerase produces transient *single-strand* breaks. As a result, the type I enzyme removes supercoils from DNA one at a time, whereas the type II enzyme removes supercoils two at a time. Although the type II topoisomerase is more efficient in removing supercoils, this enzyme requires the energy from ATP hydrolysis, but the type I topoisomerase does not. Figure 20.6 shows the untwisting of the DNA by type I topoisomerase working just ahead of the replicating DNA, introducing a nick in one of,the strands. While holding onto the cut ends, the enzyme then passes the other intact strand through the gap in the cut strand. The ends are sealed and the DNA is releaxed one full 360° turn.

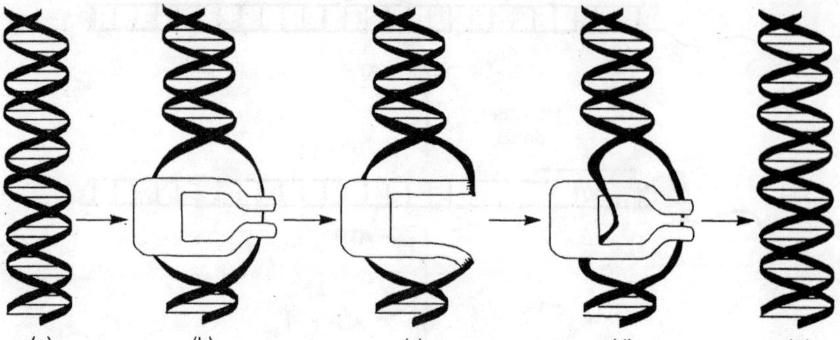

Fig. 20.6: Topoisomerase 1 binds to the double helix. It unwinds a local region (b), and introduces a nick (single-strand cut) (c). While holding onto the cut ends, the enzyme then passes the other, intact strand through the gap in the cut strand (d). The gap is sealed, and the enzyme is released from the DNA, leaving a double helix (e) which, compared to the starting DNA (a), is identical, except that the extent of negative supercoiling of the whole, circular molecule is *reduced* by one complete turn. Gyrase (topoisomerase II) works in the opposite way—it introduces negative supercoiling to compensate for the positive supercoiling in circular DNA

The topo I of both prokaryotes and eukaryotes is the type I topoisomerase. The eukaryotic topo II, bacterial gyrase and bacterial topo IV (involved in dicatenation) belong to the type II.

In eukaryotes, the topo I and topo II can remove both positive and negative supercoils while in bacteria, the topo I can remove only negative supercoils. The bacterial topo II is also called a *gyrase* and has two functions:

a. To remove positive supercoils during DNA replication.
b. To introduce negative supercoils (one supercoil for 15-20 turns of the DNA helix) so that the DNA molecule can be packed into the cell. During replication, these negative supercoils are removed by topo I.

Without topoisomerases, the DNA cannot replicate normally. Therefore, the inhibitors of topoisomerases have been used as anticancer drugs to stop the proliferation of malignant cells. However, the inhibitors may also stop the division of normal cells. Some cells (e.g. hair cells) which need to continuously divide will be most affected. This explains a noticeable side effect, the hair loss induced by such drugs.

C. Single Stranded DNA Binding Protein (SSB)

Single stranded DNA binding protein binds to the newly generated single strands of DNA at the replication fork. It serves to stabilize the separated strands and prevents DNA secondary structures from forming. Binding of one SSB promotes the binding of another SSB to the ssDNA-a phenomenon known as *co-operative binding*. Co-operative binding ensures that ssDNA is rapidly coated with SSB as it emerges from the DNA helicase. The coated

DNA is held in an elongated state that facilitates its use as a template for DNA or RNA primer synthesis. SSB interacts with ssDNA in a sequence-independent manner. SSBs primarily contact ssDNA through electrostatic interactions with the phosphate backbone and stacking interactions with the DNA bases.

D. Primase

The function of the primase is to initiate, or prime, DNA synthesis. Primase is required because DNA polymerases cannot initiate polymer synthesis; *Primase* is a specialized RNA polymerase dedicated to making short, RNA primers (5-10 nucleotides long) on a ssDNA template during replication. Since polymer synthesis is unidirectional but the DNA strands are antiparallel, primase initiates the leading strand DNA polymerization only once while it must repeatedly initiate DNA synthesis on the lagging strand.

Primase primes lagging strand DNA synthesis by synthesizing 11 $\pm$ 1 nucleotide RNA polymer once every 500-2000 nucleotides at the replication fork. Primase does not travel with the replication machinery after its function has been accomplished. It does not require specific DNA sequences to initiate synthesis of a newRNA primer. Instead it associates with the fork machinery every time a primer is initiated and dissociates when it has completed primer synthesis. In the simplest hypothesis, primase can be thought to be binding transiently either to DnaB helicase or the ssDNA bound to the helicase, as the mechanism by which helicase stimulates primer synthesis even on ssDNA. The gene that codes for primase is called DnaG.

E. DNA Polymerases

There are three types of Polymerases in *E. coli*-I, II and III involved in replication of DNA. Besides there are others that are involved in DNA repair. Table 20.2 shows the activities and functions of DNA polymerases.

i. **DNA polymerase I:** Encoded by *pol A* this was the first enzyme discovered with polymerase activity (Arther Kornberg-1957). However, this is not the primary enzyme involved with bacterial DNA replication. It is used to fill the gap between DNA fragments of the lagging strand and also during DNA repair. It has three distinct catalytic activities as shown in Figure 20.7.

   a. Template dependent 5' → 3' primer extension
   b. 3'-5' proofreading exonuclease
   c. 5'-3' error-correcting exonuclease

   The enzyme consists of a single polypeptide of 103 kD which can be cleaved proteolytically into:

   a. larger 68 kD *Klenow fragment* and
   b. smaller fragment of 35 kD.

Table 20.2: Activities and functions of DNA Polymerases

Prokaryotic (E coli)	*Number of subunits*	*Function*
Pol I	1	RNA primer removal, DNA repair
Pol II (Din A)	1	DNA repair
Pol III core	3	Chromosome replication
Pol III holoenzyme	9	Chromosome replication
Pol IV (Din B)	1	DNA repair, Translesion synthesis (TLS)
Pol V ("UmuC, UmuD'$_2$C)	3	TLS
Eukaryotic	*Number of subunits*	*Function*
Pol α	4	Primer synthesis during DNA replication
Pol β	1	Base excision repair
Pol γ	3	Mitochondrial DNA replication and repair
Pol δ	2-3	DNA replication; nucleotide and base excision repair
Pol ε	4	DNA replication; nucleotide and base excision repair
Pol θ	1	DNA repair of crosslinks
Pol ζ	1	Translesion synthesis (TLS)
Pol λ	1	Meiosis-associated DNA repair
Pol μ	1	Somatic hypermutation
Pol κ	1	TLS
Pol η	1	Relatively accurate TLS past *cis-syn*-cyclobutane dimers
Pol ι	1	TLS, somatic hypermutation
Rev 1	1	TLS

The Klenow fragment possesses the polymerase and 3'-5' exonuclease activities. The smaller fragment has a 5' → 3' exonuclease activity. It excises ~10 bases at a time and this activity is coordinated with the synthetic/proofreading activity. It is capable of synthesizing new DNA at single stranded breaks by *nick translation.*

ii. **DNA Polymerase II:** It is encoded by the pol B gene, which is involved in SOS response to DNA damage.

iii. **DNA Polymerase III:** The DNA polymerase III holoenzyme is the principal replicative enzyme in *E. coli*. This enzyme is highly processive and catalyzes polymerization at a high rate. The enzyme is a complex of 10 polypeptides (Fig. 20.8). There are two forms of the enzyme:

a. **Core enzyme:** The catalytic *core enzyme* consists only of those subunits that are required for the basic underlying enzymatic activity. It consists of 3 subunits—alpha (α), epsilon (ε) and theta (θ) and is responsible for DNA synthesis and proofreading.

- The α-subunit which is coded by *polC* gene is a 129904 kD polypeptide that contains the 5' → 3' DNA polymerase activity but no proofreading activity

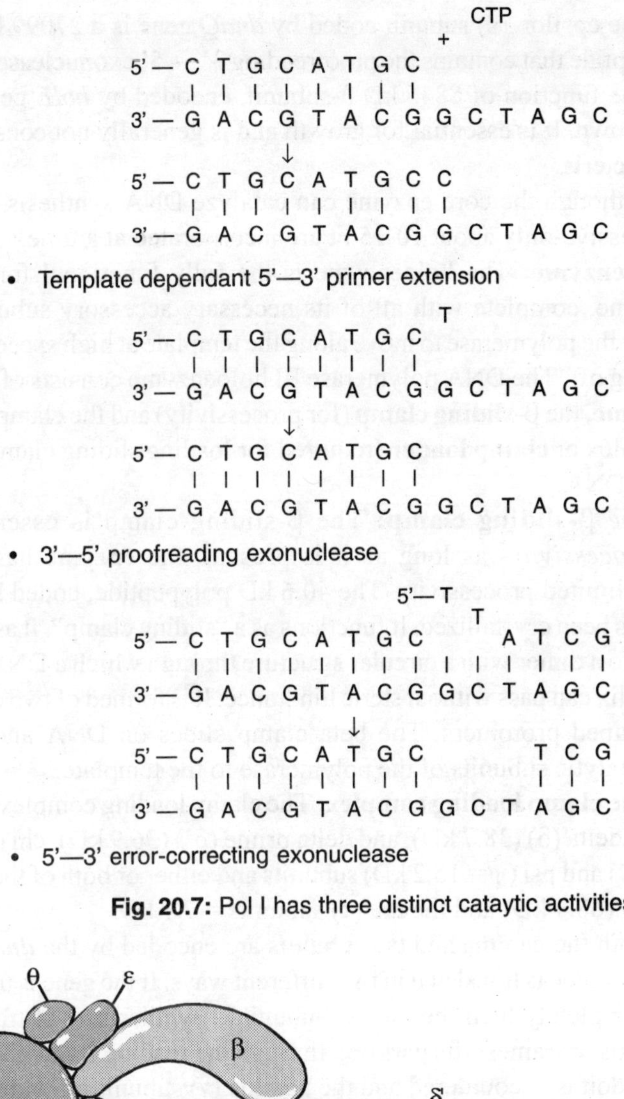

Fig. 20.7: Pol I has three distinct cataytic activities

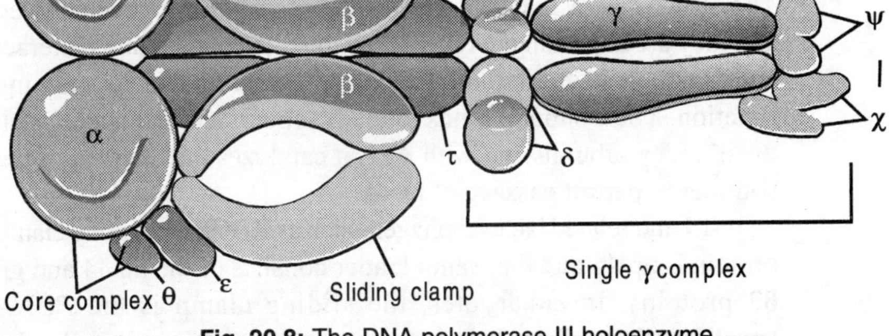

Fig. 20.8: The DNA polymerase III holoenzyme

- The epsilon (ε) subunit coded by *dnaQ* gene is a 27099 kD polypeptide that contains the proofreading 3' → 5' exonuclease activity.
- The function of 8846 kD θ-subunit, eneoded by *holE* gene is not known. It is essential for growth and is generally not conserved in bacteria.

Although the core enzyme can catalyze DNA synthesis, it is not processive-only about 10-15 nt are incorporated at a time.

b. **Holoenzyme:** The holoenzyme is the fully functional form of an enzyme, complete with all of its necessary accessory subunits that allow the polymerase to move along the template at high speed without falling off. The DNA polymerase III holoenzyme consists of the **core enzyme,** the **β-sliding clamp** (for processivity) and the clamp-loading complex or **clamp loader** (required for loading sliding clamp (beta)) onto DNA.

- *The* **β-sliding clamp:** The β-sliding clamp is essential for *processivity*- as long as it is present, the enzyme has almost unlimited processivity. The 40.6 kD polypeptide, coded by ***dnaN*** has been crystallized. It functions as a "sliding clamp". It assembles into a dimer with a circular structure through which a DNA double helix can pass without steric hindrance. It is formed of two crescent-shaped protomers. The beta clamp slides on DNA and tethers catalytic subunits of the polymerase to the template.
- The **clamp loading complex:** The clamp loading complex consists of delta (δ) (38.7 kD), and delta prime (δ') (36.9 kD), chi ($\chi = 16.6$ kD) and psi ($\psi = 15.2$ kD) subunits and either or both of the gamma (γ) (68.4 kD) and the tau (τ) subunits = 7.1 kD.

  Both the gamma and tau subunits are encoded by the ***dnaX*** gene. The gene is translated in two different ways. If the gene is translated completely then the tau (τ) subunit is synthesized. If translation slips or frameshifts part way through the reading frame then a stop codon is encountered and the gamma (γ) subunit is synthesized.

  Both the gamma (γ) and the tau (τ) subunits are 'motor' ATPases. The structures of many of these subunits have now been solved and they are beginning to reveal how the different subunits interact and function in the building of a replisome and in its ongoing function. The δ subunit binds to the β subunit and in concert with the δ' and γ subunits and with ATP, it catalyzes the opening of the β-dimer to permit passage of DNA.

  In T4 and related bacteriophages, such as RB69, the sliding clamp protein is gp 45 and the clamp loader consists of the gp 44 and gp 62 proteins. In eukaryotes, the sliding clamp is the *PCNA* **(proliferating cell nuclear antigen)** *protein* and five clamp loader subunits form a complex known as RFC (Replication factor C) Archaebacteria have two clamp loader subunits that are closely related to the eukaryotic RFC subunits as well as a clamp that is similar to eukaryotic PCNA.

IV. COLLABORATION OF PROTEINS AT THE REPLICATION FORK: THE REPLISOME

The combination of all the proteins that function at the replication fork is referred to as the **replisome** and is well illustrated in Figure 20.9. During DNA synthesis, the activities of the individual proteins are co-ordinated by interactions between them. The co-ordinated process is very similar in phages, prokaryotes and eukaryotes.

When DNA replicates, many different proteins work together to accomplish following steps:

1. The two parent strands are unwound with the help of DNA **helicases** (the hexameric dna B protein)
2. **Single stranded DNA binding proteins (SSBs)** attach to the unwound strands preventing them from winding back together.

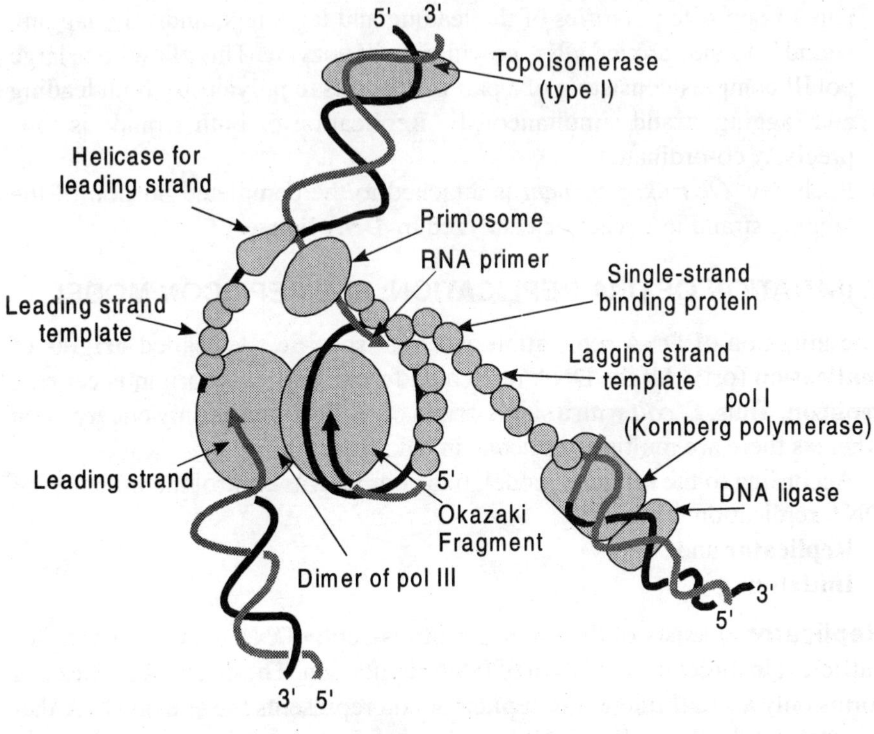

Fig. 20.9: The replisome

3. The strands are held in position, binding easily to **DNA polymerase,** which catalyzes the elongation of the leading and lagging strands.
   - Sliding clamps increase DNA processivity
   - Sliding clamps are opened and placed on DNA by clamp loaders.
4. The interaction between helicase and DNA Pol III holoenzyme (mediated by clamp loader) stimulates the activity of helicase by increasing the rate of helicase movement tenfold. The helicase slows down if it is separated from the DNA polymerase.
5. While the DNA polymerase on the leading strand can operate in a continuous fashion, RNA primer is needed repeatedly on the lagging strand to facilitate synthesis of Okazaki fragments. **DNA primase** which is one of several polypeptides bound together in a group called primosomes, helps to build the primer.
6. The primase is not tightly associated with the replication fork. Instead it associates with the helicase and SSB-coated ssDNA and synthesizes a new primer. The relatively weak interaction between the helicase and primase is important for regulating the length of Okazaki fragments. A tighter association would result in more frequent primer synthesis on the lagging strand and, therefore, shorter Okazaki fragments. A weaker interaction would result in longer Okazaki fragments.
7. Due to opposite polarities of the leading and lagging strands, the lagging strand is looped around DNA polymerase III enzyme. This allows one large pol III complex consisting of a pair of subunits to polymerize both leading and lagging strand simultaneously. Replication of both strands is thus precisely co-ordinated.
8. Each new *Okazaki fragment* is attached to the completed portion of the lagging strand in a reaction catalyzed by **DNA ligase.**

V. INITIATION OF DNA REPLICATION: THE REPLICON MODEL

The initiation of DNA replication occurs at specific sites called **origins of replication** (ori). All the DNA replicated from a particular origin is called a **replicon.** Thus, *E. coli* with a single origin of replication has only one *replicon* whereas there are multiple replicons in eukaryotes.

According to the replicon model, two components control the initiation of DNA replication. These are:

- **Replicator** and
- **Initiator.**

Replicator consists of the entire set of cis-acting DNA sequences that are sufficient to direct the initiation of DNA replication. The origin of replication forms only a small fraction of replicator and represents the site on DNA that is unwound during replication.

Initiator is a protein that specifically recognizes a DNA element in the replicator and activates initiation of replication. Initiator protein is the only sequence specific DNA binding protein involved in replication.

A. Initiation of DNA Replication in *E. coli*

1. The single replicator required for *E. coli* chromosomal replication is called the *Ori C.*
2. The initiator protein in *E. coli* is Dna A.
3. Ori C is composed of four "9-mer" Dna A binding sites and three "13-mer" repeated elements that are the site of initial DNA unwinding.
4. Initiator protein has three common functions:
   - DNA binding
   - DNA strand separation
   - Replication protein recruitment.
5. The major events in *E. coli* initiation of replication are shown in Figure 20.10.

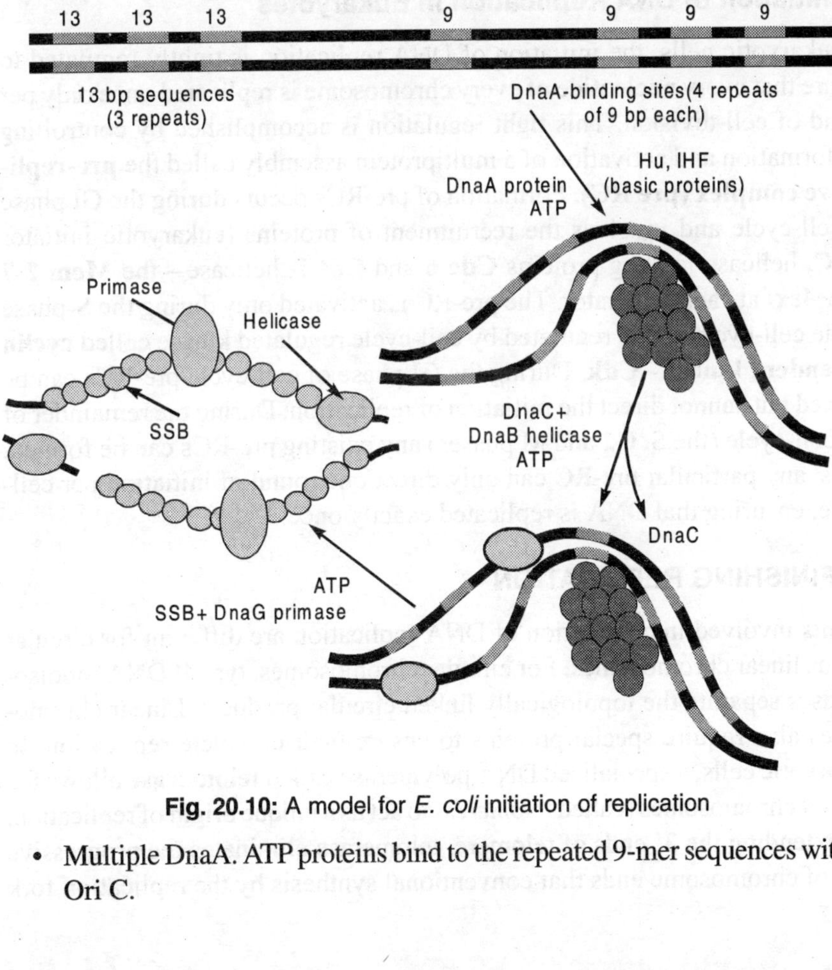

Fig. 20.10: A model for *E. coli* initiation of replication

- Multiple DnaA. ATP proteins bind to the repeated 9-mer sequences within Ori C.

- Binding of DnaA. ATP to these sequences leads to strand separation within the 13-mer repeats. This is mediated by the ssDNA binding domain in DnaA.ATP.
- DNA helicase (Dna B) and the DNA helicase loader (Dna C) associate with the DnaA bound origin.
- DNA helicase loaders catalyze the opening of the DNA helicase protein ring and placement of the ring around the ssDNA at the origin. Loading of the DNA helicase leads to the dissociation of helicase loader (Dna C) from the replicator and activates DNA helicases.
- The DNA helicases each recruit a DNA primase which synthesizes an RNA primer on each template. The movement of DNA helicases also removes any remaining Dna A bound to the replicator.
- The newly synthesized primers recruit the rest of the replicative machinery including DNA polymerase III which synthesizes the new DNA strands on both the leading and lagging strand at the replication fork.

B. Initiation of DNA Replication in Eukaryotes

In eukaryotic cells, the initiation of DNA replication is tightly regulated to ensure that every nucleotide of every chromosome is replicated once only per round of cell-division. This tight regulation is accomplished by controlling the formation and activation of a multiprotein assembly called the **pre-replicative complex (pre RC).** Formation of pre-RCs occurs during the GI phase of cell-cycle and involves the recruitment of proteins (eukaryotic initiator **ORC**, helicase loading proteins **Cdc 6** and **Cdt 1**, helicase—the **Mcm 2-7 complex**) at each replicator. The pre-RC is activated only during the S-phase of the cell-cycle and is regulated by cell-cycle regulated kinase called **cyclin dependent kinase - Cdk**. During the GI phase of cell-cycle pre-RCs can be formed but cannot direct the initiation of replication. During the remainder of the cell-cycle (the S, G2 and M phases) any existing pre-RCs can be formed. Thus, any particular pre-RC can only direct one round of initiation per cell-cycle, ensuring that DNA is replicated exactly once.

VI. FINISHING REPLICATION

Events involved in completion of DNA replication are different for circular versus linear chromosomes. For circular chromosomes, type II DNA topoisomerases separate the topologically linked circular products. Linear chromosomes also require special proteins to ensure their complete replication. In eukaryotic cells, a specialized DNA polymerase called **telomerase** allows the ends of chromosomes (called *telomeres)* to act as a unique origin of replication. By extending the 3' ends of telomere, telomerase eliminates the progressive loss of chromosome ends that conventional synthesis by the replication fork

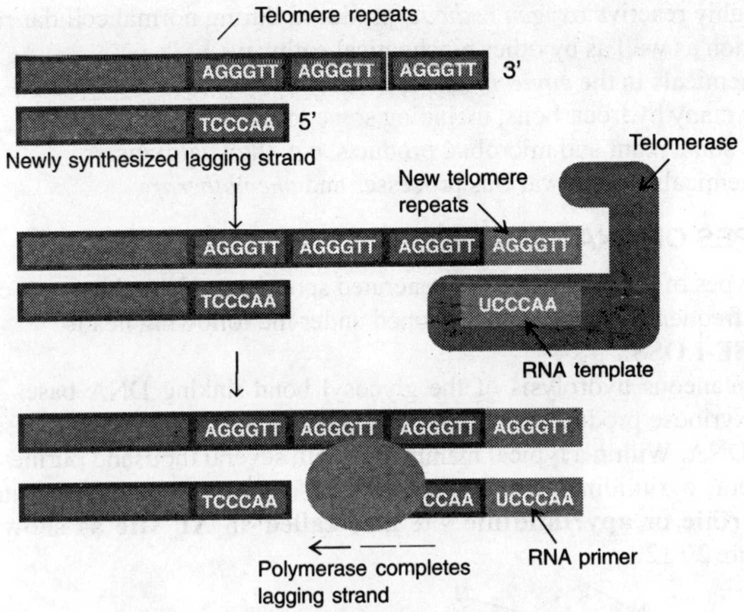

Fig. 20.11: Replication of telomeres by telomerase

machinery would cause. The various steps involved in the replication of telomeres by telomerase are shown in Figure 20.11.

DNA REPAIR

I. INTRODUCTION

DNA is the repository of hereditary information and its integrity is vital to the cell. DNA of living cells is subject to many chemical and physical alterations that would change its coding properties. If the genetic information encoded in the DNA is to remain uncorrupted, all alterations must be corrected. A failure to repair DNA produces a mutation. The importance of repair can be seen from the facts that:

- DNA is the only biomolecule that is specifically **repaired.** All others are **replaced.**
- >100 genes participate in the repair process, even in organisms with very small genomes.
- Most cancers are due to inadequacies in DNA repair.

II. AGENTS THAT DAMAGE DNA

These include:

i. Certain wavelength of *radiation* such as γ-rays, X-rays and ultraviolet rays.

ii. Highly reactive *oxygen radicals* produced during normal cellular respiration as well as by other biochemical pathways.
iii. Chemicals in the *environment*
   - many hydrocarbons, including some found in cigarette smoke.
   - some plant and microbial products, e.g. the aflatoxins.
iv. Chemicals used in various processes and *chemotherapy.*

III. TYPES OF DNA DAMAGE

Many types of DNA damage are generated spontaneously and in some cases at high frequency. They can be grouped under the following heads:

A. BASE-LOSS:

Spontaneous hydrolysis of the glycosyl bond linking DNA bases with deoxyribose produces an **abasic site** (i.e. deoxyribose lacking a base) in the DNA. Within a typical mammalian cell, several thousand purines and several pyrimidines are lost spontaneously every day. They create an **apurinic** or **apyrimidinic** site also called an **AP site** as shown in Figure 20.12.

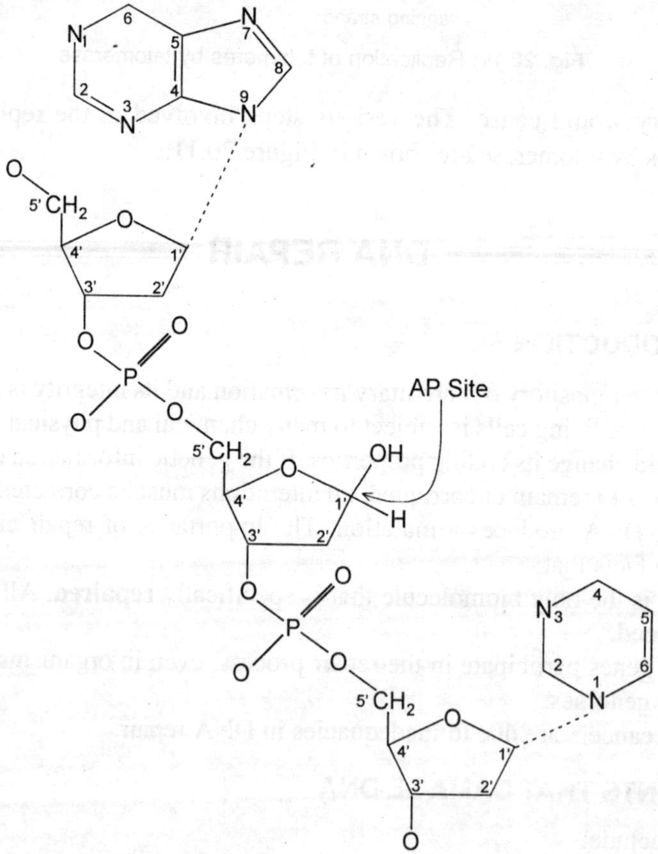

Fig. 20.13: Creation of an abasic site by hydrolysis of N-glycosyl linkage

B. BASE MODIFICATION:

(i) **Deamination:** The primary amino group of nucleic acid bases are somewhat unstable and can be converted into keto groups as shown in Figure 20.13a.

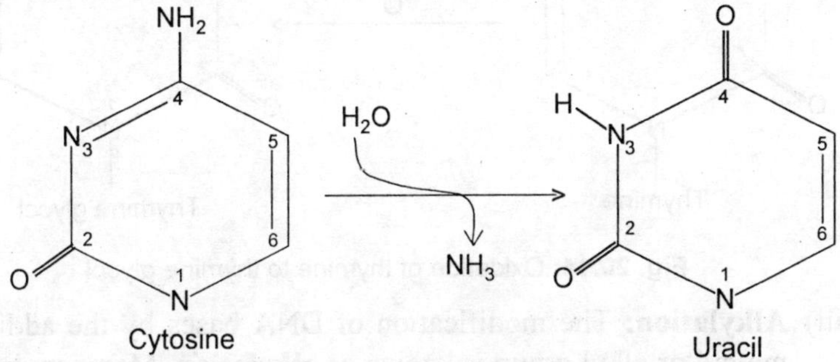

Fig. 20.13a: Deamination of cytosine creates uracil—an unnatural base for DNA

Other deamination reactions include conversion of :

adenine ⟶ hypoxanthine
guanine ⟶ xanthine
5-methyl cytosine ⟶ thymine

Deamination of 5-methyl cytosine generates the natural base thymine (Fig. 20.13b) which is not recognized as abnormal base and following a round of replication it becomes fixed as C → T transition. Hence, methylated cytosines are hotspots for mutation in vertebrate DNA.

Fig. 20.13b: Deamination of 5-methyl cytosine generates the normal base thymine

(ii) **Oxidation: Reactive oxygen species**—ROS (Singlet oxygen, peroxide radicals and hydroxyl radicals) are generated as byproducts during normal oxidative metabolism and also by ionizing radiations (X-rays, γ-rays). They can frequently modify DNA bases. A common product of thymine oxidation is thymine glycol as shown in Figure 20.14. Similarly, oxidation of guanine generates 7, 8-dihydro 8-oxoguanine or *oxoG*.The *oxoG* adduct is highly mutagenic because it can base-pair with both adenine and cytosine.

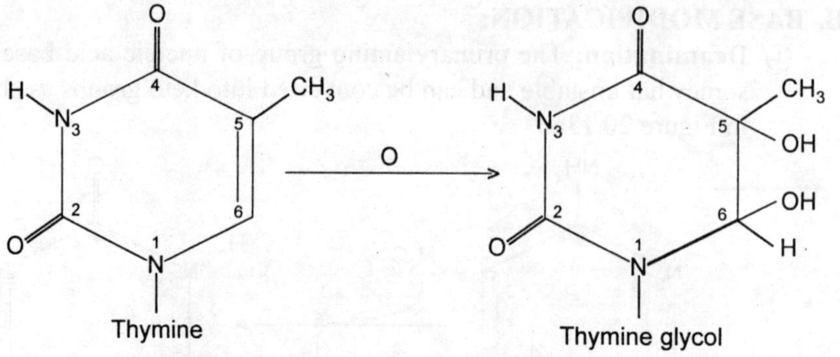

Fig. 20.14: Oxidation of thymine to thymine glycol

(iii) **Alkylation:** The modification of DNA bases by the addition of a methyl or alkyl group is known as **alkylation.** Many environmental chemicals, including "natural" ones cause alkylation of DNA bases. One of the most vulnerable sites of alkylation is the oxygen of carbon atom 6 of guanine. The product O^6-methylguanine often mispairs with thymine, resulting in the change of G : C base pair to A : T base pair when the damaged DNA is replicated. S-adenosylmethionine, the normal biological methyl donor reacts accidentally with DNA to produce several hundred alkylated bases per day. Figure 20.15 shows the alkylation of guanine into methyl guanine.

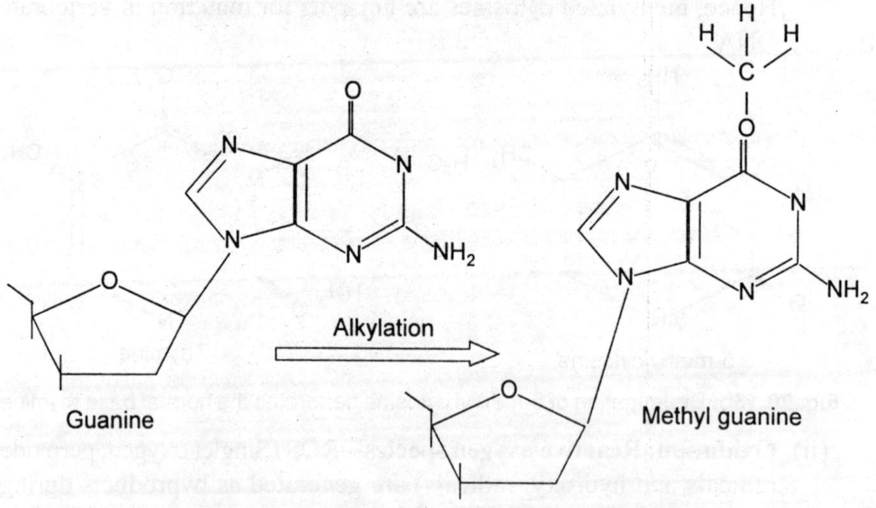

Fig. 20.15: Alkylation of guanine into methyl guanine

C. PHOTODAMAGE:

Ultraviolet light is absorbed by nucleic acid bases resulting in photochemical fusion of adjacent pyrimidines within one strand of DNA. The most common photoproducts are **cyclobutane pyrimidine dimers.** Figure 20.16 shows the **thymine dimer** where the covalent bonds are formed between carbon atoms 5 and 6 of adjacent thymines. Also shown in the Figure is a thymine-cytosine adduct in which the thymine is linked via its carbon atom 6 to carbon atom 4 of cytosine. The distortions produced by the linked bases inhibit normal base-pairing and cause the DNA polymerase to stop during replication.

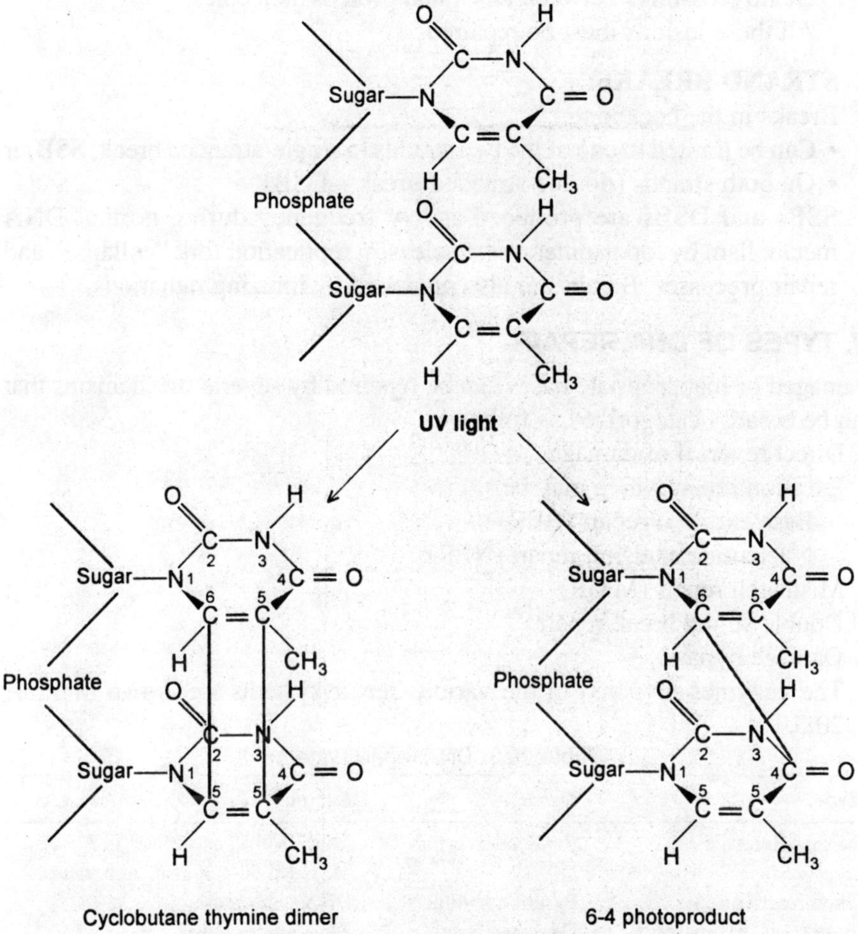

Fig. 20.16: The two forms of pyrimidine dimers

D. REPLICATION ERRORS:

Mismatches of the normal bases might occur because of failure of proof-reading function of DNA polymerase during DNA replication. One common example is the incorporation of the pyrimidine **U** (normally found only in RNA) instead of **T**. Small insertions or deletions might also occur during DNA replication. Efficient repair mechanisms correct most of these problems, as described later.

E. CROSSLINKS:

Convalent linkages may be formed between:

- Bases on the same strand ("intrastrand") or
- Bases on the opposite strands ("interstrand"). Several chemotherapeutic drugs used against cancers crosslink DNA or
- DNA and proteins. Bifunctional alkylating agents and radiation can also create crosslinks between DNA and protein molecules.

All these lesions must be repaired.

F. STRAND BREAKS:

Breaks in the backbone:

- Can be limited to one of the two strands (a single-stranded break, **SSB**)or
- On both strands (**d**ouble-**s**tranded **b**reak—DSB)

SSBs and DSBs are produced at low frequency during normal DNA metabolism by topoisomerases, nucleases, replication fork "collapse" and repair processes. Breaks are also produced by ionizing radiations.

IV. TYPES OF DNA REPAIR

Damaged or inappropriate bases can be repaired by several mechanisms that can be broadly categorized as follows:

- Direct reversal of damage
- Excision repair which may be:
  - Base excision repair (BER)
  - Nucleotide excision repair (NER)
- Mismatch repair (MMR)
- Double strand break repair
- Damage bypass.

The enzymes involved in the various repair systems are shown in Table 20.3.

Table 20.3: DNA repair system

Type	*Damage*	*Enzyme*
Mismatch repair	Replication errors	MutS, MutL, and MutH in *E coli* MSH,MLH, and PMS in humans
Photoreactivation	Pyrimidine dimers	DNA photolyase
Base excision repair	Damaged base	DNA glycosylase
Nucleotide excision repair	Pyrimidine dimer Bulky adduct on base	UvrA, UvrB, UvrC, and UvrD in *E coli* XPC, XPA, XPD, ERCCI-XPF, and XPG in humans
Double-strand break repair	Double-strand breaks	RecA and RecBCD in *E coli*
Translesion DNA synthesis	Pyrimidine dimer or apurinic site	Y-family DNa polymerases, such as UmuC in *E coli*

A. DIRECT REVERSAL OF DAMAGE:

The best studied example of direct reversal of damage is photoreversal of *photoreactivation* of cyclobutane pyrimidine dimers (CPDs) by the enzyme *photolyase*. This reaction was the first DNA repair process discovered by bacteriophage geniticists in 1949 well before the double helical structure of DNA was discovered in 1953. The enzyme reaction is diagrammed in Figure 20.17.

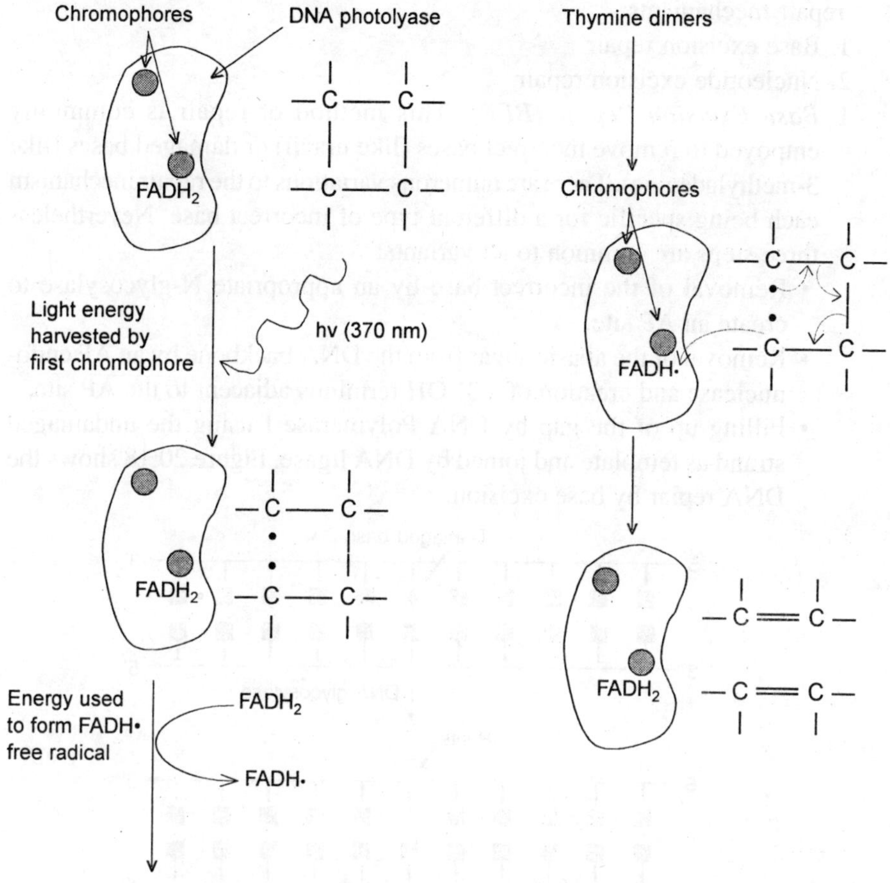

Fig. 20.17: Photoreactivation of thymine dimers

CPD *photolyase* uses visible light ($\lambda = 370$ nm) to oxidize the cyclobutane ring. The enzyme contains two chromophores that harvest the light energy to produce a free radical (FADH•). This in turn carries out the oxidation and splits the dimer. Examples of these enzymes are found in bacteria, fungi, plants and many vertebrates but not in placental mammals.

Another example of direct reversal is repair of O^6-alkyl guanine by transfer of methyl group from the DNA to a cysteine in the enzyme methyltransfrase that catalyzes this reaction.

B. EXCISION REPAIR:

The major DNA repair mechanisms take advantage of the fact that DNA is double stranded and the same information is present in both the strands. Consequently, in cases where damage is present in just one strand, the damage can be accurately repaired by cutting it out (excision) and replacing it with new DNA synthesized using the complementary strand as template. All organisms, prokaryotic and eukaryotic, employ at least two excision repair mechanisms:

1. Base excision repair
2. Nucleotide excision repair.

1. *Base Excision Repair (BER)*: This method of repair is commonly empoyed to remove incorrect bases (like uracil) or damaged bases (like 3-methyladenine). There are numerous variations to the repair mechanism each being specific for a different type of incorrect base. Nevertheless three steps are common to all variants:
   - Removal of the incorrect base by an appropriate N-glycosylase to create an AP site.
   - Removal of the abasic sugar from the DNA backbone by an AP endonuclease and creation of a 3'-OH terminus adjacent to the AP site.
   - Filling up of the gap by DNA Polymerase I using the undamaged strand as template and joined by DNA ligase. Figure 20.18 shows the DNA repiar by base excision.

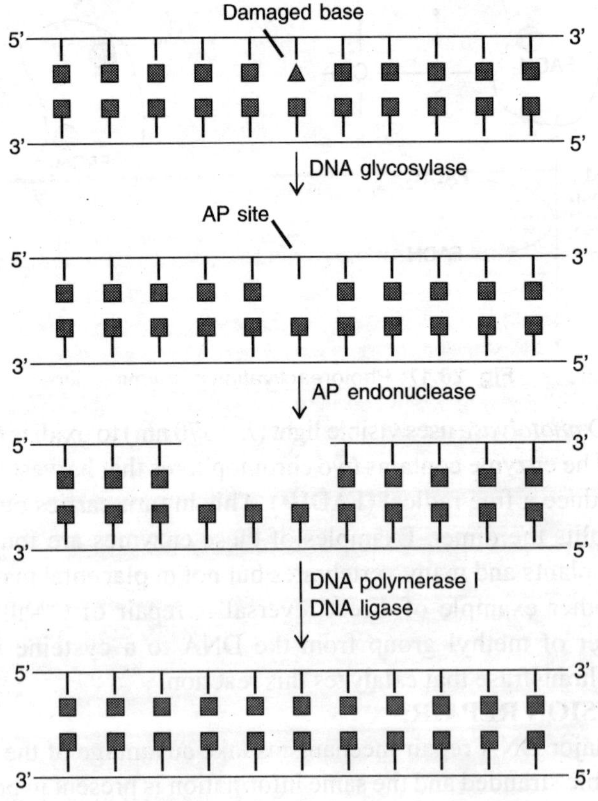

Fig. 20.18: DNA repair by base excision

DNA glycosylases are lesion specific and cells have multiple DNA glycosylases with different specificities. Thus a specific *Uracil DNA glycosylase* recognizes uracil (generated by deamination of cytosine) and another is responsible for removing oxoG (generated as a consequence of oxidation of guanine). A total of eight different DNA glycosylases have been identified in human nuclei.

2. *Nucleotide Excision Repair (NER):* NER differs from BER in several ways:
   - It uses different enzymes.
   - Enzymes are not lesion-specific, they work by recognizing distortions to the shape of the double helix such as those caused by a thymine dimer or by the presence of a bulky chemical adduct to the base.
   - Even though there may be only a single "bad" base to correct, its necleotide is removed along with many other adjacent nucleotides; i.e. NER removes a large "patch" around the damage.

   In all orgainsms, NER involves the following steps:
   1. Damage recognition.
   2. Binding of a multiprotein complex at the damaged site.
   3. Double incision of the damaged strand several nucleotides away from the damaged site, on both the 5' and 3' sides.
   4. Removal of the damage-containing oligonuleotides between the two nicks.
   5. Filling in of the resulting gap by DNA polymerase I and joining by DNA ligase.

NER in *E. coli*

In *E. coli,* proteins UvrA, UvrB, UvrC and UvrD are involved in repairing the damage (e.g. a dimer) induced by UV light. The sequence of events is shown in Figure 20.19. The complex of UvrA and UvrB scan (using ATP energy) the DNA for the presence of UV damage. Once a dimer is found, Uvr A dissociates and Uvr C joins with Uvr B to cause single strand nicks on either side of the dimer as shown. The piece of DNA containing the dimer is removed by a helicase-UvrD. The gap in the DNA is repaired by DNA polymerase I and ligase as is evident from the figure the entire process of NER is very much energy dependent. In yeast, the proteins similar to Uvr's are named RADxx (RAD stands for "radiation"), such as RAD3, RAD10 etc.

Transcription Coupled Repair (TCR) in Eukaryotes

NER is also capable of rescuing RNA polymerase that has stalled because of the presence of a lesion in the transcribed (template) strand of a gene. This process is known as *Transcription coupled repair* or *TCR.* The steps are as follows:

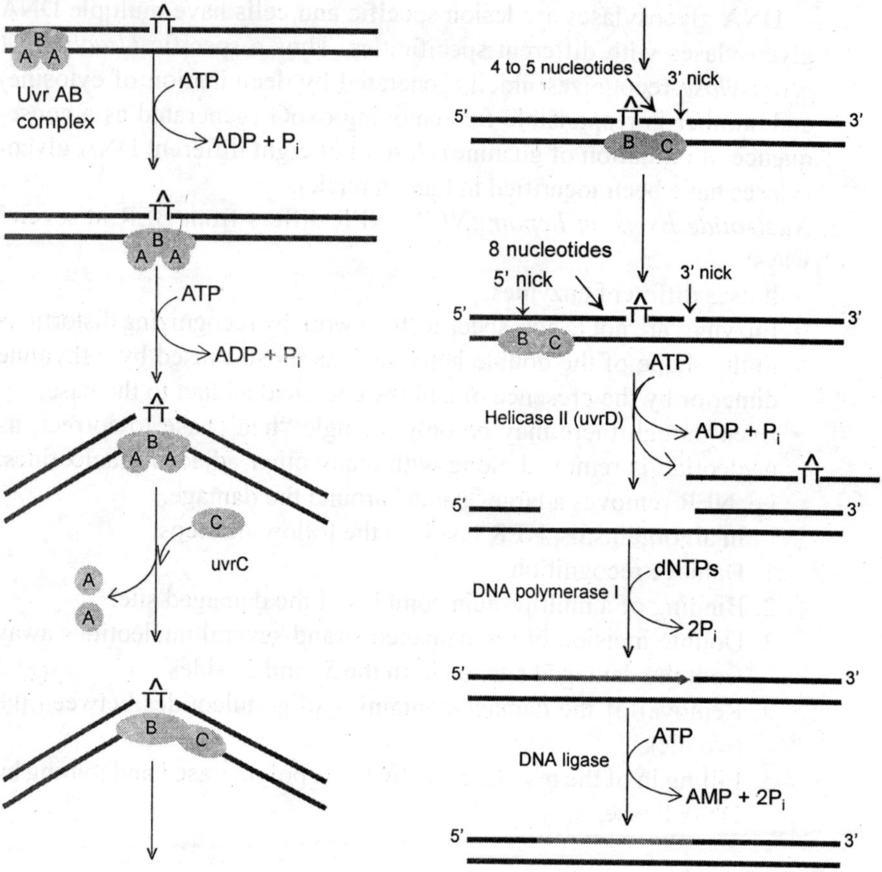

Fig. 20.19: DNA repair by nucleotide excision

1. One or more protein factors are recruited to the location of the stalled RNA polymerase. In Figure 20.20, these are the proteins CSA and CSB that are defective in the human genetic disease *Cockayne's syndrome.*
2. The proteins help to recruit the transcription factor TF II H to the damaged site and help to displace the RNA polymerase and the nascent transcript so the TF II H can access the damaged region.
3. TF II H subunits comprise of the helix-opening proteins XPA and XPD that unwind a 20-30 nucleotide stretch of DNA including the damaged region.
4. Cuts are made on both 3' side (XPG) and 5' side (ERCCI/XPF) of the dama ged area so that a tract of ~27 nucleotides containing the damage can be removed.
5. The replicative gap repair proteins (RFC, PCNA, DNA polymerase *delta* and *epsilon)* bind and synthesize a stretch of new DNA using the intact (opposite) strand as template. TF II H, XPA, XPG and XPF-ERCCI are displaced with new DNA synthesis that fills the gap.
6. The final nick is sealed by *DNA ligase I.*

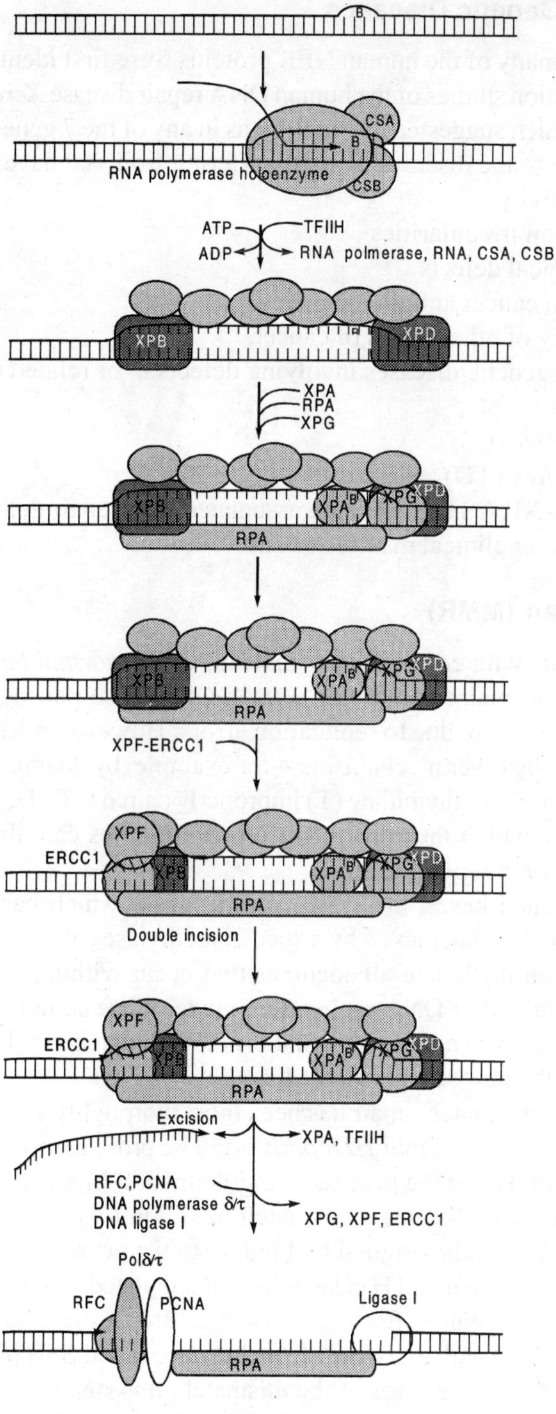

Fig. 20.20: Transcription coupled repair in eukaryotes

NER and Human Genetic Diseases

The genes encoding many of the human NER proteins were first identified in genetic complementation studies of the human DNA repair disease *Xeroderma pigmentosum*(CP) which suggested that mutations in any of the 7 genes (SPA-XPG) could give rise to the disease. Patients suffering from XP have:

- Severe light sensitivity
- Severe pigmentation irregularities
- Frequent neurological defects
- Early onset of skin cancer at hgih incidence.
- Elevated frequency of other forms of cancer.

Two other human genetic diseases involving defects in or related to NER include:

1. *Cockayne's syndrome* (CS)
2. *Trichothiodystrophy* (TTD)

All three diseases–XP, CS and TTD are associated with repair defects and have strikingly different clinical manifestations.

C. Mismatch Repair (MMR)

Mismatch repair deals with correcting mismatches of the *normal bases,* i.e. those bases that fail to maintain normal Watson-Crick base pairing (A : T, C : G). Most mismatches are due to replication errors. However, mismatches can also be produced by other mechanisms—for example, by deamination of 5-methyl cytosine to produce thymidine (T) improperly paired to G. Regardless of the mechanism by which they are produced, mismatches can always be repaired by the *mismatch repair* pathway.

To repair mismatched bases, the system has to know which base is the correct one. In *E coli*, this is achieved by a special methylase called the "*Dam methylase*", which can methylate all adenines that occur within (5') GATC sequences. Immediately after DNA replication, the template strand has been methylated, but the newly synthesized strand is not methylated yet. Thus, the template strand and the new strand can be distinguished.

The mechanism of mismatch repair has been most thoroughly studied in *E coli*. It occurs via the so called *mut HLS pathway.* The proteins involved are Mut S, Mut L and Mut H and the process in digrammed in Figure 20.21.

The must HLS complex locates a mismatch (a GT in this example) and determines which strand is the original by binding to the nearest methylated GATC sequence. Activation of mut H cleaves the unmethylated strand at GATC site. Subsequently, the segment from the cleavage site to the mismatch is removed by exonuclease (with assistance from helicase II and SSB proteins). If the cleavage occurs on the 3' side of the mismatch, this step is carried out by exonuclease I (which degrades a single strand only in the 3' → 5' direction). If the cleavage occurs on the 5' side of the mismatch, exonuclease VII or Rec

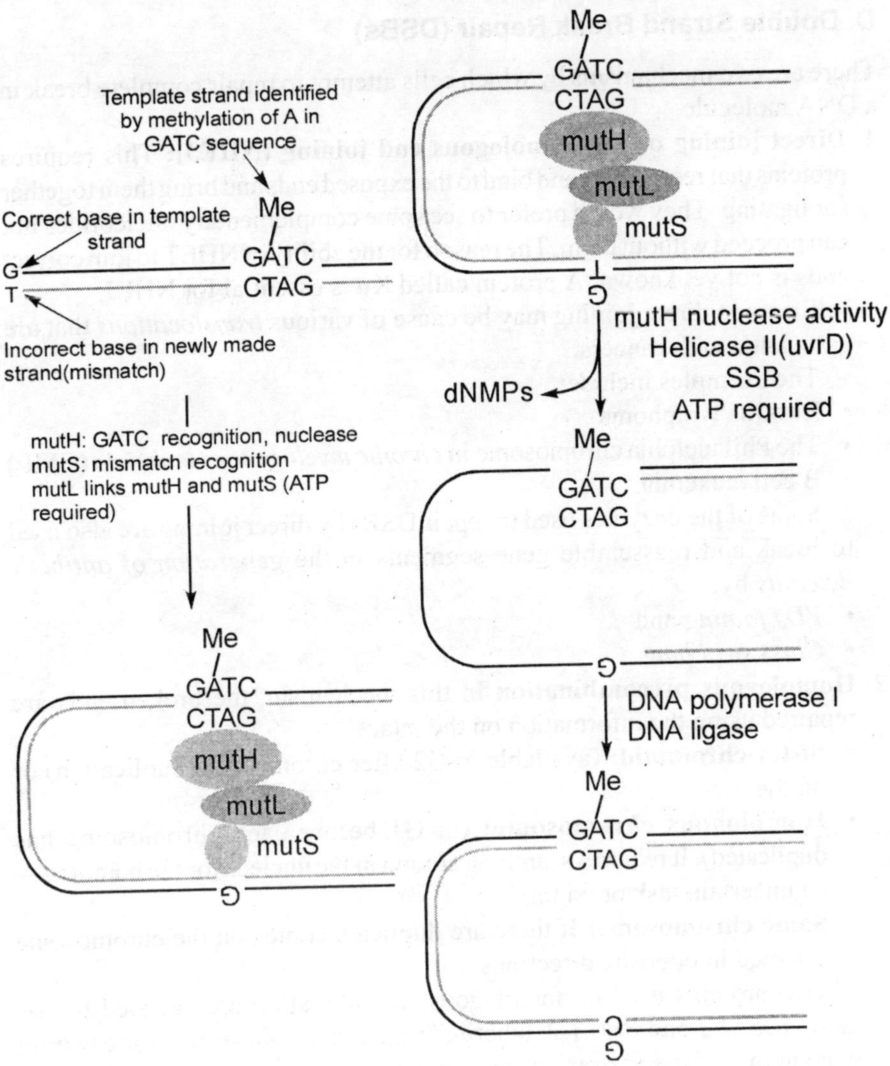

Fig. 20.21: Mismatch repair

I is used to degrade the single stranded DNA. The gap is filled by DNA polymerase III and DNA ligase.

The distance between the GATC site and the mismatch could be as long as 1000 bps. Therefore, mismatch repair is very expensive and inefficient.

Homologs of Mut S and Mut L have been identified in yeast, mammals and other eukaryotes. MSHI to MSH5 are homologous to Mut S; MLH1, PMS1 and PMS2 are homologous to Mut L. Mutations of MSH 2, PMS1 and PMS2 are related to colon cancer.

In eukaryotes, the mechanism to distinguish the template strand from the new strand is still unclear.

D. Double Strand Break Repair (DSBs)

There are two mechanisms by which cells attempt to repair complete break in a DNA molecule.

1. **Direct joining or nonhomologous end joining (NHEJ):** This requires proteins that recongnize and bind to the exposed ends and bring them together for ligating. They would prefer to see some complementary nucleotides but can proceed without them. The reason for the ability ofNHET to join correct ends is not yet known. A protein called Ku is essential for NHEJ.

   Errors in direct joining may be cause of various *translocations* that are associated with cancers.

   The examples include:
   - Burkitts lymphoma
   - The Philadelphia chromosome in *chronic myelogenous leukemia* (CML)
   - B cell leukemia.

   Some of the enzymes used to repair DSBs by direct joining are also used to break and reassemble gene segments in the *generation of antibody diversity* by:
   - *VDJ joining* and
   - *Class switching*

2. **Homologous precombination** In this mechanism, the broken ends are repaired using the information on the intact
   - **Sister chromatid:** (available in G2 after chromosome duplication) or on the
   - **Homologous chromosome:** (in G1 before each chromosome has duplicated). It requires searching around in the nucleus for the homolog—an uncertain task or on the
   - **Same chromosome:** If there are duplicate copies on the chromosome oriented in opposite directions.

   Two proteins used in homologous recombination are encoded by the genes *BRCA-1* and *BRCA-2*. Inherited mutations in these genes predispose women to breast and ovarian cancers.

   Recombination between homologous chromosomes in meiosis I also involves the formation of DSBs and their repair. So it is not surprising that this process uses the same enzymes.

   If the sequence used as a template for repairing a gene by homologous recombination differs slightly from the gene needing repair, that is it is an *allele*, the repaired gene will acquire the donor sequence. This *nonreciprocal transfer* of genetic information is called *gene conversion*. During meiosis, *gene conversions* alter the *normal Mendelian ratios*.

E. Damage Bypass

Sometimes the DNA repair mechanisms do not operate with complete efficiency and the moving replication fork encounters damage in the template strand. Such obstructions to the progression of DNA polymerase can be dealt with in two ways:

- Replication machinery must attempt to copy across the lesion or
- Cease to replicate completely.

For survival, cells opt for the first choice. Even if the lesion cannot be repaired, there is a mechanism in the cells to bypass the sites of damage. It occurs by an error-prone mechanism called **translesion synthesis** although this process may intorduce mutations, it salves the cell from a worse fate resulting from an incompletely replicated chromosome.

Translesion synthesis is catalyzed by a specialized class of DNA polymerases that synthesize DNA directly across the site of damage as shown in Figure 20.22.

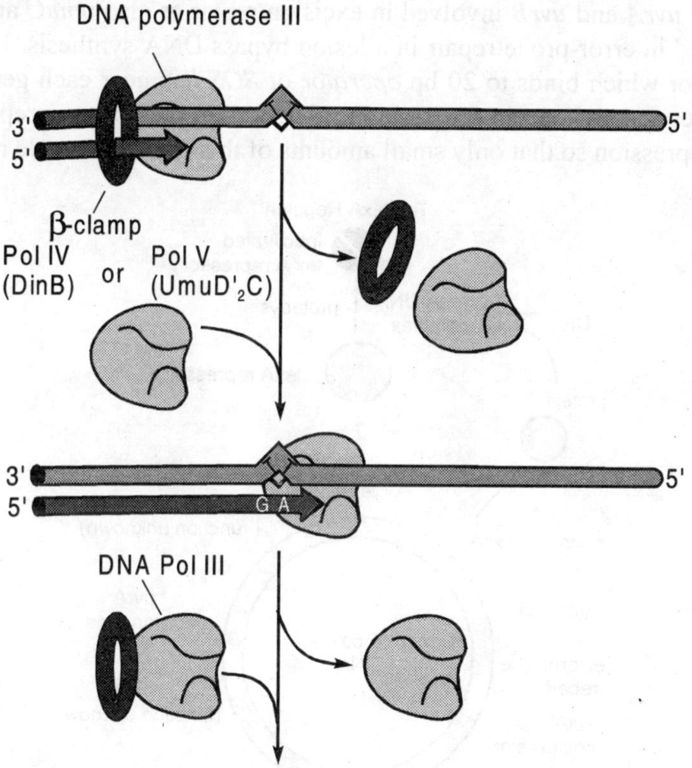

Fig. 20.22: Translesion DNA synthesis. Upon encountering a lesion in the template during replication, DNA polymerase III with its sliding clamp dissociates from the DNA and is replaced by the transleion DNA polymerase, which extends DNA synthesis across the thymine dimer on the template (upper) strand. The translesion polymerase is then replaced by the DNA polymerase III (Source: R. Woodgate)

In *E. coli* translesion, synthesis is carried out by a complex of proteins UmuC and UmuD that constitute a distinct family of polymerases found in many organisms and known as the **γ-family** of DNA polymerases. Although these are template dependent they incorporate nucleotides in a manner that is independent of base pairing.

The SOS Response

While most DNA repair mechanisms are constitutive (i.e. active at all times) a few are activated in response to a signal. This may take the form of a damage of DNA that inhibits replication such that it threatens the survival of the cell. In this case, an emergency response comes into play the so, called *SOS repair* or *post-replication repair* which involves the expression of many genes whose products include repair functions. In *E coli,* the expression of these genes is governed by a complex interplay of the products of two genes *lexA* and *recA*. As shown in Figure 20.23a some of the genes regulated by the lexA repressor include *uvrA* and *uvrB* involved in excision repair as also *umuC* and *umuD* involved in error-prone repair in a lesion bypass DNA synthesis. LexA is a repressor which binds to 20 bp *operator* or *SOS box* near each gene that it regulates and prevents its transcription. The lexA repressor even represses its own expression so that only small amounts of this protein is made routinely.

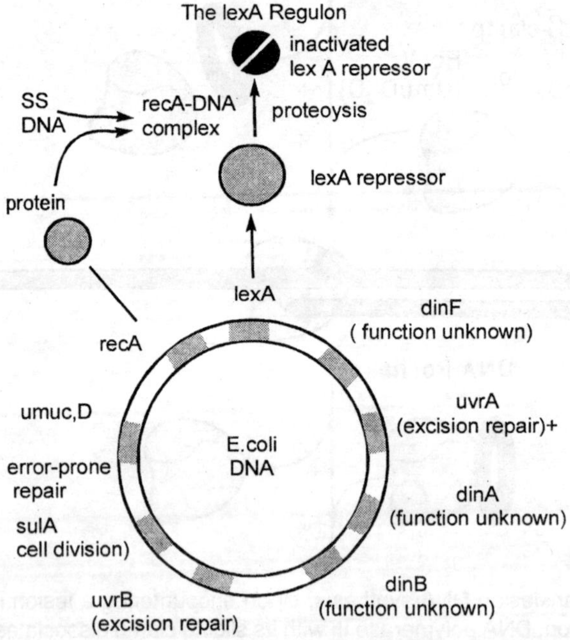

Fig. 20.23a: All operators regulated by lexA repressor

Induction of SOS repair system is initiated by activated RecA protein. The RecA protein is activated by binding to ssDNA at the damaged site and interacts with the LexA repressor protein stimulating the lexA repressor to inactivate itself by proteolytic cleavage. With the repressor inactivated, all the genes involved in SOS response can be expressed and the enzymes needed for DNA repair are synthesized. The various steps involved in SOS response are shown in Figure 20.23b.

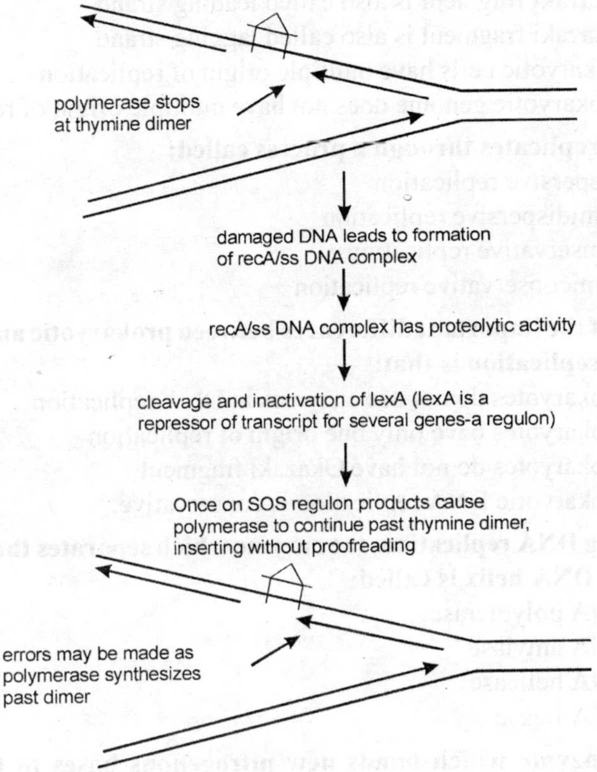

Fig. 20.23b: SOS repair

The SOS repair system is error-prone and is "switched on" as a last-ditch effort to allow the cell to survive and hence the name SOS response. The enhanced mutagenesis that results may be seen as a benefit in that it may result in progeny that through mutation are better adapted to live in the noxious environment that led to the induction of SOS system.

SHORT QUESTIONS

1. Briefly describe the process of DNA replication.
2. Write short notes on the following
   A. Topoisomerase
   B. β–sliding clamp
   C. Telomerase
   D. The replicon model
   E. Mismatch repair

MCQs

Select one most appropriate answer given for each of the following questions:

1. **Which of the following statements is false?**
   A. Okazaki fragment is also called leading strand
   B. Okazaki fragment is also called lagging strand
   C. Eukaryotic cells have multiple origin of replication
   D. Prokaryotic genome does not have multiple origin of replication.
2. **DNA replicates through a process called:**
   A. Dispersive replication
   B. Semidispersive replication
   C. Conservative replication
   D. Semiconservative replication
3. **One of the important differences between prokaryotic and eukaryotic DNA replication is that:**
   A. Prokaryotes do not use enzymes in DNA replication
   B. Prokaryotes have only one origin of replication
   C. Prokaryotes do not have Okazaki fragment
   D. Prokaryotic DNA replication is conservative.
4. **During DNA replication, the enzyme which separates the two strands of the DNA helix is called:**
   A. DNA polymerase
   B. DNA amylase
   C. DNA helicase
   D. DNA ligase
5. **The enzyme which bonds new nitrogenous bases to those on the original DNA strand is called:**
   A. DNA amylase
   B. DNA polymerase
   C. DNA helicase
   D. DNA ligase
6. **Okazaki fragments are formed because:**
   A. DNA polymerase can only create a new strand of DNA from the 5' end to the 3' end.
   B. DNA polymerase can only create a new strand of DNA from 3' end to 5' end.
   C. Newly formed DNA strand tends to break apart into fragments.
   D. DNA helicase sometimes inadvertently breaks the DNA
7. **Which one of the following enzymes unwinds the helical DNA by breaking the hydrogen bond:**
   A. DNA polymerase
   B. DNA amylase

C. DNA helicase
D. DNA ligase

8. Which one of the following enzymes adds DNA nucleotides to RNA primer?
A. DNA polymerase II
B. DNA helicase
C. DNA polymerase III
D. DNA ligase

9. Which one of the following enzymes synthesizes a short RNA primer at the origin of replication?
A. DNA ligase
B. DNA primase
C. DNA polymerase
D. DNA helicase

10. Which one of the following enzymes links the DNA fragments of lagging strands together?
A. DNA primase
B. DNA ligase
C. DNA polymerase II
D. DNA helicase

---

1 A 2 D 3 B 4 C 5 B 6 A 7 C 8 C
9 B 10 B

21

Transcription

I. INTRODUCTION

The process by which the genetic information stored in the DNA is passed onto RNA is known as transcription. Only one strand of the DNA is used as a **template** (sometimes called the **antisense strand**) and its sequence is complementary to the sequence of the RNA being produced. The other strand called the **coding strand** or **sense strand** has a sequence identical to the RNA transcript except that it has a 'T' in place of 'U' in the RNA. The enzyme responsible for the synthesis of RNA is the **DNA dependent RNA polymerase** or simply RNA polymerase. RNA synthesis is very similar to DNA synthesis except that in transcription:

- The new strand is made up of ribonucleotides.
- The RNA product does not need a primer.
- The RNA product does not remain base-paired to the template DNA.
- The process is less accurate than replication (one mistake occurs in 10,000 nucleotides added, compared to one in 10^7 nucleotides for replication).

Besides, transcription selectively copies only certain parts of the genome and makes several copies of a given section. In contrast replication must copy the entire genome and do so only once every cell division.

II. FEATURES OF ENZYMATIC SYNTHESIS OF RNA

The essential features of the synthesis of RNA are following:

1. The precursors in the synthesis of RNA are the four ribonucleoside-5'-triphosphates (rNTP)-rATP, rGTP, rCTP and rUTP. The ribose of each ribonucleotide has 2-OH groups one each on the 2' and 3' carbon atoms.
2. As shown in Figure 21.1 in the polymerization reaction the 3'-OH terminus of the RNA being synthesized reacts with the 5' triphosphate of a precursor rNTP resulting in the formation of a phosphodiester bond with the release of a pyrophosphate (PPi).

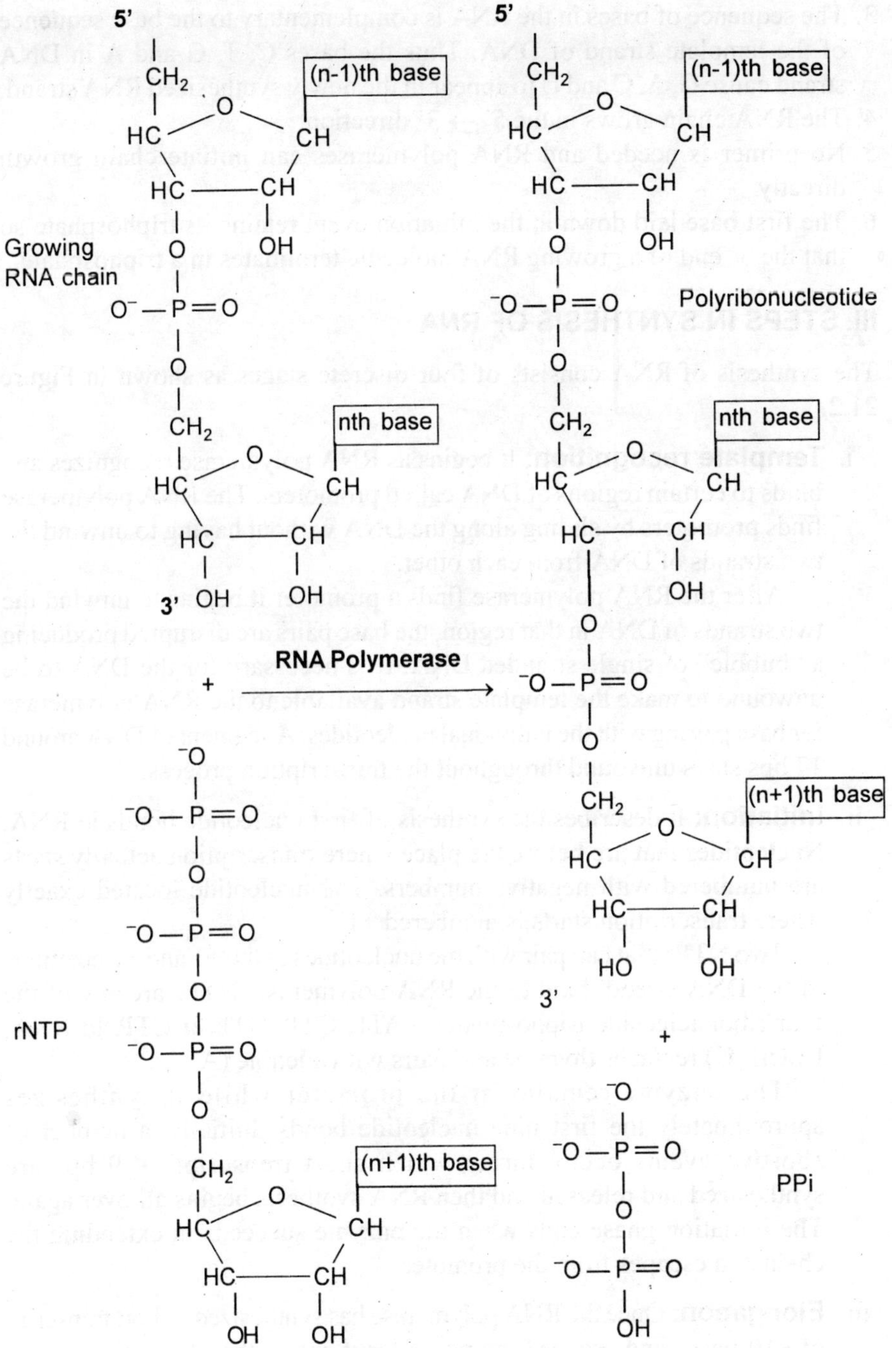

Fig. 21.1: Mechanism of chain elongation reaction catalyzed by RNA polymerase. The nucleotides used to extend a growing RNA chain are ribonucleotide triphosphates (rNTPs). Two phosphate groups are released as pyrophosphate (PPi) during the reaction.

3. The sequence of bases in the RNA is complementary to the base sequence of the template strand of DNA. Thus the bases C, T, G and A in DNA strand cause G, A, C and U to appear in the newly synthesized RNA strand.
4. The RNA chain grows in the 5' → 3' direction.
5. No primer is needed and RNA polymerases can initiate chain growth directly.
6. The first base laid down in the initiation event retains its triphosphate so that the 5' end of a growing RNA molecule terminates in a triphosphate.

III. STEPS IN SYNTHESIS OF RNA

The synthesis of RNA consists of four discrete stages as shown in Figure 21.2.

i. **Template recognition:** It begins as RNA polymerase recognizes and binds to certain regions of DNA called promoters. The RNA polymerase finds promoters by sliding along the DNA without having to unwind the two strands of DNA from each other.

   After the RNA polymerase finds a promoter it begins to unwind the two strands of DNA in that region, the base pairs are disrupted producing a "bubble" of single stranded DNA. It is necessary for the DNA to be unwound to make the template strand available to the RNA polymerase for base pairing with the individual nucleotides. A segment of DNA around 17 bps stays unwound throughout the transcription process.

ii. **Initiation:** It describes the synthesis of first nucleotide bonds in RNA. Nucleotides that are before the place where transcription actually starts are numbered with negative numbers. The nucleotide located exactly where transcription starts is numbered +1.

   Two NTPs that can pair with the nucleotides in the +1 and +2 positions of the DNA strand, bind to the RNA polymerase. NTPs are any of the four ribonucleoside triphosphates—ATP, GTP, UTP or CTP. In RNA, Uracil (U) replaces thymine and pairs with adenine (A).

   The enzyme remains at the promoter while it synthesizes approximately the first nine nucleotide bonds. Initially, a number of abortive events occur during which short transcripts < 9 bps are synthesized and released and then RNA synthesis begins all over again. The initiation phase ends when the enzyme succeeds in extending the chain and escapes from the promoter.

iii. **Elongation:** Once the RNA polymerase has synthesized a short transcript of ~10 bases and escaped the promoter it enters the *elongation phase.* There is a change in the conformation of polymerase and it grips the template more firmly. During elongation the enzyme:

   a. moves along the DNA and extends the RNA chain in the 5' → 3' direction.

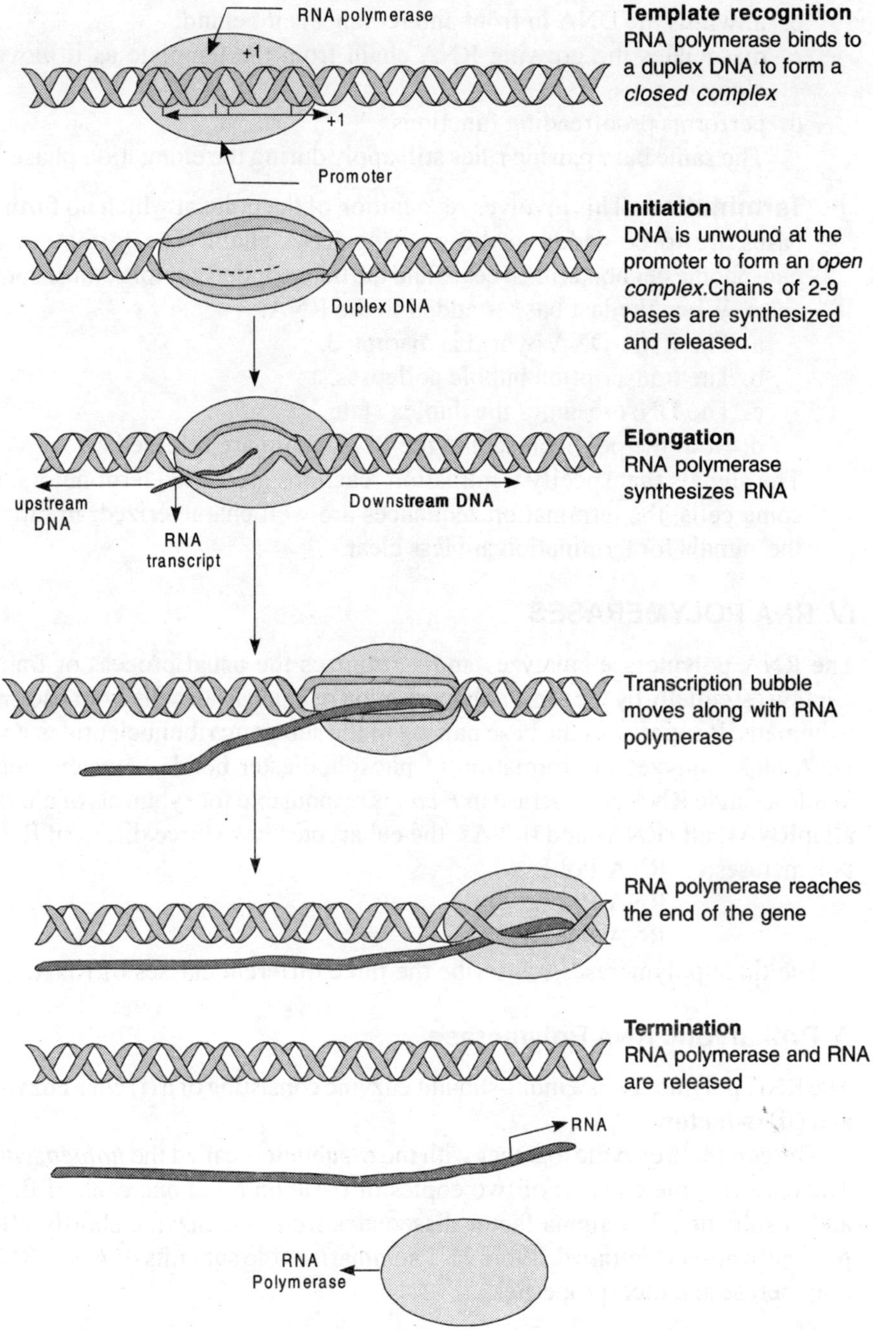

Fig. 21.2: The phases of transcription cycle: template recognition, initiation elongation and termination. The features shown hold true both for prokaryotes and eukaryotes. Transcription start site is at +1 position

b. unwinds the DNA in front and re-anneals it behind.
c. dissociates the growing RNA chain from the template as it moves along.
d. performs proofreading functions.

The same base pairing rules still apply during the elongation phase.

iv. **Termination:** This involves recognition of the point at which no further bases are to be added to the growing RNA chain. The formation of phosphodiester bonds must cease and the transcription complex must come apart. When the last base is added to the RNA:
   a. The RNA-DNA hybrid is disrupted.
   b. The transcription bubble collapses.
   c. The DNA resumes the duplex state.
   d. Both the polymerase and the RNA chain are released.

The signals that specify termination reactions are called terminators. In some cells, the termination sequences are well characterized; in others the signals for termination are less clear.

IV. RNA POLYMERASES

The RNA polymerase catalyzes and scrutinizes the usual process of transcription essentially by the same polymerization reaction in all cells from bacteria to humans. It supervises the base pairing of the substrate ribonucleotides with DNA and catalyzes the formation of phosphodiester bonds between them. While a single RNA polymerase in *E coli* is responsible for synthesis of almost all mRNAs, all rRNAs and tRNAs, the eukaryotes have three different RNA polymerases: RNA Pol I
RNA Pol II
RNA Pol III

The three polymerases transcribe the three different classes of RNA.

A. Prokaryotic RNA Polymerase

The RNA polymerase is a multi-subunit enzyme consisting of a (i) **core enzyme** and (ii) **σ-factor.**

The complete enzyme together with the σ-subunit is called the ***holoenzyme.*** The core enzyme consists of two copies of α-subunit and one each of β, β' and ω subunits. The sigma factor dissociates from the enzyme shortly after polymerization is initiated. Table 21.1 summarizes the subunits of *E coli* RNA polymerase and their properties.

Core Enzyme

In the core enzyme the omega (ω) subunit was for many years considered a curiosity since no function could be ascribed to it and it did not appear to be necessary in function. However, it is now known that omega is necessary to

Table 21.1: Subunit composition of *E. coli* RNA polymerase

Subunit	*Size (amino acids)*	*Size (daltons)*	*Gene*	*Function*
Alpha (α) (2 copies)	329	36,511	rpoA	Required for the assembly of the core enzyme, interacts with some regulatory proteins and is also involved in catalysis.
Beta (β)	1342	150,616	rpoB	Forms the catalytic center and is involved in chain initiation and elongation.
Beta' (β')	1407	155,159	rpoC	Is also a part of the catalytic center and binds to the DNA template.
Sigma (σ)	613	70,263	rpoD	Directs enzyme to the promoter.
Omega (ω)	91	10,237	rpoZ	Required to restore denatured RNA polymerase *in vitro* to its fully functional form.

restore denatured RNA polymerase *in vitro* to its fully functional form. It may function by binding simultaneously to the N-terminus and the C-terminus of the β-subunit. The ω-subunit is a part of the *Thermus aquaticus* polymerase enzyme whose structure was determined in May 2002.

The **core RNA polymerase** can bind to DNA and catalyze synthesis of RNA but it has no specificity. This form of enzyme will bind quite well to non-specific DNA but it cannot recognize promoters.

The Holoenzyme

Only the holoenzyme can initiate transcription. Sigma factor ensures that bacterial RNA polymerase binds in a stable manner to DNA only at promoter. The σ-factor does two things:

- It reduces the affinity of the enzyme for non-specific DNA.
- It greatly increases the affinity of the enzyme for promoters.

The Sigma Subunit

Sigma is a specificity factor. It directs RNA polymerase to the promoter and ensures that transcription is initiated only when it is supposed to be initiated.

The principal sigma factor in *E coli* is σ^{70} because the protein is 70 kD in size. Table 21.2 depicts the *six* alternative sigma factors that are used in special circumstances.

Sigma factor has a helix-turn-helix DNA binding motif by which it interacts with the DNA. One of the helices inserts into the major groove and interacts with the bases in the -35 region while the other helix makes contact with the DNA backbone. The interaction of σ-factor with -10 region is less well-understood.

Table 21.2: The different kinds of σ-factors found in bacterial cells

Sigma factor	*Gene*	*Function*
σ^{70}	rpoD	Principal sigma factor (regulates expression of most genes)
σ^{54}	rpoN	Required for expression of many genes involved in nitrogen metabolism
σ^{32}	rpoH	Required for expression of heat-shock genes
σ^{S}	rpoS	Gene expression in stationary phase cells
σ^{F} (σ^{28})	rpoF	Expression of flagellar operons
σ^{E} (σ^{32})	rpoE	Involved in heat-shock and oxidative stress responses, regulates expression of extracytoplasmic proteins
$\sigma^{Fec\ 1}$	fec1	Regulates the *fec* genes for iron dicitrate transport

RNA polymerase is one of the largest enzymes known and can easily be seen by electron microscopy. There are about 7000 RNA polymerase molecules in a single *E coli* cell. Depending upon growth conditions ~2000-5000 are engaged in transcription at any given time.

B. Eukaryotic RNA Polymerases

The three different eukaryotic polymerases transcribe different classes of RNA

- RNA polymerase I - transcribes rRNA
- RNA polymerase II - transcribes mRNA
- RNA polymerase III - transcribes tRNA and other small RNAs.

Table 21.3 summarizes the properties and products of the three polymerases.

Table 21.3: Properties and products of the three eukaryotic polymerases

	RNA Polymerase I	*RNA Polymerase II*	*RNA Polymerase III*
1. Location	Nucleolus	Nucleoplasm	Nucleoplasm
2. RNA product	Ribosomal RNA (rRNA) except the 5s rRNA	Nuclear pre-messenger RNA-(pre-mRNA) also called heterogenous nuclear RNA or hnRNA; some small nuclear RNAs (snRNA)	Transfer RNA (tRNA), 5s rRNA and other small RNAs
3. Sensitivity to α-aminitin (toxin of poisonous	Resistant	50% inhibition at 0.02 μg/ml	50% inhibition at 20 μg/ml. Response to α-aminitin is not well conserved; in animal cells it is inhibited by high levels mushroom) but in yeast and insects it is not inhibited.
4. Sensitivity to actinomycin D	Very sensitive	Slightly sensitive	Slightly sensitive

All eukaryotic RNA polymerases are large proteins appearing as aggregates of 500 kD or more and are more complex than bacterial polymerases. They typically have 8-14 subunits. The yeast *S. cerevesiae* RNA polymerase II is the most well-studied eukaryotic RNA polymerase and the three largest subunits show similarity to the bacterial RNA polymerase. The subunit composition of the eukaryotic RNA polymerase II is shown in Table 21.4.

Table 21.4: Subunits of Eukaryotic RNA polymerase II

Gene	*Protein mass*	*Phenotype of deletion mutant*
RPB1	190	Non-viable
RPB2	140	Non-viable
RPB3	35	Non-viable
RPB4	25	Conditional
RPB5	25	Non-viable
RPB6	18	Non-viable
RPB7	19	Non-viable
RPB8	17	Non-viable
RPB9	14	Conditional
RPB10	8.3	Non-viable
RPB11	14	Non-viable
RPB12	7.7	Non-viable

The term "conditional" in the above table means that deletion mutant of that gene have a problem only under certain conditions, and therefore that protein is not absolutely required. RPB stands for "RNA Polymerase B" because before the nomenclature was changed RNA Polymerases were called as PolA, PolB and PolC rather than Pol I, Pol II and Pol III.

The largest subunit has a *carboxy terminal domain-CTD* which consists of multiple repeats of consensus sequence of 7 amino acids YSPTSPS. This sequence is unique to RNA polymerase II and the number of repeats may vary from ~26 in yeast to 52 in mouse.

The subunits can be classified into three categories.

a. Core-subunits (related to α, β and β' of prokaryotes)
b. Common subunits (found in all three polymerases)
c. Non-essential subunits.

a. **Core subunits:** Three core subunits have been identified. These are:
   - RPB1
   - RPB2 and
   - RPB3

   RPB1 and RPB2 are homologous to the *E coli* β' and β subunits respectively. RPB3 has homologies to the prokaryotic α-subunit.

b. **Common subunits:** RPB5, RPB6, RPB8, RPB10 and RPB12 are the common subunits. Very little is known about their function except that they are found in all three RNA polymerases.

c. **Non-essential subunits:** RPB4 and RPB9 appear to be essential only under certain conditions. Deletion mutants of these two functions well at normal temperatures but fail to grow at either higher or lower temperature. RPB4 seems to be required to keep RPB7 associated with the polymerase.

d. **Other subunits:** Two other proteins present in the enzyme are RPB7 and RPB11. Deletion mutants of these two are non-viable so they are essential. RPB7 has a σ-like role being responsible for correct initiation of transcription. However, this is not exactly like σ and initiation requires much more than this protein.

C. Structure of RNA Polymerases

The structures of bacterial and yeast RNA polymerase II core enzymes are very similar. The overall shape of each enzyme resembles a crab claw, the pincers being made up of the largest subunits, β and β' in case of bacterial enzyme and RPB1 and RPB2 for the eukaryotic enzyme. The 'active site' made up of regions from both these subunits is found at the base of the pincers. The active site can bind two Mg^{++} ions. There are various channels that allow DNA, RNA and ribonucleotides into and out of the enzyme's active center cleft.

V. PROMOTERS

The region of the DNA where the RNA polymerase binds specifically to initiate transcription is called a **Promoter.** It functions by being **recognized**— the information for promoter function is provided directly by the DNA sequence.

A. Prokaryotic Promoters

The promoter recognized by *E coli* RNA polymerase has following features:

- The start point designated as +1.
- A conserved sequence of 6 nucleotides 10 bps upstream of the start point referred to as the *–10* **sequence** or **element** or the **Pribnow box.**
- A conserved sequence of 6 nucleotides 35 bps upstream of the start point called as the *–35* **sequence** or **element**.
- The separation between the -10 and -35 regions consisting of a non-specific stretch of 17-19 nucleotides.

The features of *E. coli* promoters have been summarized in Figure 21.3.

The DNA sequences recognized by RNA polymerase are not always exactly the same. By comparing many different (~300) promoters and ascertaining the most common base found in each position in the –35 and –10 hexamer, **consensus sequences** were derived for these regions.

The –10 consensus sequence is T A T A A T

The –35 consensus sequence is T T G A C A

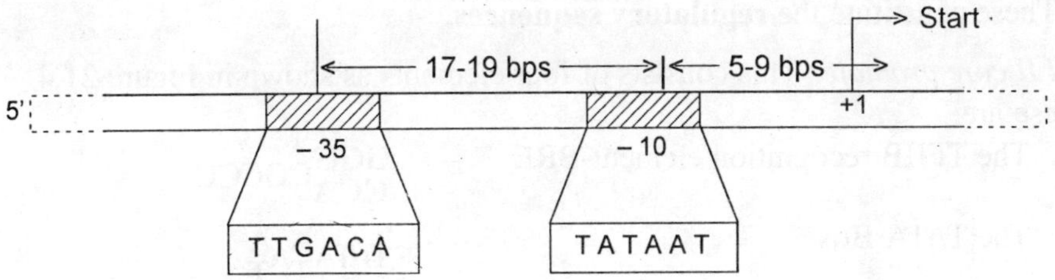

Fig. 21.3: Features of bacterial promoters

Promoters with sequences closer to the consensus are generally "stronger" than those that match less well. The strength of a promoter indicates the number of transcripts it initiates in a given time. Transcription from strong promoter can start as frequently as every two seconds and as little as every 10 seconds in weak promoters in *E coli.*

Although the actual sequence in the intervening region is unimportant, the distance is critical in holding the two sites at the appropriate separation for the geometry of RNA polymerase.

B. Eukaryotic Promoters

In eukaryotes, three different RNA polymerases fulfill the different transcriptional tasks and accordingly their promoter elements are more diversified as compared to the bacterial promoters. The promoters for RNA polymerases I and II are mostly upstream of the start point but some promoters for RNA polymerase III lie downstream of the start point. Each promoter contains characteristic sets of short conserved sequences that are recognized by the appropriate class of factors. RNA Pol I and Pol III each recognize a relatively restricted set of promoters and rely upon a small number of accessory factors.

Promoters utilized by RNA Pol II show more variation. The promoter contains several short (<10 bp) sequence elements that bind transcription factors dispersed over > 200 bp. Some of these elements and the factors that recognize them are common and are found in a variety of promoters. They are used constitutively. Others are specific — they identify particular classes of genes and their use is regulated.

RNA Polymerase II Promoter

A promoter for RNA Pol II consists of two types of regions:
- The region around the start point itself defined by the initiator and the TATA box close by. This forms the **core-promoter.**
- Several short cis-acting elements, farther upstream of the TATA box govern the efficiency and the specificity by which the promoter is recognized.

These constitute the **regulatory sequences.**

Pol II core promoter This consists of four elements as shown in Figure 21.4. These are:

i. The TFIIB recognition element-BRE — $^{GGG}_{CCA}$CGCCC

ii. The TATA Box — TATA$^{A}_{T}$A$^{A}_{T}$

iii. Initiator or Inr. — $^{CC}_{TT}$AN$^{TCC}_{ATT}$

iv. Downstream promoter element or DPE — $^{A}_{G}$G$^{A}_{T}$CGTG

In a given promoter, usually only two of these four elements are present. The minority of promoters that do not contain a TATA element are called **TATA-less promoters.**

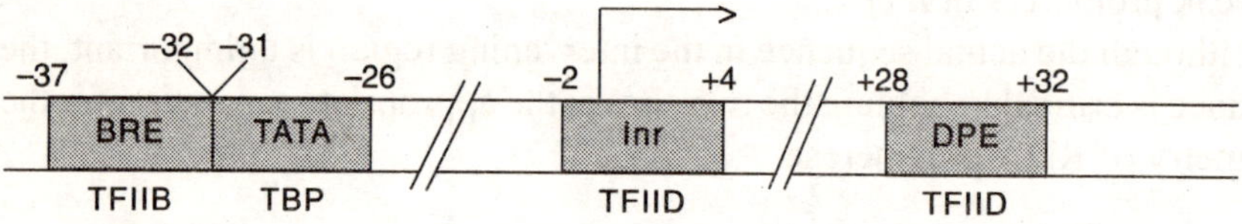

Fig. 21.4: Pol II core promoter: The positions of the various promoter elements and the transcription factors recognizing them are shown

Beyond and upstream of the core promoter are the **regulatory sequences** required for efficient transcription. These could be promoter proximal elements, upstream activation sequences, enhancers, silencers, boundary elements, insulators, etc. All these DNA elements bind regulatory proteins.

Often at –75 position is found the **CAAT Box.** It can function in either orientation and has a strong role to play in determining the efficiency of the promoter.

The **GC Box** is at position –90 and contains the sequence GGG CGG. Often multiple copies are present in the promoter and they occur in either orientation. It too is a relatively common promoter element.

DNA sequences called **enhancers** also help stimulate transcription in eukaryotes. Enhancers are not well understood yet, but they are known to be able to increase transcription rates even when they are located thousands of base pairs away from the transcription start site.

RNA Polymerase I Promoter

RNA Pol I is required for the transcription of only one gene that codes for the rRNA precursor and that too from a distinct single type of promoter. The promoter for RNA Pol I has been best characterized in human cells.

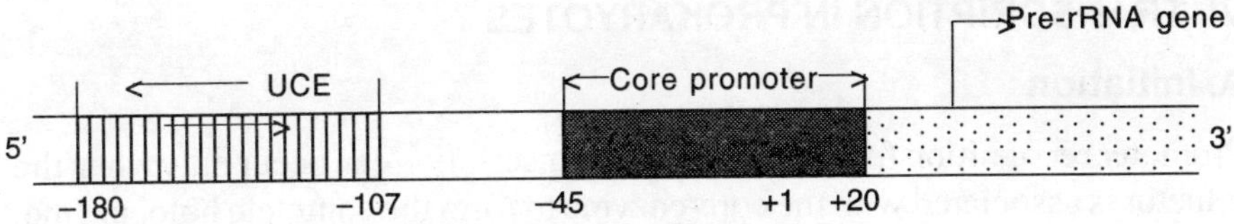

Fig. 21.5: Pol I promoter region

Thc promoter is bipartite and comprises two parts as shown in Figure 21.5.

i. The core promoter extending from – 45 to +20

ii. The upstream control element –UCE extending from –180 to –107.

Transcription can be initiated from the core-promoter alone but the efficiency of transcription is greatly increased by the presence of UCE. Both regions are unusual in that they are G:C rich and are ~85% identical.

For initiation of transcription Pol I requires two other factors:

i. UBF

ii. SLI — This comprises of TBP and three TAFs specific for Pol I transcription.

RNA Polymerase III Promoter

The promoters recognized by RNA polymerase III are of two types being recognized in different ways by different groups of factors. The promoters for tRNA and 5s-rRNA genes are **internal** and lie downstream of the transcription start site. The promoters for snRNA (small nuclear RNA) genes lie upstream of the start point. The individual elements that are necessary for promoter function consist exclusively of sequences recognized by transcription factors, which in turn direct the binding of RNA polymerase. The structures of three types of RNA polymerase III promoters are summarized in Figure 21.6.

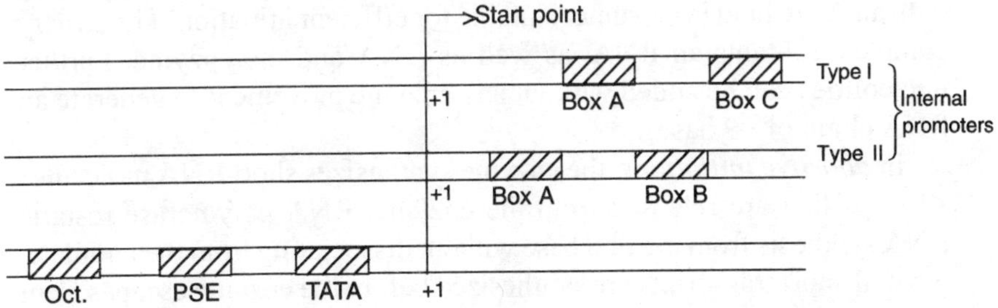

Fig. 21.6: Promoters for RNA polymerase III may consist of bipartite sequences downstream of start point, with box A separated from either box C or box B. Or they may consist of separated sequences upstream of start point (Oct, PSE, TATA)

The distance between box A and box B in a type 2 promoter can vary extensively, but the boxes usually cannot be brought too close together without abolishing function.

VI. TRANSCRIPTION IN PROKARYOTES

A. Initiation

Promoter recognition for the initiation of transcription can occur only when the σ factor is associated with the core-enzyme to form the complete holoenzyme. The core enzyme can synthesize RNA on any region of the DNA template but it cannot distinguish between promoter and other sequences of DNA. Sigma factor ensures that the bacterial RNA polymerase binds in a stable manner to DNA only at promoters. Transcription initiation involves 3 steps.

i. Formation of a **closed binary complex:** Here the DNA remains a duplex after the holoenzyme is bound to the promoter. When RNA polymerase holoenzyme initially binds to DNA, it covers some 75-80 bps extending from –55 to +20. The sigma factor has specific regions that make contact with the –35 and –10 promoter elements. The –35 region provides a secure binding to the enzyme. It is in the –10 region that the melting of the DNA is initiated.

ii. Formation of an **open complex:** The binding of the enzyme is "tight" so that the polymerase is more intimately engaged with the promoter. The DNA "melts" so that the strands separate over a distance of about 14 bps (between –11 and +3) at the start site to form the transcription "bubble". For strong promoters formation of an open complex is an irreversible reaction. The transition from closed to open complex involves structural changes in the enzyme.

iii. Formation of a **ternary complex***:* The opening of the DNA frees the template strand. The first two ribonucleotides are brought into the active site of the enzyme, aligned on the template strand and the phosphodiester bond forms between them forming a ternary complex. Thus transcription is initiated by the RNA polymerase without the need for a primer. RNA polymerase starts most transcripts with an A and only chains beginning with an A are held in a manner suitable for efficient initiation. The ternary complex contains an RNA, as well as DNA and the enzyme. Further nucleotides can be added without any enzyme movement to generate an RNA chain of ~9 bases.

In *abortive initiations,* the enzyme synthesizes short RNA molecules <10 bps that are released from the enzyme. RNA polymerase restarts RNA synthesis from the first base without dissociating from the template. Several short transcripts are synthesized before the enzyme escapes from the promoter and enters into the elongation phase which continues until polymerase is instructed to terminate transcription by specific sequences downstream of the gene.

B. Elongation

After transcription initiation, the sigma factor dissociates from the holoenzyme and elongation of the RNA chain is brought about by core enzyme and Nus proteins. Elongation of RNA chain occurs in the 5' → 3' direction and is rather a monotonous process in which the enzyme moves forward 1bp along DNA for every nucleotide added to the RNA chain. The overall reaction rate is ~40 nucleotides/second at 37°C. The RNA polymerase performs the following functions during elongation:

i. Facilitates the unwinding and rewinding of DNA
ii. Holds the separated strands of DNA and RNA product
iii. Catalyzes the addition of ribonucleoside to growing RNA chain
iv. Proofreading
v. Adjusts to difficulties in progressing by cleaving the RNA product and restarting RNA synthesis.

For the performance of all these functions, the enzyme has different channels:

i. NTP-uptake channel	:	Allows ribonucleotides to enter the active center.
ii. RNA-exit channel	:	Allows growing RNA chain to leave the enzyme.
iii. Downstream DNA channel	:	Allows the DNA ahead of the enzyme, yet to be transcribed, to enter the active center cleft between the pincers of the enzymes.
iv. NT channel	:	Allows the non-template strand to exit the active center and travels across the surface of the enzyme.
v. T channel	:	This is the path for the template strand. Only 8-9 nucleotides of the growing chain are base paired to the DNA template at any given time. The remainder of the RNA chain is directed out through the RNA exit channel.

The proofreading function of RNA polymerase is performed in two ways:

i. **Pyrophosphorolytic editing:** RNA polymerase pauses when there is a mismatched base, goes back to catalyze the removal of incorrectly inserted nucleotide by reincorporation of PPi. Now a fresh ribonucleotide can be incorporated in the growing RAN chain.
ii. **Hydrolytic editing:** In this mechanism, the polymerase backtracks by one or more nucleotides and cleaves the RNA product, removing the error-containing sequence. Hydrolytic editing is stimulated by *Gre-factors* that release the RNA polymerase from elongation arrest at sequences difficult to transcribe and ensures that the polymerase elongates efficiently.

C. Termination

Bacterial RNA polymerase terminates transcription at two types of termination signals depending upon the presence or absence of additional protein factors.

These terminators are therefore of two types:

i. **Rho independent**
ii. **Rho dependent**

Rho Independent Termination

Core-enzyme can terminate *in vitro* at certain sites in the absence of any other factor. Such *intrinsic terminator* sequences signal the RNA polymerase terminator to stop transcribing.

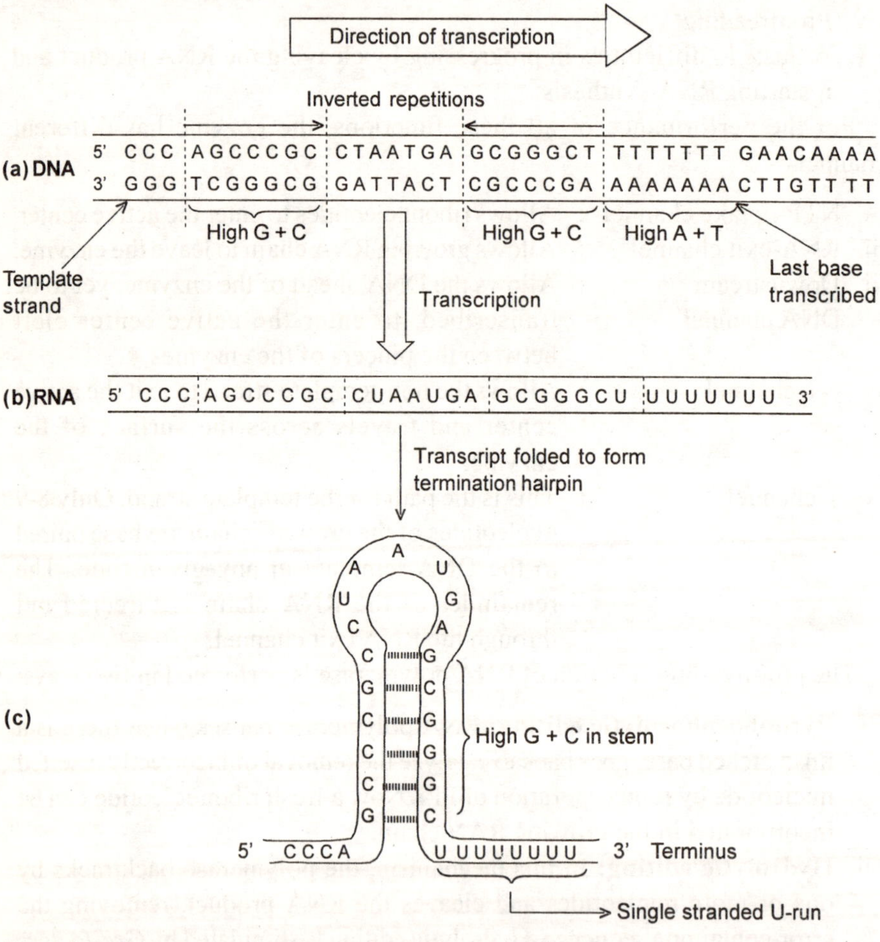

Fig. 21.7: Sequence of a Rho-independent terminator (a) base sequence of the DNA of *E. coli* trp. operon at which transcription termination occurs, (b) sequence of RNA and (c) structure of the terminator hairpin

The **intrinsic terminators** have following features as shown in Figure 21.7.

a. An inverted-repeat base sequence containing a central non-repeating segment. This sequence is capable of intrastrand base-pairing forming a stem and loop (hairpin) configuration of about 7-20 bp in the RNA transcript.

b. A G + C rich region near the loop end of putative stem.
c. A region with a stretch of about eight A: T base pairs that yield in the RNA a sequence of 6-8 uracil residues often followed by an adenine.

The stem and loop hairpin structure causes the RNA polymerase to pause. When this is followed by a string of U residues in the right location the weak rU. dA bonds in the RNA-DNA hybrid are easily disrupted and the polymerase dissociates. If, however, the pause site does not correspond to a terminator, the enzyme moves on again to continue transcription. Both the sequence of hairpin and the length of U-run influence the efficiency of termination.

Rho Dependent Termination

Rho is a protein of ~46 kD and is active as a hexamer. It acts by binding to the single stranded nascent RNA as it exits the polymerase. It also has an ATPase activity. It uses the energy derived from the hydrolysis of ATP to forcibly pull the RNA away from its contact with the DNA in the transcription-elongation complex with RNA polymerase. The sequences required for Rho-dependent termination are 50-90 bases long and lie upstream of the terminator. It binds to the so-called *Rho utilization* sites or *Rut sites* that are rich in C residues and poor in G-residues. Also they do not fold into secondary structures. The "hot pursuit" model of Rho action is shown in Figure 21.8.

At first Rho attaches to some specific site on nascent RNA. It then translocates along the mRNA faster than the polymerase moves along the template DNA. The RNA polymerase pauses when it reaches a terminator and termination occurs if Rho catches it there. Rho has a 5'-3' helicase action. It directly gains access to the stretch of RNA-DNA hybrid in the transcription bubble and causes it to unwind. Termination occurs with the release of Rho and RNA polymerase from the nucleic acids.

Rho fails to bind any transcript that is being translated presumably because of the presence of the ribosomes that are translating the RNA. In bacteria transcription and translation are tightly coupled so Rho terminates only those transcripts that are still being transcribed beyond the end of the gene or the operon.

VII. TRANSCRIPTION IN EUKARYOTES

Most of the basic features of transcription are the same in both prokaryotes and eukaryotes. However, transcription in eukaryotes has following differences:

1. There are three RNA polymerases in contrast to the single RNA polymerase in prokaryotes.
2. Several initiation factors called the **general transcription factors** or **GTFs** are required for efficient, promoter specific initiation of transcription in eukaryotes. This is in contrast to the single initiation factor σ required by the single RNA polymerase in bacteria.

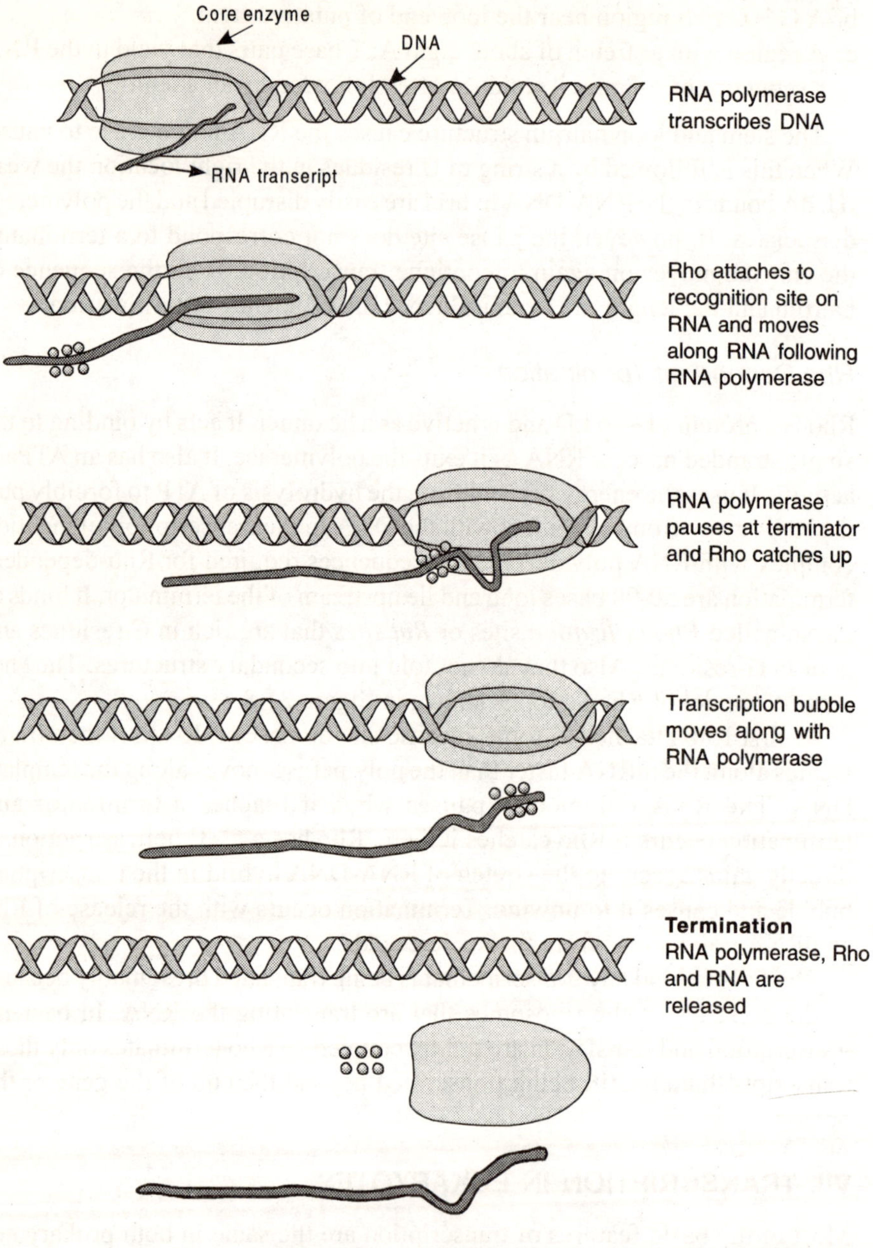

Fig. 21.8: Rho dependent termination

3. Many mRNA molecules are very long-lived.
4. Both 5' and 3' termini of mRNA are modified. The 5' end has a complex 'cap' while the 3' end has a long stretch (upto 250 nucleotides) of polyA residues.

5. The primary transcripts are much longer than the mRNA. During processing intervening sequences specified by the intergenic regions or introns are excised and the fragments rejoined.
6. All eukaryotic mRNA molecules are monocistronic and termination occurs beyond the coding region.

Besides, eukaryotic DNA is organized into nucleosomes. Therefore, in order to initiate transcription *in vivo* several additional factors are required like the DNA binding regulatory proteins and chromatin modifying enzymes.

A. Initiation

None of the three polymerases recognize their promoters directly. A feature common to all three is that it is the transcription factors that recognize specific sequence elements of any promoter and they, in turn serve to bind the RNA polymerase and position it correctly at the start point. The complete set of GTFs and polymerase bound together at the promoter is called **preinitiation complex (PIC).** The preinitiation complex formed by each of the three RNA polymerases consists of a factor TBP (Tata Binding Protein) that binds in the vicinity of the start point.

Formation of PIC by RNA Polymerase II

The TATA box (if there is one) and the initiator region immediately at the start point, are responsible for the selection of the exact start point at promoters for RNA polymerase II. The sequence of events in the formation of PIC on a naked DNA template *in vitro* is as follows and is shown in Figure 21.9.

i. The general transcription factor TFIID recognizes the TATA element and binds to a region extending upstream of the TATA element. TFIID is a multi-subunit complex of ~800 kD containing TBP and eleven TAFs varying in mass from 30-250 kD. TBP is the **Tata Binding Protein** and binds to the TATA DNA sequence. TAFs are tissue-specific and have various functions.
ii. TBP binds the DNA in an unusual way. It uses a b-sheet that is inserted in the minor groove of the DNA. This not only widens the minor groove but also bends the DNA at the TATA element by ~80°.
iii. Distortion of the DNA allows for a closer association of the TAFs and RNA polymerase than would be possible in linear DNA.
iv. In *vitro,* the proteins assemble in the following order TFIIA followed by TFIIB, TFIIF RNA polymerase and then TFIIE and TFIIH. TFIIF is a two-subunit factor that binds to RNA polymerase. TFIIH controls the ATP-dependent transition of the PIC to the open complex. It is the helicase-like activity of TFIIH that stimulates the unwinding of promoter DNA.

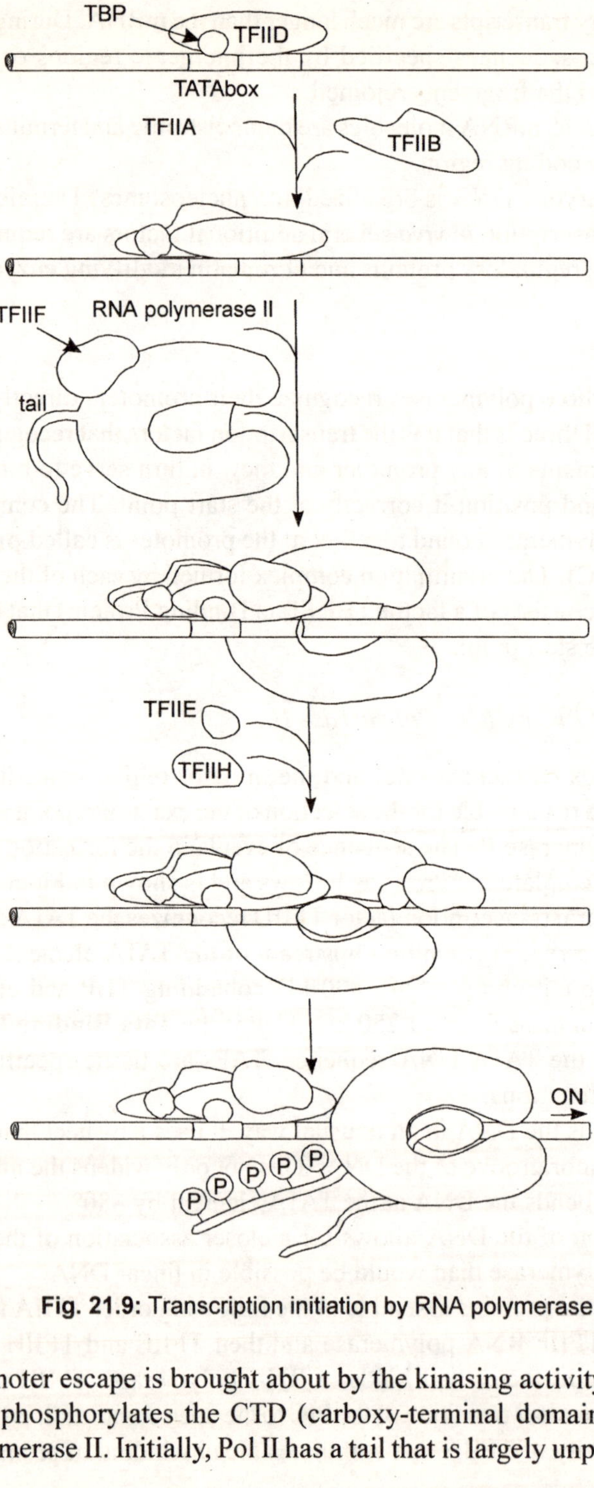

Fig. 21.9: Transcription initiation by RNA polymerase II

v. Promoter escape is brought about by the kinasing activity of helicase H that phosphorylates the CTD (carboxy-terminal domain) tail of RNA polymerase II. Initially, Pol II has a tail that is largely unphosphorylated.

The tail contains a series of repeats (27 in yeast and 52 in humans) of a heptapeptide sequence Tyr-Ser-Pro-Thr-Ser-Pro-Ser each of which contains specific sites for phosphorylation.

vi. Most of the TFII factors are released before RNA polymerase II leaves the promoter.

In *vivo* the DNA is packaged into nucleosomes and chromatin. Transcription initiation, therefore, requires additional proteins as shown in Figure 21.10.

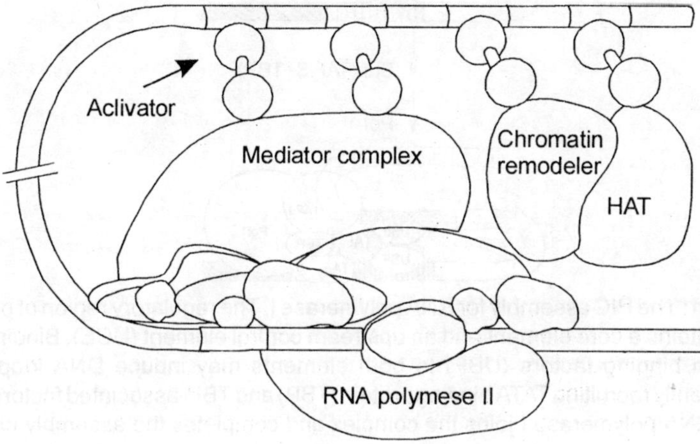

Fig. 21.10: Assembly of the preinitiation complex in presence of mediators

The proteins involved in the formation of PIC are:

i. **Activators**
ii. **Mediator complex**
iii. **Nucleosome modifiers** and **remodellers**.

Formation of Preinitiation Complex by RNA Polymerase I

RNA polymerase 1 is devoted to the synthesis of pre-rRNA. Like RNA polymerase II a *preinitiation complex* (PIC) has to be formed before RNA polymerase I can exert its function. The order of PIC assembly is illustrated in Figure 21.11.

Formation of Pre-initiation Complex by RNA Polymerase III

The genes transcribed by RNA polymerase III include those for tRNA, 5s rRNA genes. Figures 21.12 and 21.13 depict the assembly of preinitiation complex for tRNA and 5s rRNA genes respectively.

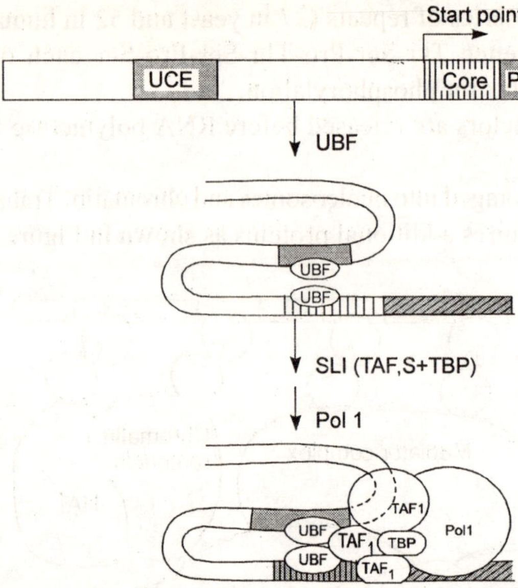

Fig. 21.11: The PIC assembly for RNA polymerase I. The regulatory region of pre-rRNA gene contains a core element and an upstream control element (UCE). Binding of two upstream binding factors (UBF) to both elements may induce DNA looping and subsequently recruiting TATA binding protein (TBP) and TBP-associated factors (TAF_1). Finally, RNA polymerase I joins the complex and completes the assembly process

a. *Transcription of tRNA Gene*
b. *Transcription of 5s rRNA Gene*

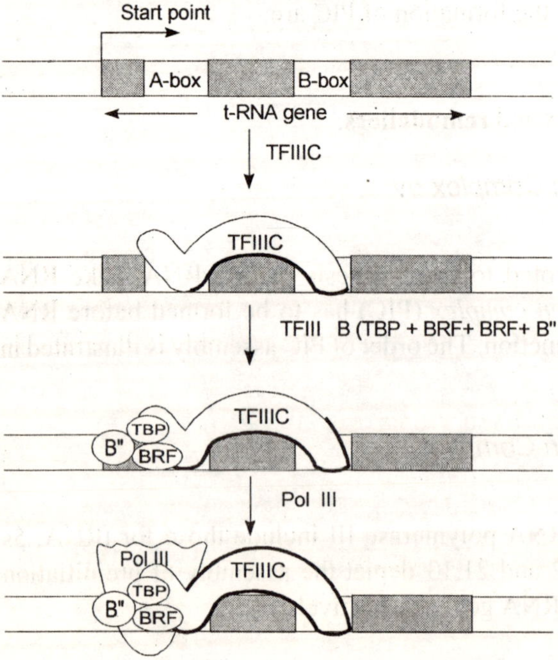

Fig. 21.12: The PIC assembly for the transcription of tRNA gene. Initiation via the internal Pol III promoter involves the assembly factor TFIIIC, the initiation factor TFIIIB and RNA poly merase III.

The regulatory region of 5s-rRNA gene contains a C-box also located inside the transcription unit. The PIC assembly begins with the binding of TFIIIA to the C-box as shown in Figure 21.13.

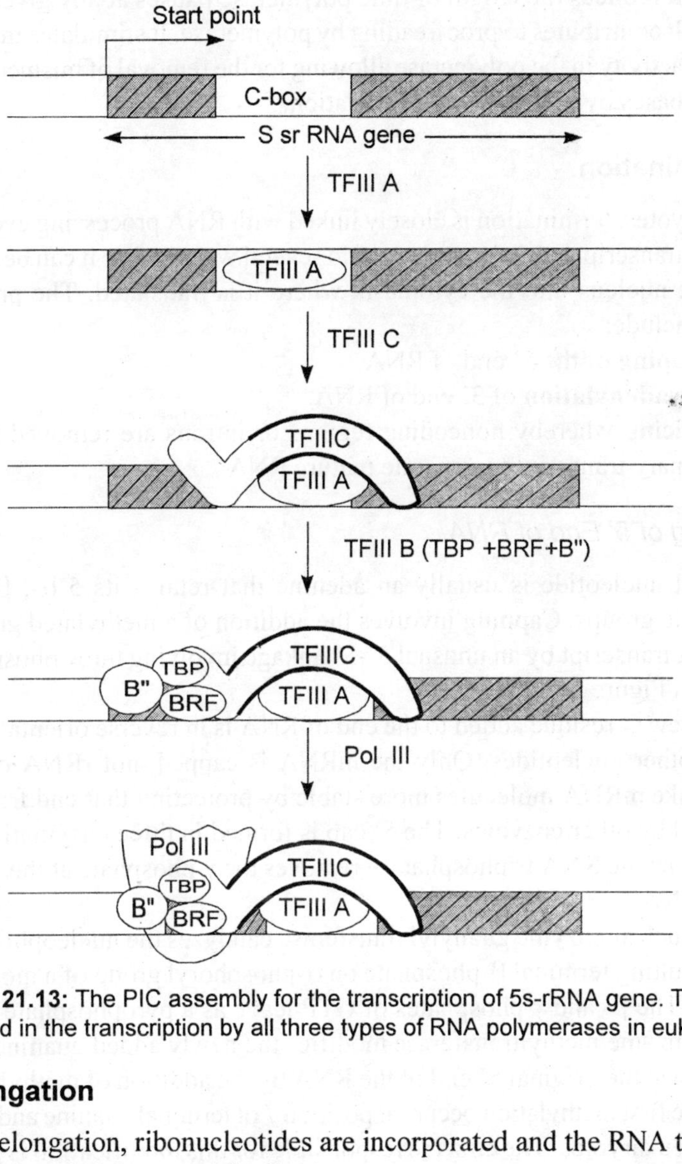

Fig. 21.13: The PIC assembly for the transcription of 5s-rRNA gene. TBP is involved in the transcription by all three types of RNA polymerases in eukaryotes

B. Elongation

During elongation, ribonucleotides are incorporated and the RNA transcript grows in length. More specifically:

a. Pol II enzyme sheds most of the initiation factors
b. Elongation factors like TFIIS, hSPT5, P-TEFb are recruited to the phosphorylated CTD tail of RNA polymerase II.
c. P-TEFb functions in 3 ways:
   – It phosphorylates the serine residue at position 2 of CTD tail.

- It phosphorylates and activates another protein hSPT5.
- It recruits another elongation factor TAT-SF1.

d. Elongation factor TFIIS functions in two ways:
- It reduces the length of time polymerase pauses at any given spot
- It contributes to proofreading by polymerase. It stimulates the RNAse activity in the polymerase allowing for the removal of misincorporated bases by limited local degradation.

C. Termination

In eukaryotes, termination is closely linked with RNA processing events. The primary transcript has to be processed in several ways before it can be exported from the nucleus into the cytoplasm where it is translated. The processing events include:

i. **Capping** of the 5' end of RNA.
ii. **Polyadenylation** of 3' end of RNA.
iii. **Splicing** whereby noncoding regions or introns are removed from the primary transcript to generate mature RNA.

Capping of 5' End of RNA

The first nucleotide is usually an adenine that retains its 5'α-, β- and γ-phosphate groups. Capping involves the addition of a methylated guanine to the RNA transcript by an unusual 5'-5' linkage involving three phosphates as shown in Figure 21.14.

The new G residue added to the end of RNA is in reverse orientation from all the other nucleotides. Only the mRNA is capped, not rRNA or tRNA. Caps make mRNA molecules more stable by protecting that end from being degraded by other enzymes. The 5' cap is formed in three enzymatic steps:

- The enzyme RNA triphosphatase removes the γ-phosphate at the 5' end of the RNA.
- The nuclear enzyme guanylyl transferase catalyzes the nucleophilic attack of resulting terminal β-phosphate on α-phosphoryl group of a molecule of GTP. The β- and γ-phosphates of GTP leave as a pyrophosphate (PPi).
- The enzyme methyltransferase modifies the newly added guanine and the purine at the original 5' end of the RNA by the addition of methyl groups.

  The first methylation occurs at position 7 of terminal guanine and is found in all eukaryotes. The enzyme responsible for this modification is guanine-7-methyltransferase.

Polyadenylation of 3' End of RNA

The polyadenylation of 3' end of mRNA is intimately linked with termination of transcription. The poly (A) sequence is not coded in the DNA but is added

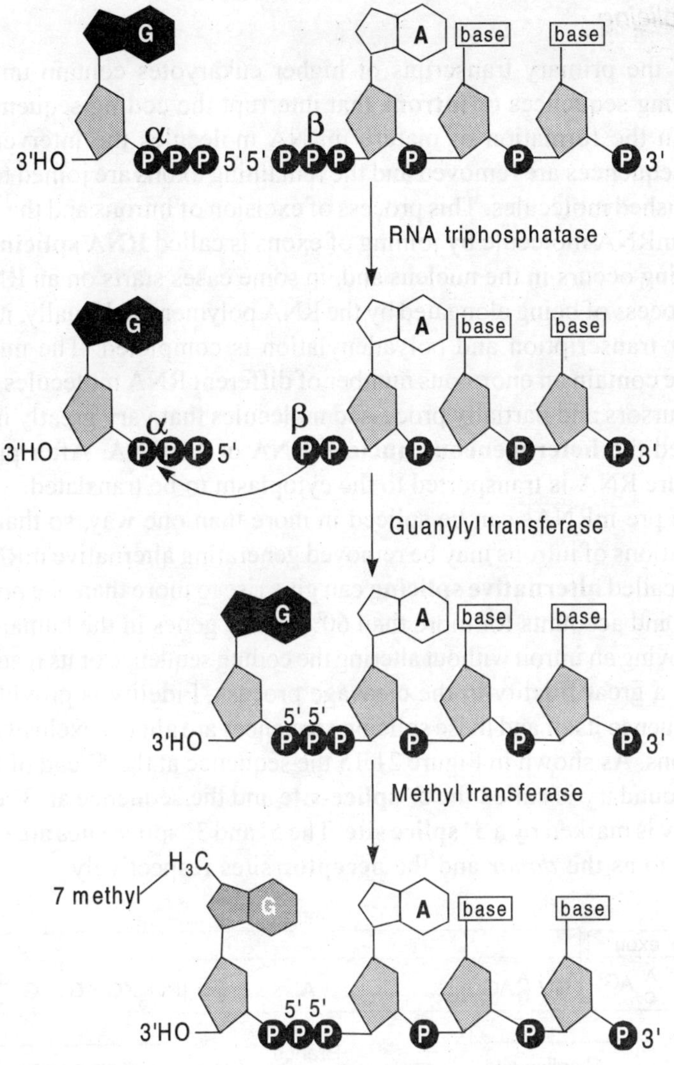

Fig. 21.14: The structure and formation of the 5' RNA cap

to the RNA in the nucleus after transcription. The enzyme responsible for the addition of poly (A) tail is the **poly-A polymerase.** It uses ATP as precursor. The adenylate residues are not added to the 3' terminus of the primary transcript. Transcription continues through the site where poly (A) residues are to be added, so an endonucleolytic cleavage occurs before a poly (A) is added. The sequence AAUAAA 10-25 bases upstream from the poly (A) site signals polyadenylation. The poly (A) tail is associated with a **poly** (A) **binding protein** (PABP) and confers stability on the mRNA. It is not known what links polyadenylation to termination but it is clear that polyadenylation signal is required for termination.

RNA Splicing

Most of the primary transcripts of higher eukaryotes contain untranslated intervening sequences or **introns** that interrupt the coding sequences or the **exons**. In the formation of mature mRNA molecules the intervening non-coding sequences are removed and the remaining exons are joined together to form finished molecules. This process of excision of introns and the formation of final mRNA molecule by joining of exons is called **RNA splicing**.

Splicing occurs in the nucleus and, in some cases starts on an RNA that is in the process of being elongated by the RNA polymerase. Usually, it is carried out after transcription and polyadenylation is completed. The nucleus can therefore contain an enormous number of different RNA molecules including the precursors and partially processed molecules that vary greatly in size and are called the **heterogeneous nuclear RNA** or **HnRNA**. After processing, the mature RNA is transported to the cytoplasm to be translated.

Some pre-mRNAs can be spliced in more than one way, so that different combinations of introns may be removed generating alternative mRNAs. This process called **alternative splicing** can give rise to more than one polypeptide product and accounts for more than 60% of the genes in the human genome.

Removing an intron without altering the coding sequence or its reading frame requires a great fidelity in the cleavage process. Fidelity is provided by the base sequence itself and these splicing sequences are almost exclusively within the introns. As shown in Figure 21.15 the sequence at the 5' end of the exon-intron boundary is called the *5' **splice-site*** and the sequence at 3' end of the boundary is marked by a *3'* **splice site**. The 5' and 3' splice sites are sometimes referred to as the *donor* and the **acceptor** sites respectively.

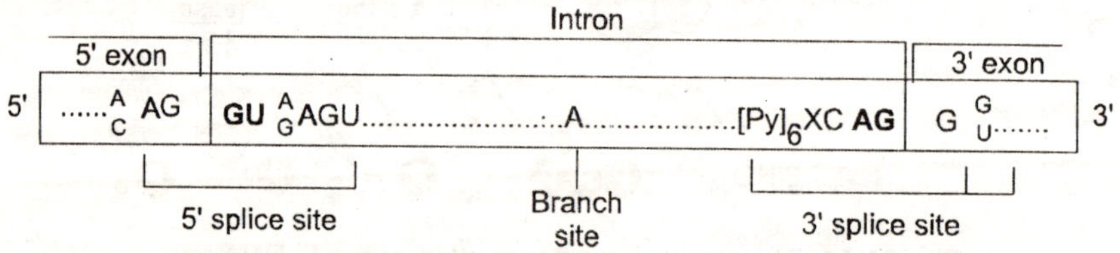

Fig. 21.15: Sequences at the exon-intron boundary

Py = pyrimidine; x = any base. The underlined bases are the same in all introns, i.e. they are conserved. Also conserved is A at the branch site. Where two alternative bases are similarly favoured they are both indicated at the same position

Two transesterification reactions bring about the removal of introns. At first, an A in the branch site attacks a G in the 5' splice site. Then, the liberated 5'exon attacks the 3' splice site. As a result of these reactions the exons are fused and the intron is released in the form of a branched structure called a lariat.

The Spliceosome: Splicing of nucleosomal pre-mRNAs requires a large complex of proteins and RNA called ***spliceosomes***. Spliceosomes actually form on the splice sites through interactions with a number of small nuclear ribonucleoproteins called ***snRNPs.*** There are five snRNPs U1, U2, U4, U5 and U6. Each of these has at least one RNA molecule, i.e. **snRNA** and a number of proteins. Each snRNP has a separate function. The snRNPs bring the 3' and 5' splice junctions into close proximity at the spliceosome, which has enzymatic activities and preform the intricate process of cutting out the intron sequences and joining the two exon segments.

Self-splicing RNAs: There are a few rare introns that can remove themselves from within RNA molecules by a process known as self-splicing. The sclf-splicing introns come in two classes-group I introns and group II introns. The intron itself folds into a specific conformation within the precursor RNA and catalyzes the chemistry of its own release.

RNA Editing

This is another mechanism that allows an RNA to be changed after transcription so as to encode a different protein from that encoded by the gene. Two mechanisms for editing are:

a. Enzymatic modification of bases generating forms that alter how they are read by tRNAs.
b. Insertion or deletion of multiple U nucleotides within the message.

MCQs

Select one most appropriate answer given for each questions below:

1. **Which of the following molecules functions to transfer information from the nucleus to the cytoplasm?**
   A. DNA
   B. mRNA
   C. tRNA
   D. Proteins
2. **From what substrates is RNA made:**
   A. Ribonucleoside 5' – triphosphates
   B. Ribonucleoside 5' – diphosphates
   C. Deoxyribonucleotides
   D. Ribonucleotide monophosphates
3. **The initiation of mRNA is brought about by which of the following enzymes:**
   A. RNA polymerase holoenzyme
   B. RNA polymerase core-enzyme
   C. DNA polymerase
   D. Klenow fragment

4. **Which of the following is a feature of prokaryotic mRNA?**
   A. Intron
   B. Poly A trail at 3' end
   C. Polycistronic mRNA
   D. 5' cap structure
5. **RNAs that catalyze biological reactions such as self-splicing introns are known as:**
   A. Enzymes
   B. Ribozymes
   C. Spliceosome
   D. Lariats
6. **Which of the following statements is false about RNA synthesis?**
   A. The new strand is made up of ribonucleotides.
   B. The RNA product requires a primer
   C. The process is less accurate than replication
   D. The RNA product does not remain base paired to the template DNA.

---

1 B 2 A 3 A 4 C 5 B 6 B

22

Translation

I. INTRODUCTION

Translation is the RNA directed synthesis of polypeptides. The process has two important features:

i. The genetic information contained in the order of nucleotides in mRNA that constitute the **genetic code.**
ii. Generation of a linear sequence of amino acids in proteins by decoding the information in mRNA by enzymatic protein synthesis.

The entire process requires

i. All three classes of RNAs namely, mRNA, tRNA and rRNA
ii. Aminoacyl tRNA synthetases
iii. Ribosomes.

The template for correct addition of individual amino acids is the mRNA, yet both tRNA and rRNAs are involved in the process. The tRNAs carry activated amino acids into the ribosome which is composed of rRNA and proteins. The ribosome is associated with the mRNA ensuring correct access of activated tRNAs and containing the necessary enzymatic activities to catalyze peptide bond formation.

II. THE GENETIC CODE

DNA transfers information to mRNA in the form of a code defined by a sequence of nucleotide bases. During protein synthesis, ribosomes move along the mRNA molecule and read its sequence, three nucleotides at a time—the **codon** from the 5'—3' end. Each amino acid is specified by the mRNA's codon which pairs with a sequence of three complementary nucleotides—the **anticodon** carried by a particular tRNA. The genetic code could be read in two's, threes or fours. A double codon with four nucleotides would be far too less of the requirement of 20 amino acids while a codon with 4 nucleotides would be far in excess of the number of amino acids. So, a triplet codon with 64 possible permutations forms the **genetic code** for the transfer of information from DNA→RNA→amino acids Table 22.1 lists all 64 permutations.

Table 22.1: The genetic code

first position (5' end)	U		C		A		G		third position (3' end)
	second position								
U	UUU	Phe	UCU	Ser	UAU	Tyr	UGU	Cys	U
	UUC		UCC		UAC		UGC		C
	UUA	Leu	UCA		UAA*	stop	UGA*	stop	A
	UUG		UCG		UAG*	stop	UGG	trp	G
C	CUU	LEu	CCU	Pro	CAU	His	CGU	Arg	U
	CUC		CCC		CAC		CGC		C
	CUA		CCA		CAA	Gln	CGA		A
	CUG		CCG		CAG		CGG		G
A	AUU	LLe	ACU	Thr	AAU	Asn	AGU	Ser	U
	AUC		ACC		AAG		AGC		C
	AUA		ACA		AAA	Lys	AGA	Agr	A
	AUG	Met	ACG		AAG		AGG		G
G	GUU	Val	GCU	Ala	GAU	Asp	GGU	Gly	U
	GUC		GCC		GAC		GGC		C
	GUA		GCA		GAA	Glu	GAA		A
	GUG		GCG		GAG		GGG		G

* Chain-terminating or "nonsense" codons

† Also used in bacteria to specify the initiator formyl-Met-$tRNA^{fMet}$

A. Deciphering the Genetic Code

1. The first step to solving the genetic code depended on the development of a cell free *in vitro* translation system by Paul Zamecnik. This system which consisted of a **membrane-free cell supernatant, ATP, GTP, radioactively labelled amino acids** and **RNA** was capable of directing the synthesis of radioactively labelled protein.
2. In 1961, **Marshall Nirenberg** and **Heinrich Matthaei** were using such a system to investigate the synthesis of viral proteins. They used the TMV RNA as their experimental template. As a control RNA template they used

the **homopolymer poly** (U)—which they synthesized from UDP using polynucleotide phosphorylase. They did not expect that this template would code for or direct protein synthesis. However, Nirenberg and Matthaei showed that the only amino acid that was incorporated into a polypeptide when **poly (U)** was the RNA template was **phenylalanine**.

The use of **poly (A)** and **poly (C)** as templates similarly showed that AAA was a codon for lysine and that CCC was a codon for proline. However, poly (G) did not work at all in the system.

The use of homopolymers is quite limited. The use of **random mixed copolymers** helped to extend the utility of the system and the information obtained from it.

3. **Nirenberg** and **Phil Leder** showed that aminoacylated tRNAs could be bound to ribosomes, if the ribosomes contained trinucleotides acting as mRNA.
4. **Khorana** showed that tri-and tetra-nucleotides could be polymerized into polymers with repeating sequences that could be used in cell-free *in vitro* translation assays.

   In the case of trinucleotides, three polypeptides will be synthesized, each of which is homopolymer of a single amino acid.

   The synthetic polynucleotide $(AAG)_n$ can be read in three different frames.

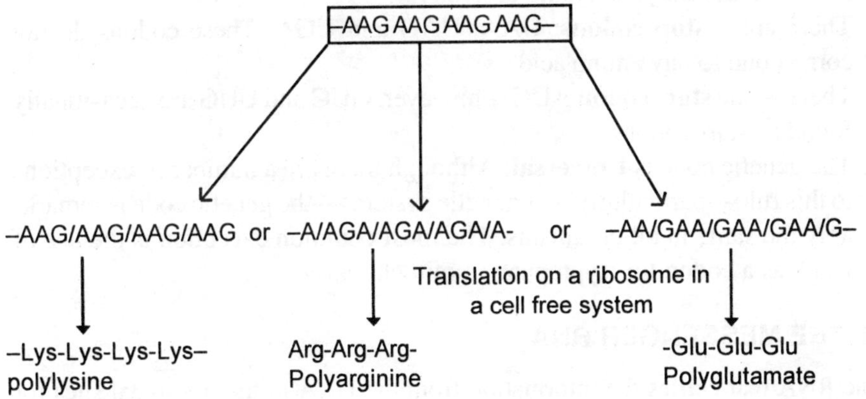

In case of tetranucleotides, a single polypeptide usually will be synthesized which contains a repeating amino acid sequence.

B. Features of the Genetic Code

1. The code is **triplet, unpunctuated** and **nonoverlapping**. Three bases are required to specify each amino acid. There are no gaps between codons. Codons do not overlap.

2. The code is **degenerate:** Most amino acids are specified by more than one codon. In fact, only *Met* and *Trp* are specified by a single codon. Table 22.2 lists the amino acids and the no. of codonos specifying them. Degeneracy is found only in the third nucleotide of the codon.

Table 22.2: The degeneracy of codons

No. of codons	*Amino acids*
1	Met, Trp
2	Asn, Asp, Cys, Gln, Glu, His, Lys, Phe, Tyr.
3	Ile
4	Ala, Gly, Pro, Thr, Val
6	Arg, Leu, Ser

3. The genetic code is **unambiguous:** In general no codon specifies more than one amino acid. The exceptions so far are AUG, UGA and UAG.
   AUG : The start codon used for initiating protein synthesis in bacteria specifies both **Methionine** and **N-formyl methionine**
   UGA : Specifies the twenty first amino acid **selenocysteine** as well as being a stop codon.
   UAG : Specifies the twenty second amino acid (the most recent to be added to the list) **Pyrrolysine** (an amideo-linked 4-substituted pyrroline-5-carboxylate lysine derivative).
4. There are 3 **stop codons**: UAA, UAG and UGA. These codons do not correspond to any amino acid.
5. There is one **start codon** AUG—however, GUG and UUG are occasionally found as start codons.
6. The genetic code is **Universal**: Although there are a number of exceptions to this rule—particularly in organelle systems—the genetic code is remarkably the same in all organisms. The most common exception is the use of UGA as a codon for tryptophan in mitochondria.

III. THE MESSENGER RNA

The RNA that carries the information from the DNA to the ribosomal sites for protein synthesis is called **messenger RNA.** It contains the information in the form of three nucleotide codons each of which specifies an amino acid. The protein coding region has a continuous stretch of non-overlapping codons that form an **open reading frame** (ORF) with a **start codon** and a **termination codon**. The start codon specifies the **reading frame** for all subsequent condons. The mRNA is longer than the coding region. In monocistronic RNAs, it has an untranslated sequence before the start codon at 5' end called the **leader sequence** and another untranslated region after the stop codon at 3' end called the **trailer sequence.**

Messenger RNAs containing multiple ORFs as found in bacteria are said to be **polycistronic** whereas those with a single ORF as found in eurkaryotes are said to be **monocistronic** (Fig. 22.1a and b).

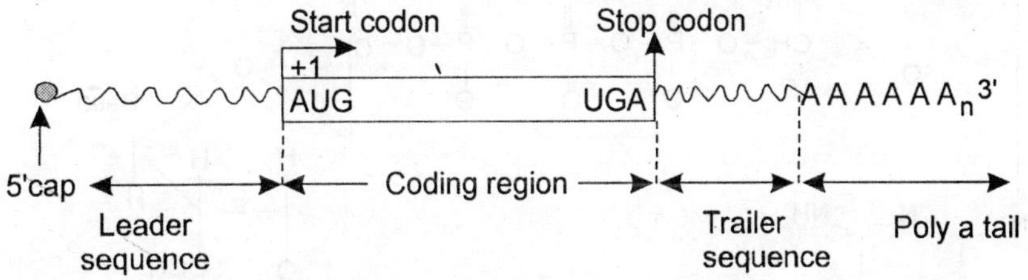

Fig. 22.1a: A monocistronic mRNA with a single ORF in eukaryotes

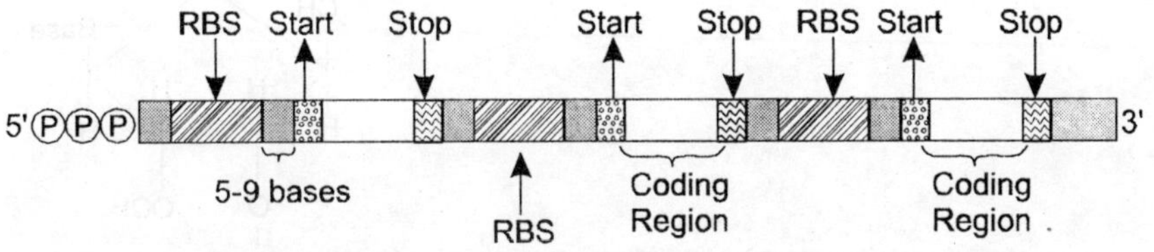

Fig. 22.1b: A polycistronic mRNA with multiple ORFs in prokaryotes

The prokaryotic mRNA has a ribosome binding site (RBS) that recruits the ribosomes to the start site. It consists of a **Shine Dalgarno sequence** 5'-A G G A G G–3' located 5-9 bps upstream of the start codon AUG. This sequence is complementary to a short sequence 5'-CCUCCU-3' located at 3' end of a 16s rRNA. The interaction between the ribosomes and RBS on mRNA serves to distinguish the start AUG from any of the AUGs that may be present internally within the coding region of the mRNA. The extent of complementarity and spacing between the RBS and the start codon has an influence on the efficiency of translation.

The eukaryotic mRNAs differ from the prokaryotic mRNA in three important respects.

—Presence of a **5' cap** in eukaryotes

—Presence of a **Kozak sequence** near the start codon in eukaryotes

—Presence of a **Poly A** tail at the 3' end in eukaryotes.

The 5' cap is a 7-methylguanosine residue linked to a triphosphate at 5' end of mRNA by an unusual 5'→5' linkage instead of the regular 5'→3' linkage (Fig. 22.2). The cap recruits the ribosome to mRNA. Once bound, the ribosomes slide along the mRNA in a 5'→3' direction till they encounter the start codon. The 5' untranslated region at the cap end of the eukaryotic mRNA is folded into a secondary structure held together by base pairing. This secondary structure inhibits

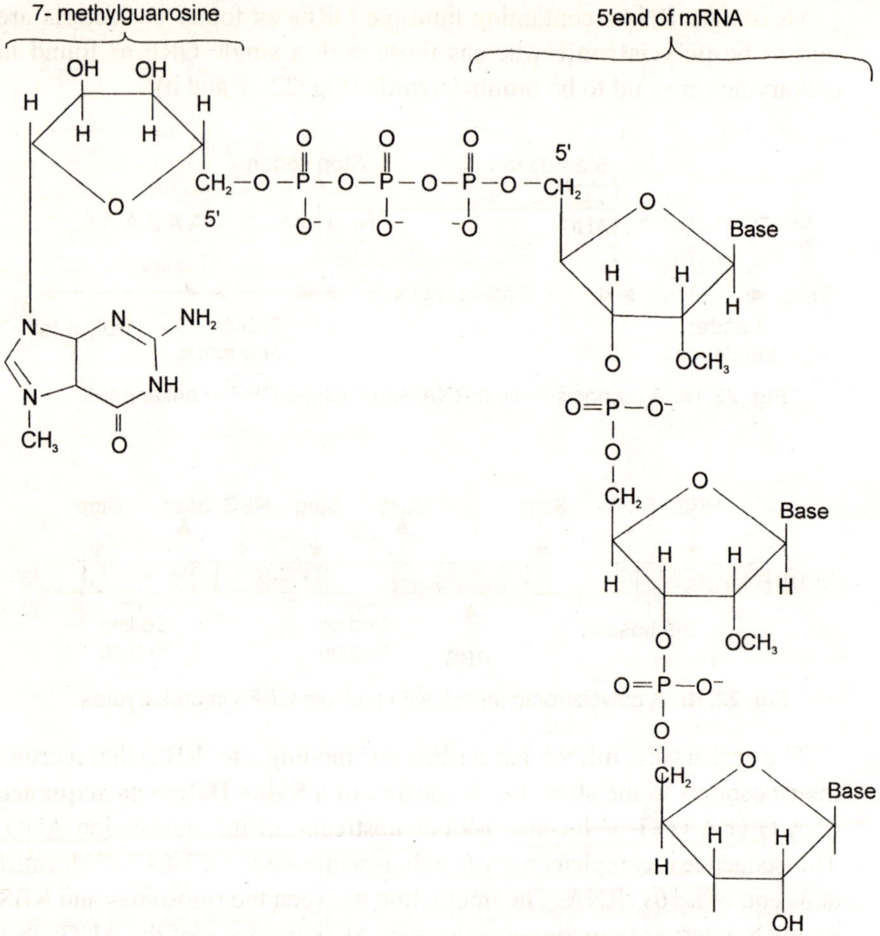

Fig. 22.2: Structure of the cap at 5' of eukaryotic mRNA

efficient translation. Specific proteins called cap binding proteins bind to the cap and unwind the secondary structure at the 5' end of mRNA. This facilitates efficient translation.

A. The Kozak Sequence

In some mRNAs in eukaryotes is also present a purine base three bases upstream of the start codon AUG and a guanine immediately downstream. This sequence i.e. 5' **G/A** NNAUGG-3' is called the Kozak sequence and its presence increases the efficiency of translation. These bases are thought to interact with the initiator tRNA and not rRNA.

B. The PolyA-Tail

A stretch of ~200 adenine residues is added to the 3' end of the eukaryotic mRNA by poly A-polymerase after transcription. The poly-A tail contributes to the stability of mRNA molecule and enhances the level of translation by promoting efficient recycling of ribosomes.

IV. CHARACTERISTICS OF tRNAs

The process by which the genetic information contained within RNA molecule is decoded and translated into an amino acid sequence is accomplished by the tRNA molecules. The tRNA serves as an adaptor between the codons and the amino acids and has two crucial features:

- It represents a single amino acid to which it is covalently linked.
- It contains a trinucleotide sequence, the anticodon which is complementary to the codon representing its amino acid.

The tRNA molecules are small, single-stranded nucleic acids ranging in size from 75-95 nucleotides. They have a 3'-OH terminus, but the opposite end terminates in a 5' monophosphate rather than a 5' triphosphate. Due to pairing of complementary base sequences, short double stranded regions form causing the molecule to assume a secondary structure in which open loops are connected to one another by double-stranded stems

(i) Secondary Structure of tRNA

The secondary structure is in the form of a planar cloverleaf (Fig. 22.3) that has the following four major arms:

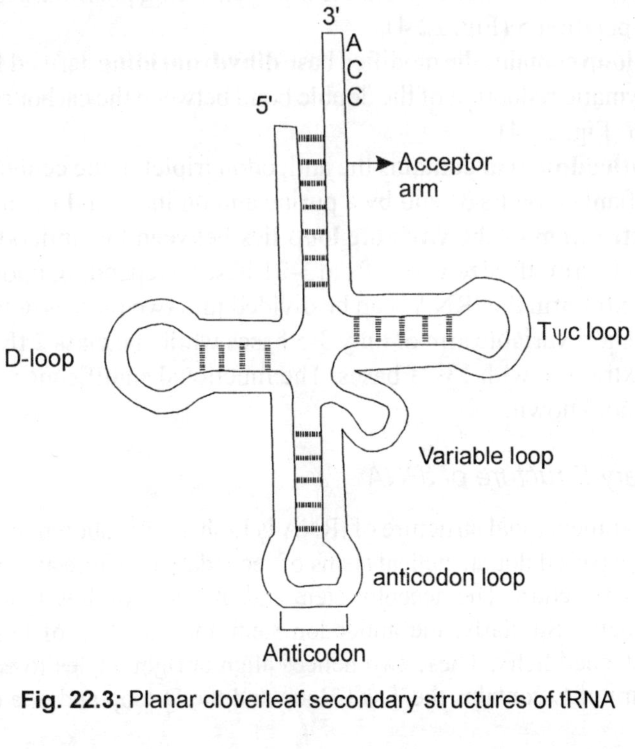

Fig. 22.3: Planar cloverleaf secondary structures of tRNA

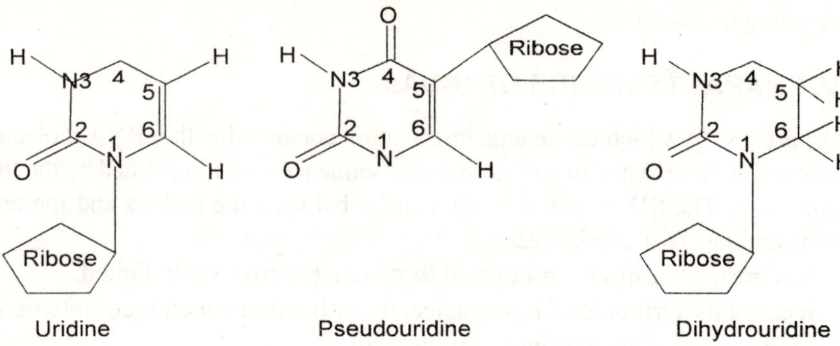

Fig. 22.4: Modified bases found in tRNA

1. The **acceptor arm**: It is the site for the attachment of amino acid. It is formed by base pairing between the 5' and 3' ends of tRNA molecule. The 3' end of the molecule has a 5'-CCA-3' sequence that protrudes from the double stranded stem.
2. The **TψC arm** that has the unusual base ψ = **pseudouridine** that often occurs within the sequence 5' - TψC-3'. Pseudouridine is derived from uridine by an isomerization in which the site of attachment of the uracil base to ribose is switched from the nitrogen at ring position 1 to the carbon at ring position 5 (Fig. 22.4).
3. The **D-loop** contains the modified base **dihydrouridine** derived from uridine by enzymatic reduction of the double bond between the carbons at positions 5 and 6 (Fig. 22.4).
4. The **anticodon arm** contains the anticodon triplet in the center of the loop and is flanked on its 3' end by a purine and on its 5' end by uracil.
5. The **extra-arm** or the **variable loop** lies between the anticodon arm and the T ψ C arm. Its size varies from 3-21 bases. Depending upon the nature of the extra-arm the tRNAs can be divided into two classes. Class I tRNAs have a short variable arm of only 3-5 bases while the class 2 tRNAs have a large extra-arm with 13-21 bases. The functional significance of the extra-arm is not known.

(ii.) Tertiary Structure of tRNA

The three-dimensional structure of tRNA is L-shaped as shown in Figure 22.5.

The base paired double helical stems of secondary structure are maintained in the tertiary structure. The acceptor stem and the stem of T ψ C loop form on extended helix. Similarly, the anticodon stem and the stem of D-loop form a second extended helix. These two helices align at right angles to each other so that the turn of L-contains the T ψ C loop and the D-loop and one extremity of

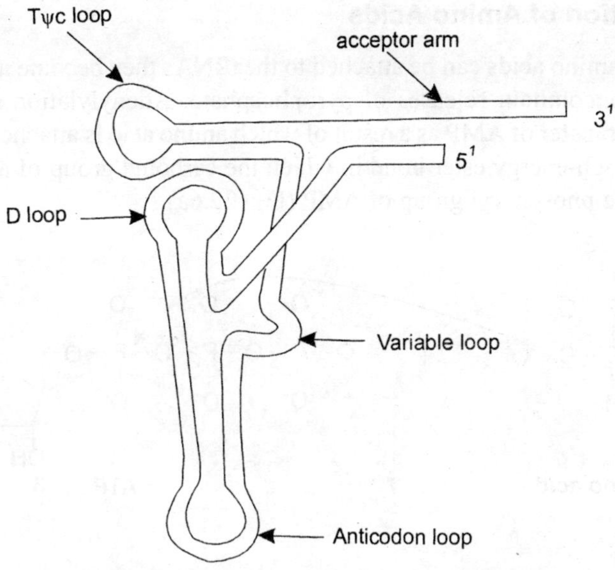

Fig. 22.5: Schematic representation of three-dimensional structure of tRNA

the L-shaped structure contain the amino acid residues while the other extremity is formed by the anticodon loop.

V. ACTIVATION OF AMINO ACIDS AND CHARGING OF tRNAs

The tRNA molecules are charged with their respective amino acids at the 3' end via a high energy acyl linkage between the carboxyl group of the amino acid and the 2' or 3' hydroxyl group of the adenosine nucleotide that protrudes from the acceptor arm. The enzymes responsible for charging the amino acids are called **aminoacyl tRNA synthetase** and the charged tRNA molecule is called as the **aminoacyl tRNA**. During protein synthesis, the hydrolysis of the high-energy acyl bond releases the energy that drives the formation of peptide bonds between the amino acids of the polypeptide chain.

A. Aminoacyl tRNA Synthetase

The aminoacyl tRNA synthetases covalently attach the amino acid to tRNA. Only one tRNA synthetase attaches each amino acid to all of the appropriate tRNAs. Thus there are at least 20 different tRNA synthetases for the 20 different amino-acids. The tRNA synthetases recognize unique structural features of cognate tRNAs as well as the correct amino acids. The aminoacyl tRNA formation is very accurate and the frequency of mischarging is very low. There are two classes of tRNA synthetases. Class I tRNA synthetases attach the amino acid to 2'-OH of tRNA whereas class II enzymes attach the amino acid to 3' OH of tRNA.

B. Activation of Amino Acids

Before the amino acids can be attached to the tRNAs they become **adenylylated** with the concomitant releases of pyrophosphate. Adenylylation refers to the enzymatic transfer of AMP as a result of which amino acid is attached to adenylic acid via a high-energy ester bond in which the carbonyl group of amino acid is joined to the phosphoryl group of AMP (Fig. 22.6a).

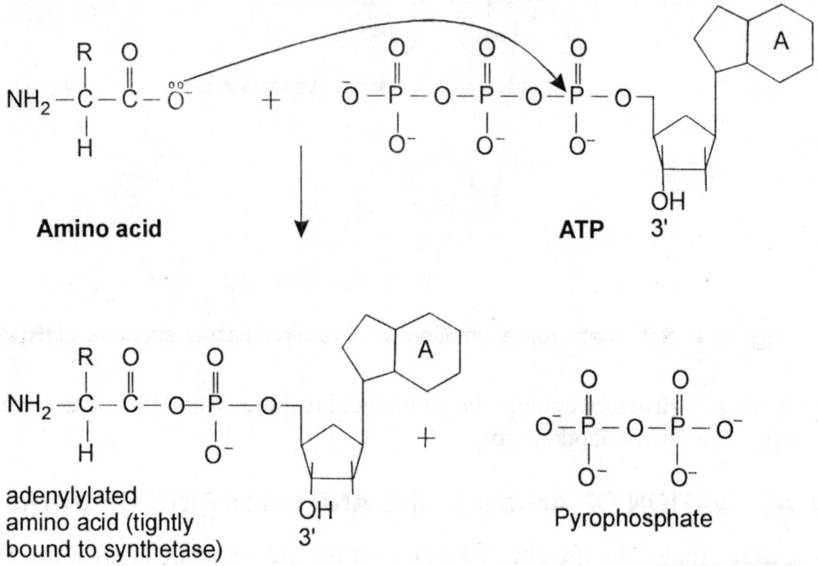

Fig. 22.6a: Adenylylation of amino acid

C. Charging of tRNA

In this step, the adenylylated amino acid which remains tightly bound to the synthetase reacts with tRNA as a result of which the amino acid is transferred to the 2' or 3' hydroxyl group with the concomitant release of AMP as shown in Figure 22.6b.

VI. THE WOBBLE HYPOTHESIS

The genetic code is degenerate, i.e., many amino acids are encoded by more than one codon. For example, CCU, CCA, CCC and CCG all encode proline. In some of these cases, different tRNAs recognize the different codons, but certain tRNAs recognize several different codons. In 1966, Francis Crick proposed the **wobble hypothesis** that explains how some tRNA molecules respond to several codons and also provides insight into the pattern of redundancy in the code. Crick suggested that the base at the 5' end of the anticodon does not have as strict base-pairing requirements as the other two base pairs, allowing it to form hydrogen bonds with several bases at 3' end of the codon. He proposed that since the anticodon

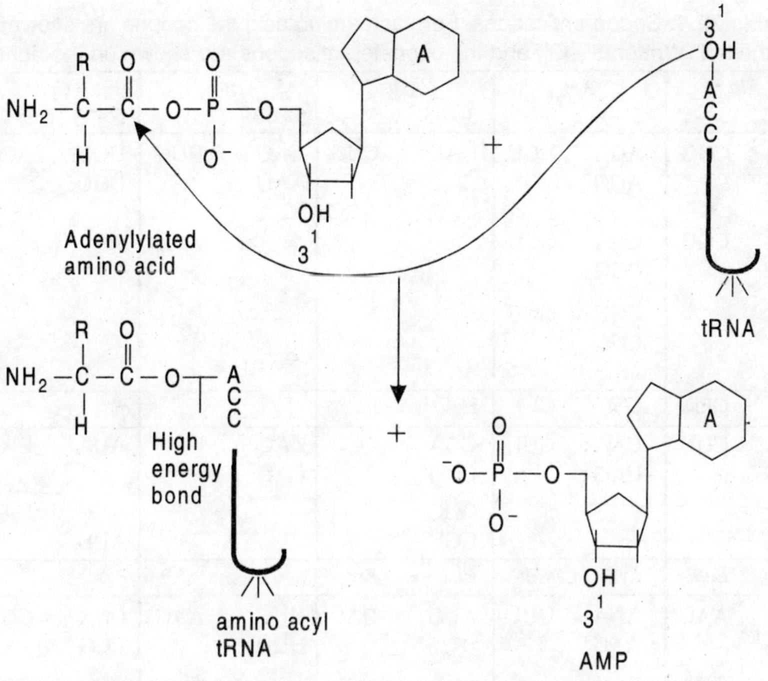

Fig. 22.6b: Transfer of adenylylated aminoacid to tRNA

is located within a single stranded RNA loop, the codon-anticodon interaction might not be constrained to the same extent as in DNA-double helix. He showed that the steric requirements were less stringent at the 3rd position of the codon. This is especially true when the base at 5' end of the anticodon is Inosine (derived from adenine whose C6 is deaminated to give 6-keto group of inosine). Inosine is found at 5' end of the anticodon in many tRNA molecules. Unlike the standard base pairing rules, the base pairing rules for this "wobble" position are shown in Table 22.3.

Table 22.3: Allowed pairings according to the wobble hypothesis

5' base in anticodon (first position in tRNA)	*3' base in codon (third position in mRNA)*
G	U or C
C	G
A	U
U	A or G
I	A, U or C

Because of the wobble hypothesis, fewer than 61 tRNAs are required to recognize all of the possible sense codons. For example, based upon the wobble base pairing rules and the codon table, the minimal number of tRNAs required can be predicted as shown in the Table 22.4.

Table 22.4: Codon/anticodons. For each amino acid the codons are shown to the left (Written 5'→3') and the cognate anticodons are shown on the right

Ala		*Arg*		*Asp*		*Asn*		*Cys*	
5'→3'	3←5'	5'→3'	3←5'	5'→3'	3'←5'	5'→3'	3'←5'	5'→3'	3←5'
GCC	CGG	AGA	UCU	GAC	CUG	AAC	UUG	UGC	ACG
GCU		AGG		GAU		AAU		UGU	
GCA	CGU	CGA	GCU						
GCG		CGG							
		CGU	GCG						
		CGC							
	Glu	Gln	Gly	His	Ile				
GAA	CUU	CAA	GUU	GGA	CCU	CAC	GUG	AUA	UAU
GAG		CAG		GGG		CAU			
				GGC	CCG			AUC	UAG
				GGU				AUU	
	Leu	*Lys*	*Met*	*Phe*	*Pro*				
UUA	AAU	AAA	UUU	AUG	UAC	UUC	AAG	CCA	GGU
UUG		AAG				UUU		CCG	
CUA	GUA							CCU	GGG
CUG								CCC	
CUU	GUG								
CUC									
	Ser	*Thr*	*Trp*	*Tyr*	*Val*				
AGC	UCG	ACA	UGU	UGG	ACC	UAC	AUG	GUA	CAU
AGU		ACG				UAC		GUG	
UCA	AUG	ACU	UGA					GUC	CAG
UCG								GUU	
UCC	AGG								
UCU									
Amber suppressor		Ochre suppressor		Opal suppressor					
UAG	AUC	UAA	AUU	UGA	ACU				
		(UAG)		(UGG)					

Note: The first two bases of codon and anticodon interact by standard Watson-Crick base pairing rules, but the third base of the codon can pair by wobble rules.

Based upon Table 22.4, according to wobble rule a minimum of 32 tRNAs are needed to recognize all of the sense codons in mRNA. The pairings permitted by the wobble rules are those that give ribose—ribose distances close to that of the standard A: U or G: C base pairs. Purine-purine (with the exception of I:A pairs) or pyrimidine-pyrimidine pairs would give ribose-ribose distances that are too long or too short respectively.

The wobble rules do not permit any single tRNA molecule to recognize four different codons. Three codons can be recognized only when inosine occupies the first (5') position of the anticodon.

VII. THE RIBOSOME

Ribosomes are the large ribonucleoprotein particles containing several enzymes needed for protein synthesis. They provide the environment for controlling the interaction between mRNA and aminoacyl tRNA. In this process, the mRNA containing the codons and the aminoacyl tRNA molecules containing the respective anticodons are brought-together into proper position and orientation so that the corresponding series of amino acids is incorporated into a polypeptide chain. Together with its accessory factors, the ribosome provides the full range of activities required for all the steps in protein synthesis.

A. Subunit Structure

Both the prokaryotic and eukaryotic ribosomes have two subunits each known as the large and the small subunits. They are described in terms of their sedimentation velocity in an ultracentrifuge referred to as Svedbergs or S-value. The S-value is dependent upon both the shape and mass of the molecule and is not simply a measure of the molecular mass. Bacterial ribosomes sediment at ~ 70S while the eurkaryotic ribosomes are larger, usually sidementing at ~ 80S.

Figure 22.7 shows the two subunits of *E. coli* 70S ribosome. The 70S *E. coli* ribosome consists of a large 50S subunit and a small 30S subunit. The two subunits dissociate *in vitro* when the concentration of Mg^{++} ions is reduced. In *vivo*, the ribosomal subunits assemble and disassemble during each round of protein

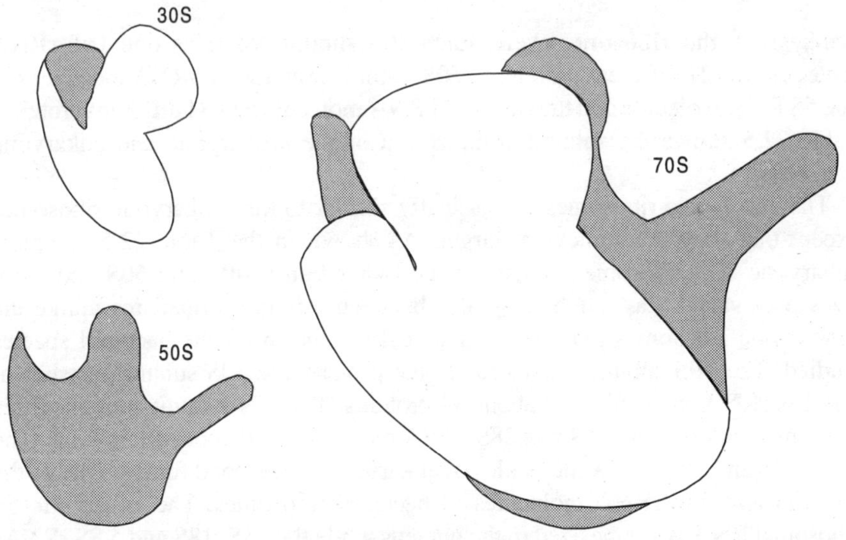

Fig. 22.7: Three-dimensional model of the 70S ribosome. The small subunit is light and the large subunit is dark

Table 22.5: Subunit composition of prokaryotic and eukaryotic ribosomes. The rRNA and protein composition of the different subunits are shown

	Prokaryotes	*Eukaryotes*
Small subunit	30S (MW = 900,000) 21 proteins 16S rRNA (1, 540 nucleotides)	40S (MW = 1, 400, 000) ~33 proteins 18S rRNA (1,900 nucleotides)
Large subunit	50S (MW = 1,600,000) ~34 porteins 23S rRNA (2,900 nucleotides) 5S rRNA (120 nucleotides)	60S (MW = 2,800,00) ~ 49 proteins 28S rRNA (4,700 nucleotides) 5S rRNA (120 nucleotides) 5.8S rRNA (160 nucleotides)
Whole ribosome	70S	80S

synthesis in the **ribosome cycle**. Each 30S subunit contains one 16S rRNA molecule and 21 different proteins; a 50S subunit contains two RNA molecules—one 5S RNA molecule and the other 23S RNA molecule and 34 different proteins. Table 22.5 shows the subunit composition of the prokaryotic and eukaryotic ribosomes.

The eukaryotic ribosomes are basically similar to the prokaryotic ribosomes except that they are somewhat larger. As shown in the Table 22.5 a typical eukaryotic 80S ribosome consists of the two subunits 40S and 60S and these sizes may vary by as much as $\pm$ 10% between different organisms unlike the prokaryotic ribosomes that are all practically same in all the bacterial species studied. The 40S subunit analogous to the prokaryotic 30S subunit consists of one 18s rRNA molecule and about 33 proteins. The 60s subunit contains three RNA molecules—5S, 5.8S and 28S rRNA molecules and about 49 proteins. The 5.8S, 18S and 28S rRNA molecules of eukaryotes correspond functionally to the 5S, 16S and 23S rRNA molecules of bacterial ribosomes. The eukaryotic 5S ribosomal RNA is synthesized in the nucleus while the 28S, 18S and 5.8S rRNAs are transcribed in the nucleolus, as a 45S precursor and then processed into 3 RNA species.

B. tRNA Binding Sites of Ribosome

The ribosome contains three tRNA binding sites called

- A-site — Binding site for the aminoacylated tRNA
- P-site — Binding site for peptidyl tRNA
- E-site — Binding site for the exit of the tRNA that is released after the growing polypeptide chain has been transferred to the aminoacyl tRNA.

Figure 22.8 shows the tRNA binding sites of the ribosome.

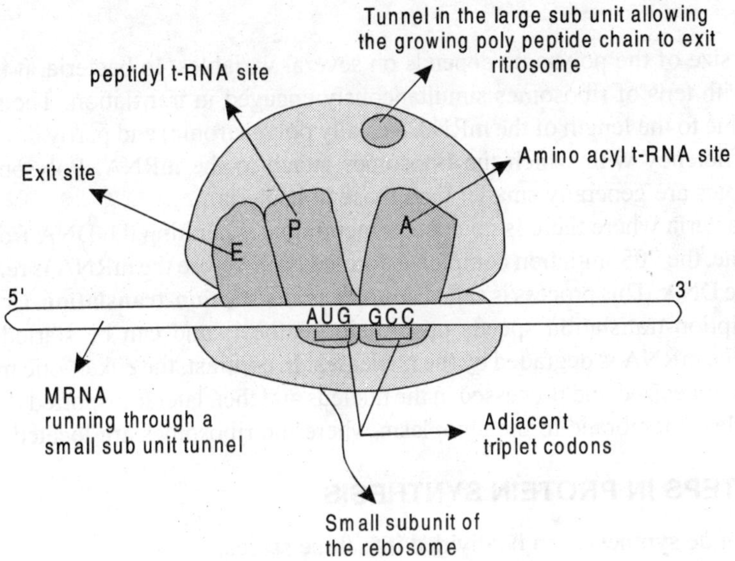

Fig. 22.8: Schematic representation of tRNA binding sites in the ribosome

C. The Ribosome Cycle

Bacterial ribosomes engaged in elongating the polypeptide chain exist as a 70S particle. At termination they are released from the mRNA to enter a pool of free ribosomes. Ribosomes in free pool can dissociate into separate subunits; so free 70S ribosomes are in dynamic equilibrium with free 30S and 50S subunits. The sequence of association and dissociation of ribosomal subunits is known as the **ribosome cycle.**

Although a ribosome can synthesize only one polypeptide at a time, each mRNA can be translated simultaneously by multiple ribosomes. An mRNA bearing multiple ribosomes is known as a **polyribosome** or a **polysome** (Fig. 22.9).

A single ribosome makes contact with approximately 30 nucleotides of mRNA but because of its bulk they occur at a density of one ribosome for every 80 nucleotides of mRNA. The occurrence of polysomes increases the overall rate of protein synthesis compared to the rate that would occur if there were no polysomes.

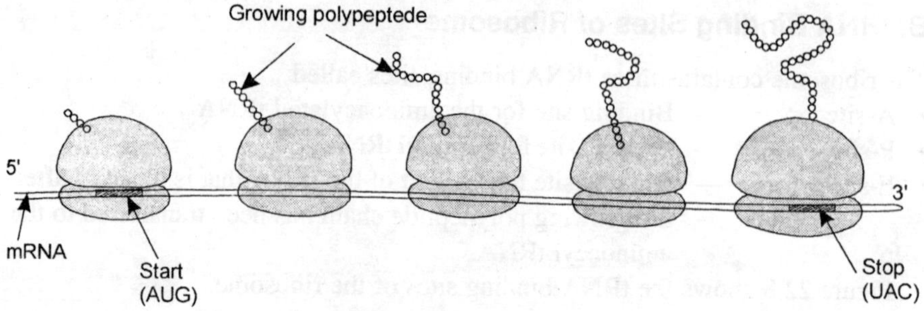

Fig. 22.9: A polyribosome

The size of the polysome depends on several variables. In bacteria, it is very large with tens of ribosomes simultaneously engaged in translation. The size is partly due to the length of the mRNA (usually polycistronic) and partly due to the high efficiency with which the ribosomes attach to the mRNA. Polysomes in eukaryotes are generally smaller than those in bacteria.

In bacteria where there is no nuclear membrane separating the DNA from the ribosome, the 70S initiation complex is formed even before the mRNA is released from the DNA. This process is called **coupled transcription-translation.** Coupled transcription-translation speeds up protein synthesis and can be started even before the mRNA is degraded by the nucleases. In contrast, the eukaryotic mRNA is first synthesized and processed in the nucleus and then later transported through the nuclear membrane to the cytoplasm where the ribosomes are located.

VIII. STEPS IN PROTEIN SYNTHESIS

Polypeptide synthesis can be divided into three stages:

i. Initiation
ii. Elongation
iii. Termination

The overall mechanism of protein synthesis is the same in prokaryotes and eukaryotes, however, there are differences in detail.

A. Translation Initiation in Prokaryotes

The following events occur during the initiation of protein synthesis.

- The ribosome binds to mRNA.
- The initiating tRNA charged with amino acid is placed in the P-site of the ribosome.
- The ribosome is precisely positioned over the start codon.

Initiation is aided by **translation initiation factor**

a. ***Formation of a 30S preinitiation complex:***

At first a 30S preinitiation complex is formed. This requires:

- The 30S subunit (not the entire 70S ribosome)
- An mRNA molecule
- f-Mel-$tRNA^{met}$ (initiator tRNA)
- Three initiation factors IF-1, IF-2 and IF-3
- Guanosine 5' triphosphate (GTP)

i. Initiation factors IF-1, IF-2 and IF-3 bind to the 30S subunit in order to prepare it for binding to mRNA and initiator tRNA. First IF-3 binds to the 30S subunit and prevents the association of large subunit. IF-3 must be released before 50S subunit can bind. Thus IF -3 occupies the part of subunit that will become the E-site. Next, the IF-2 GTP and IF-1 bind to the 30S subunit IF-1 binds the portion of small subunit that will become the A-site. IF-2 is a GTPase and interacts with IF-1 and guides fMet tRNA to the P-site.

ii. An important feature of initiation of polypeptide synthesis is the use of a specific initiating tRNA molecule. The methionine in $tRNA^{met}$ is formylated to yield N-formyl methionine tRNA designated as fMet $tRNA^{fmet}$ (Fig. 22.10). Both $tRNA^{Met}$ and $tRNA^{fMet}$ recognise the codon AUG but only $tRNA^{fMet}$ is used for initiation. The enzyme needed for formylation of methionine is found only in prokaryotes and not in eukaryotes

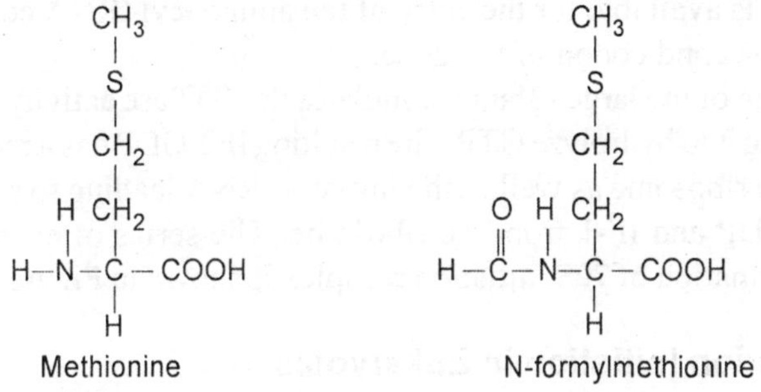

Fig. 22.10: Methionine and N-formyl methionine

During initiation the fMet $tRNA^{fmet}$ enters the P-site directly.

iii. Binding of the ribosome to mRNA in prokaryotes occurs at a special sequence on mRNA called the **ribosome binding site**. It consists of a short sequence AGGAGGU that precedes the start codon AUG and is called the **Shine Dalgarno sequence**. The association of the 30S subunit with mRNA is mediated by complementary base pairing between the 3' terminus of 16S rRNA and the ribosome binding sequence on mRNA as shown in Figure 22.11. The ideal spacing between the ribosome binding site and the initiating codon AUG is approximately 10 bases and places the AUG in the region of P-site of the ribosome.

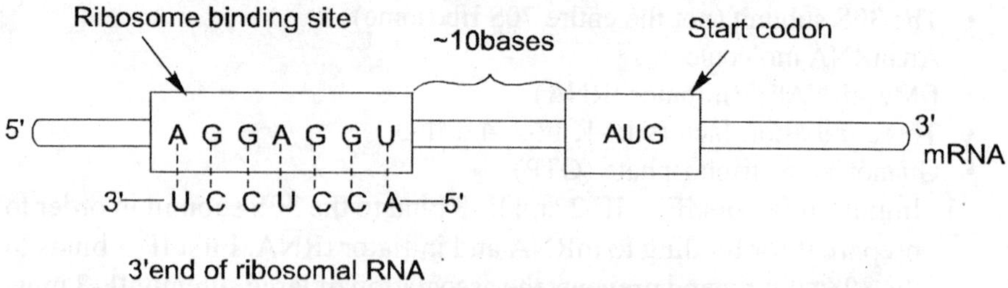

Fig. 22.11: 16S rRNA interacts with the ribosome binding site to initiate the binding of the 30S subunit to mRNA

With the binding of the 30S subunit to the mRNA, the start codon AUG and the f-Met- tRNA^{fMet} anticodon are aligned to start translation.

b. ***Formation of 70S initiation complex:***

Base pairing of mRNA start codon and fmet-tRNA^{fMet} anticodon initiates the reaction cascade to form the 70S initiation complex.

- The small subunit undergoes a change in conformation resulting in the release of IF-3.
- In the absence of IF-3, the 50S subunit binds to the 30S initiating complex such that the initiation fMet-tRNA^{fMet} lies in the now intact P-site and the A-site is available for the entry of the aminoacyl tRNA complementary to the second codon of the gene.
- Binding of the large subunit stimulates the GTPase activity of IF-2 - GTP causing it to hydrolyze GTP. The resulting IF2-GDP has a reduced affinity for the ribosome as well as the initiator tRNA leading to release of both IF2-GDP and IF-1 from the ribosome. The series of events leading to the formation of 70S initiation complex is shown in Figure 22.12.

B. Translation Initiation in Eukaryotes

Translation initiation involves the formation of a 43S preinitiation complex similar to the 30S pre-initiation complex. However, there are important differences:

1. Eukaryotic mRNAs are generally monocistronic, recognition of an internal start codon is not required.
2. There is no Shine-Dalgrno sequence in the mRNA of eukaryotes.
3. Recognition of the initiation codon occurs after the recognition of the methylated cap that is found at the 5' end of the RNA.
4. The initiator tRNA is charged with methionine which is not formylated.
5. Larger number (> 30 polypeptides) of initiation factors are involved.

a. Formation of 43S Preinitiation Complex

Analogous to the prokaryotic initiation factors two eukaryotic GTP-binding proteins eIF-2 and eIF 5B mediate the binding of Met-tRNA^{met} to the P site of the

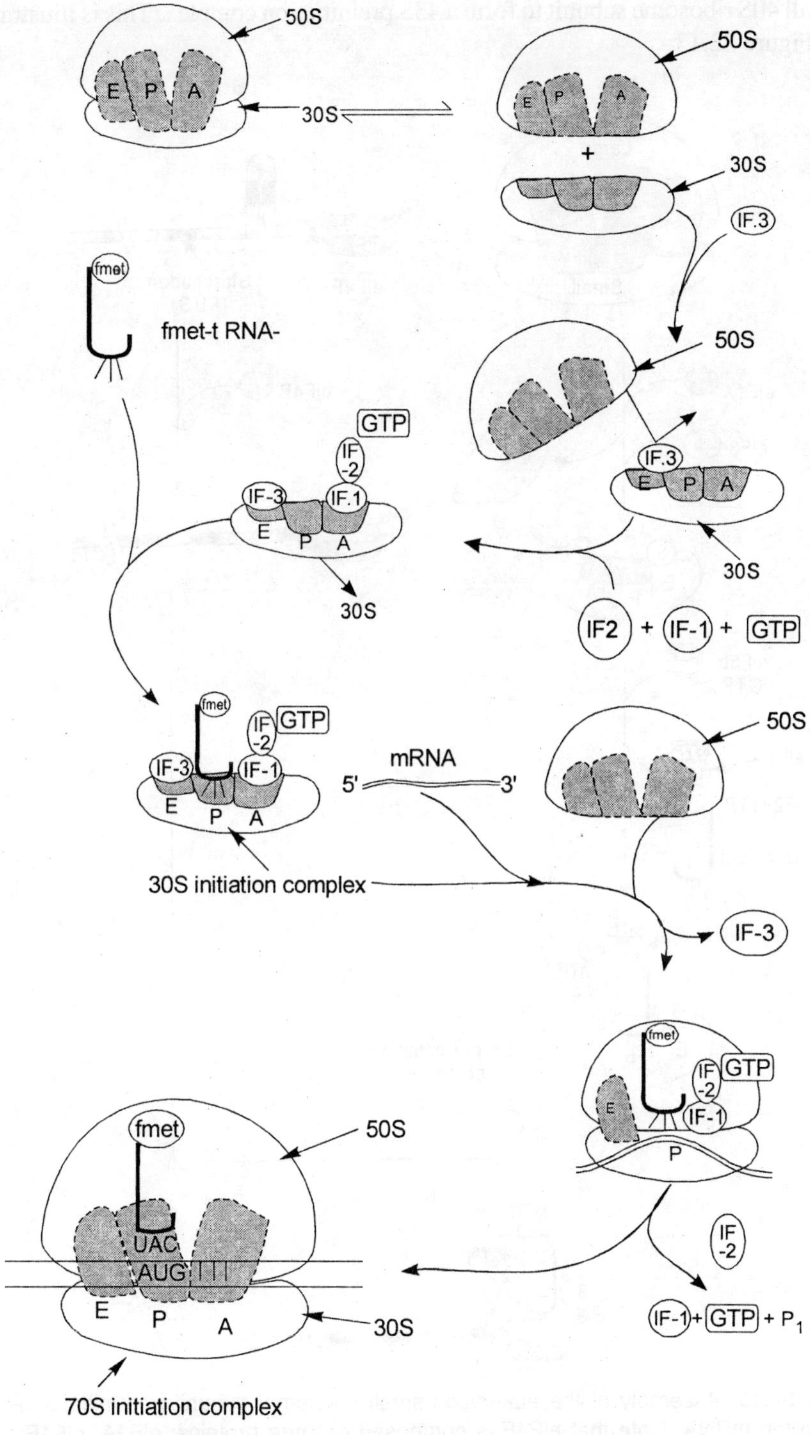

Fig. 22.12: Summary of translation initiation in prokaryotes

small 40S ribosome subunit to form a 43S preinitiation complex. This is illustrated in Figure 22.13.

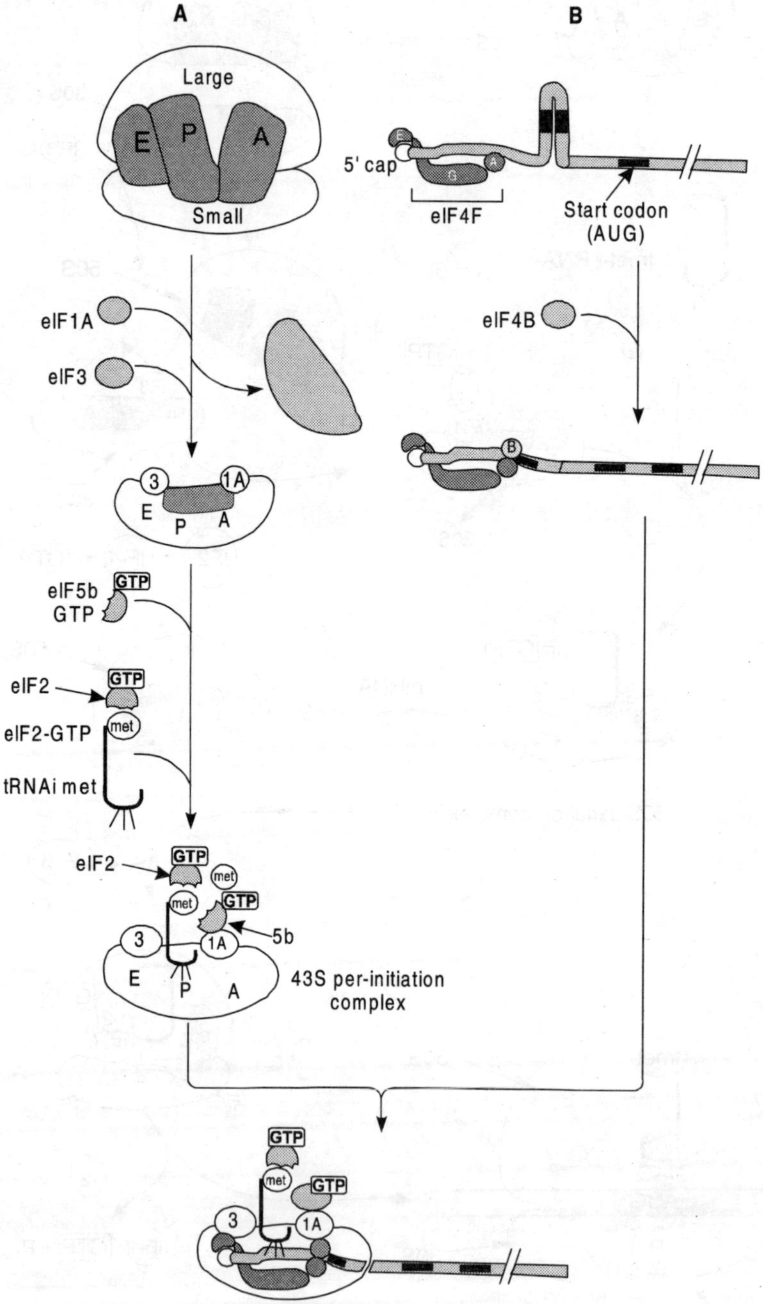

Fig. 22.13: Assembly of the eukaryotic small ribosomal subunit and initiator tRNA onto the mRNA. Note that eIF4F is composed of three proteins; eIF4A, eIF4E and eIF4G. eIF4E directly binds the 5' cap, tethering the other two proteins to the end of the mRNA

b. Binding of 43S Preinitiation Complex to mRNA

The first step in the recognition of mRNA is the recognition of 5' methylated cap found at the end of most eukaryotic RNAs. This process is mediated by eIF 4F which is a three subunit protein—One subunit binds directly to the cap while the other two bind non-specifically to the mRNA. RNA helicase activity of one of the subunits is activated by another protein eIF 4B. The helicase unwinds the hairpin structures found at 5' end of the RNA. The 43S preinitiation complex binds to the eIF 4F/eIF 4FB bound mRNA through interaction between eIF 4F and eIF-3.

c. Formation of 80S Initiation Complex

After the recognition of 5' cap the 43S initiation complex migrates along the mRNA in the 5→3' direction **scanning** it for the presence of a start codon AUG. This is an ATP-dependent process driven by the eIF 4F-associated helicase. The start codon is recognized in an optimal sequence context which is usually GC$^{A}_{G}$ CC **AUG** G. Correct base pairing between the intiator tRNA and start codon triggers the release of eIF-2 and eIF-3 and the binding of large 60S subunit to form the 80S initiation complex. GTP-hydrolysis by eIF 5B leads to the release of other factors as well and the Met-tRNA^{met} and AUG are placed in the P-site of 80S ribosome. The events leading to the formation of 80S initiation complex are summarized in Figure 22.14.

C. Elongation

Once the complete ribosome is formed at the initiation codon, the next step in protein synthesis, i.e. elongation of the polypeptide chain by the incorporation of new amino acids corresponding to the codons on mRNA can start. Elongation requires:

- Placement of charged tRNA in the **A-site**
- Formation of peptide bond by **peptidyl transferase.**
- **Translocation**
- Elongation factors EF-Tu, EF-G and EF-Ts in prokaryotes and eEFs in eukaryotes. Elongation of polypeptides occurs in a cyclic manner such that at the end of one complete round of amino acid addition the A site will be empty and ready to accept the aminoacyl t-RNA detected by the next codon of the mRNA. This means that not only does incoming amino acid need to be attached to the peptide chain but the ribosome must move down the mRNA to the next codon.

  The summary of steps in elongation is shown in Figure 22.15.

(i) Placement of Charged tRNA in the A-site

Any aminoacyl tRNA except the initiator tRNA can enter the A-site. Its entry is mediated by the elongation factor EF-Tu (in bacteria). EF-Tu can bind to an aminoacyl tRNA only when it is associated with GTP. When the GTP is hydrolyzed

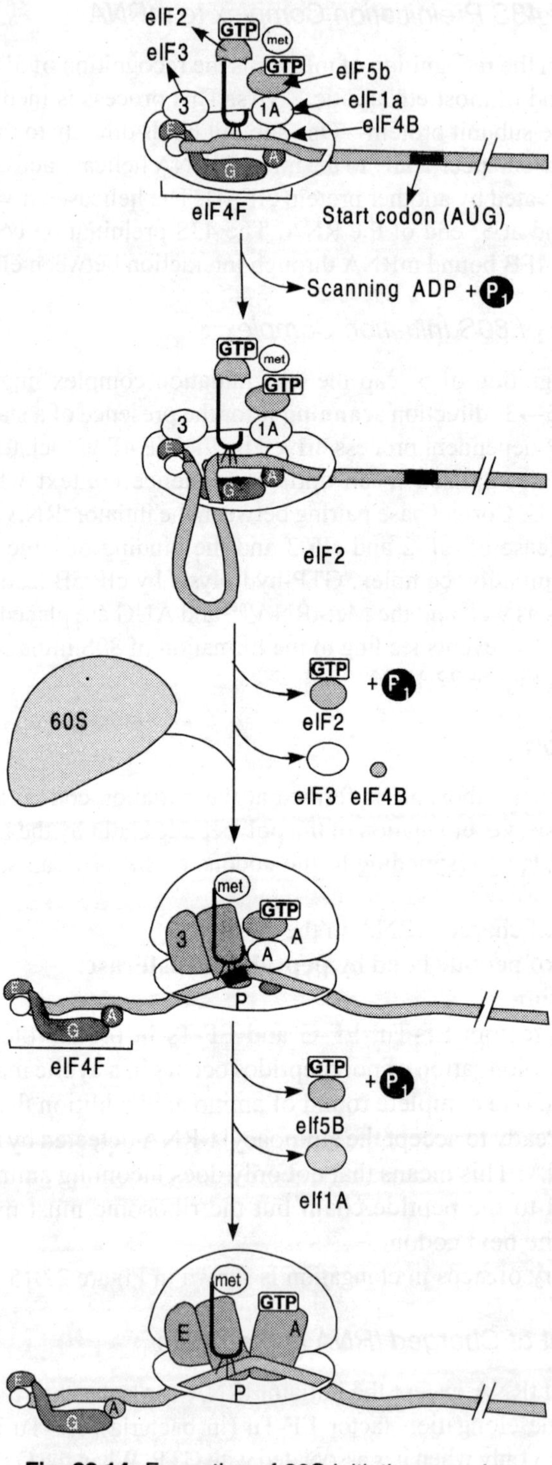

Fig. 22.14: Formation of 80S initiation complex

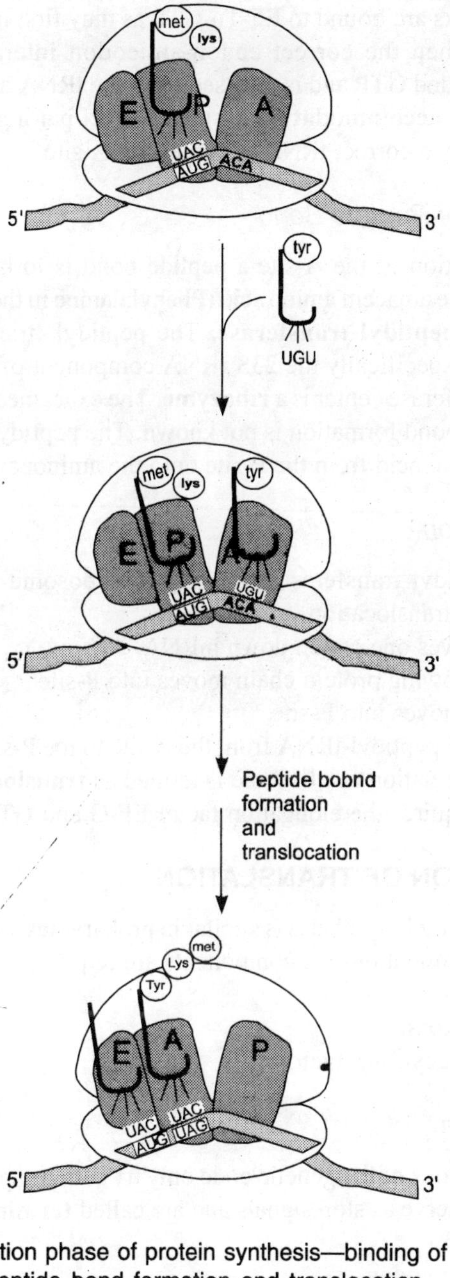

Fig. 22.15: Elongation phase of protein synthesis—binding of charged tRNA, peptide bond formation and translocation.

to GDP the factor becomes inactive and has little affinity for charged tRNAs. Another elongation factor EF-Ts mediates the regeneration of the used form of EF-Tu GDP into the active form EF-Tu GTP.

Charged tRNAs are bound to EF-Tu-GTP as they first interact with the A site of ribosome. When the correct codon-anticodon interaction occurs EF-Tu hydrolyzes its bound GTP and is released from the tRNA and the ribosome. This process of **tRNA accommodation** ensures correct pairing between mRNA and tRNA so that only a correct tRNA occupies the A-site.

(ii) Formation of Peptide Bond

After the occupation of the A-site a peptide bond is formed between initiatior methionine and the adjacent amino acid (Phenylalanine in the figure) by an enzyme complex called **peptidyl transferase.** The peptidyl transferase is composed entirely of RNA specifically the 23S rRNA component of the large subunit, i.e. the peptidyl transferase center is a **ribozyme**. The exact mechanism by which 23S rRNA catalyzes bond formation is not known. The peptidyl transferase reaction transfers the amino acid from the P-site onto the aminoacyl tRNA in the A-site.

(iii) Translocation

During the peptidyl transferase reaction, the ribosome remains in its place, however, during translocation.

- Ribosome moves one codon down mRNA.
- tRNA with growing protein chain moves into P-site
- Spent tRNA moves into E-site.

The movement of peptidyl-tRNA from the A site to the P-site, and the movement of the mRNA in relation to ribosome is termed as **translocation.**
Translocation requires the elongation factor EF-G and GTP.

1. TERMINATION OF TRANSLATION

Termination of protein synthesis is similar in prokaryotes and eukaryotes. In order to accomplish termination two components are required.

i. Stop codons
ii. Release factors
iii. Ribosome recycling factor

A. Stop Codon

Out of the 64 codons of the genetic code only 61 codons specify amino acids, the remaining three serve as stop signals and are called **termination codons** or **stop codons.** These are

UAG or amber codon
UAA or ochre codon
UGA or opal codon

When a point mutation creates a termination codon (nonsense-mutation) it leads to premature termination of the protein at the mutated codon. Non-sense mutations shows that any one of the three codons UAG, UAA or UGA is sufficient to terminate protein synthesis. In bacteria, UAA is the most commonly used termination codon. Next comes UGA and then UAG. Frequently, there appear to be errors reading UGA and protein synthesis continues till the next stop codon is encountered. The termination events are shown in Figure 22.16. There are no tRNAs represented by the stop codons. Instead they are directly recognized by protein factors called release factors.

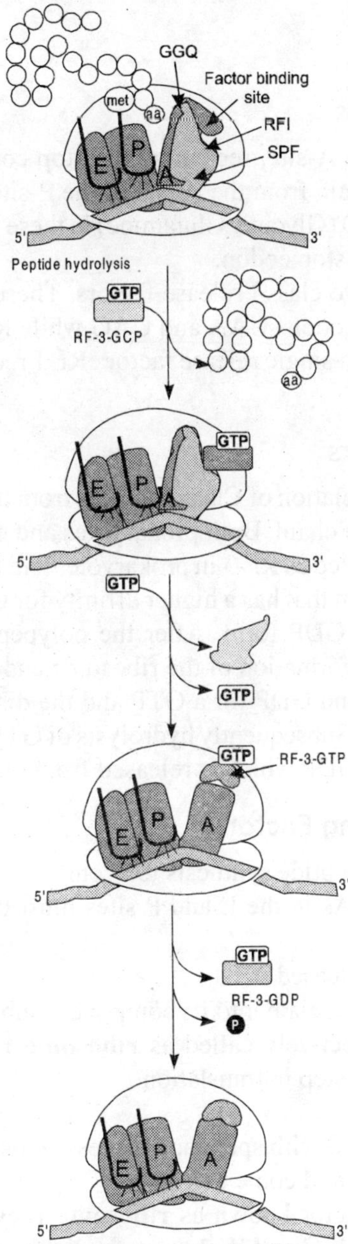

Fig. 22.16: Termination of protein synthesis p9olypeptide release is catalyzed by two release factors

B. Release Factors

Release factors are proteins that activate the hydrolysis of the polypeptide chain from the peptidyl-tRNA upon recognition of stop codons as shown in Figure 22.16

There are two classes of release factors

a. Class I RFs
b. Class II RFs.

Class I Release Factors

These factors bind to the A-site, recognize the stop codons and trigger the release of polypeptide chain from the tRNA in the P-site. A conserved three amino acid sequence GGQ (Glycine, Glutamine) in these factors is responsible for the recognition of the stop codon.

In *E. coli,* there are two class I release factors. These are RF 1 and RF 2. RF1 recognizes the stop codons UAA and UAG while RF2 recognizes UAA and UGA. In eukaryotes, a single release factor eRF 1 recognizes all the three stop codons.

Class II Release Factors

These stimulate the dissociation of Class I factors from the ribosome after the release of the polypeptide chain. Both prokaryotes and eukaryotes have only one class II RF. This is called as RF-3 in prokaryotes and ERF-3 in eukaryotes. It is a GTP-binding protein that has a higher affinity for GDP than GTP. Thus free RF 3 is in an RF-3-GDP form. After the polypeptide chain has been released, a change in conformation of the ribosome and RF1/RF2 stimulates RF-3 to exchange its bound GDP for a GTP and the displacement of class I factor from the ribosome. Subsequently hydrolysis of GTP in RF3-GTP results in the formation of RF3-GDP which is released from the ribosome.

C. Ribosome Recycling Factor

For a new round of polypeptide synthesis to occur

- The deacetylated tRNAs in the E and P sites must be removed from the ribosome.
- The mRNA must be released.
- The ribosome must dissociate into its component subunits.

These events are collectively called as **ribosome recycling**. Ribosome recycling is a critical last step in translation.

Without recycling:

- Cell would be littered with spent and useless ribosomes.
- Protein synthesis would come to a halt.
- In prokaryotes, a factor known as **ribosome recycling factor** (RRF) acts together with EF-G and IF-3 to cause dissociation of 30S and 50S subunits after polypeptide release.

IX. PROTEIN SYNTHESIS INHIBITORS

Many of the antibiotics utilized for the treatment of bacterial infections as well as certain toxins function through the inhibition of translation. Inhibition can be effected at all stages of translation from initiation to elongation to termination. Some of the antibiotic inhibitors of protein synthesis are shown in Table 22.6.

Table 22.6: Several antibiotics and their role in inhibition of translation

Inhibitor	*Comments*
Chloramphenicol	Inhibits prokaryotic *peptidyl transferase*
Streptomycin	Inhibits prokaryotic peptide chain initiation, also induces mRNA misreading
Tetracycline	Inhibits prokaryotic aminoacyl-tRNA binding to the ribosome small subunit
Neomycin	Similar in activity to streptomycin
Erythromycin	Inhibits prokaryotic translocation through the ribosome large subunit
Fusidic acid	Similar to erythromycin only by preventing EF-G from dissociation from the large subunit.
Puromycin	Resembles an aminoacyl-tRNA, interface with peptide transfer resulting in premature termination in both prokaryotes and eukaryotes
Diptheria toxin	Catalyzes ADP-ribosylatin of and inactivation of eEF-2
Ricin	Found in coaster beans, catalase elevate of the eukaryotic large subunit rRNA.
Cycloheximide	Inhibits eukaryotic *peptidyltransferase*

STUDY QUESTIONS

1. Describe the process of translation initiation in prokaryotes with a suitable diagram.
2. Write short notes on the following
   a. Wobble hypothesis
   b. tRNA structure
   c. Polyribosomes
   d. Antibiotic inhibitors of protein synthesis.
   e. Features of genetic code.

MCQs

Select one most appropriate answer given for each of the questions below:

1. The first step in the elongation stage of protein synthesis is:

A. Translocation
B. The binding of aminocyl t–RNA to A–site.
C. Peptide bond formation
D. Formation of initiation complex

2. Which of the following structures consisting of 10–20 ribosomes simultaneously translates 1 mRNA molecule:

A. Polyribosomes
B. Synthetase
C. Signal recognition particle
D. Polymerase

3. Which of the following is the normal cause of chain termination in protein synthesis:

A. The tRNA corresponding to a chain termination triplet cannot bind an amino acid
B. mRNA synthesis stops at chain termination triplet
C. There is no tRNA within the anticodon corresponding to a chain termination triplet
D. Ribosomes fall off automatically

4. During protein synthesis an anticodon on tRNA pairs with:

A. Other tRNA nucleotide bases
B. DNA nucleotide bases
C. rRNA nucleotide bases
D. mRNA nucleotide bases

---

1 B 2 A 3 C 4 D

23

Regulation of Gene Expression

I. INTRODUCTION

The activity of gene is embedded in each cell, however, pattern of expression of a particular gene differs from cell to cell depending upon their functions. For example, haemoglobin gene is expressed only in RBC while immunoglobulin genes are expressed in B-cells. In many cases, gene activity is regulated at the level of transcription either through signals originating within the cell itself or in response to environmental conditions. Certain gene products are needed only in special circumstances e.g., gamma interferon is produced by T cells during viral infection. Thus gamma interferon gene operates in *on* and *off* manner that enable T lymphocytes to produce it when viral infection sets in because gamma interferon is required to combat viral infection. The flow of genetic informations is regulated in other ways as well. The control points of gene expression include:

1. *Transcriptional regulation* of the synthesis of RNA transcripts by controlling initiation or termination.
2. *RNA processing*, or regulation through RNA splicing or alternative pattern of splicing.
3. *Translational control* of polypeptide synthesis.
4. *Stability of mRNA,* because mRNA that persists in the cell have long-lasting effects.
5. *Post-translational control*, i.e. enzyme activity, activation and stability, etc.
6. *DNA rearrangements* in which gene expression changes depending on the position of DNA sequences.

The regulatory system significantly differs in prokaryotic and eukaryotic systems, mainly because prokaryotes are generally free-living unicellular organisms that grow and divide indefinitely under suitable environmental conditions. In contrast, in multicellular organisms division of most cells ceases in adulthood and each type of cell needs to maintain its identity in due time.

II. TRANSCRIPTIONAL REGULATION IN PROKARYOTES

In bacteria, on and off gene activity is often controlled through transcription. When several enzymes act in sequence in a single metabolic pathway, usually all or none of the enzymes are produced. Such *coordinate regulation* results from control of the synthesis of one or more polycistronic mRNA molecules encoding all of the gene products that function in the same metabolic pathway. This type of regulation is not found in eukaryotes.

A. Inducible and Repressible Systems of Negative Regulation

The molecular mechanisms of regulation usually fall into two broad categories: (a) *Negative regulation* and (b) *Positive regulation.*

a. Negative Regulation

In negative regulation, the default state is "on" and a transcription takes place until it is turned "off" by a *repressor* protein that binds to DNA upstream from the transcription start site (Fig. 23.1). A negatively regulated system may be either *inducible* or *repressible* depending upon how the active repressor is formed.

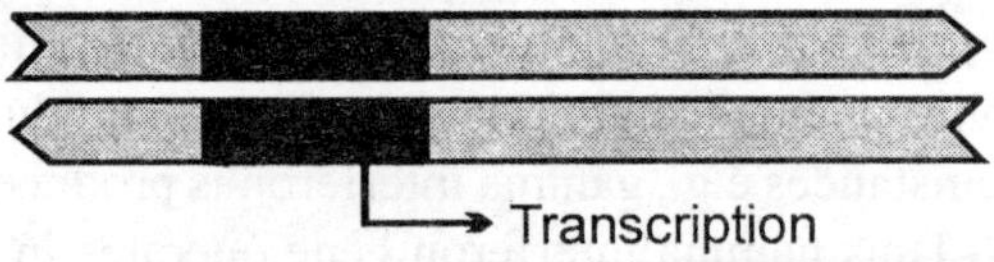

Fig. 23.1: Negative regulation of transcription

i. *Inducible transcription:* In inducible transcription, a repressor DNA-binding protein normally keeps transcription in "off" state. In the presence of a *inducer* (a small molecule), the repressor binds preferentially with the inducer and in turn loses its DNAbinding capability, allowing transcription to occur (Fig. 23.2). Many *catabolic pathways* are inducible and use the initial substrate of a degradative enzyme as inducer.

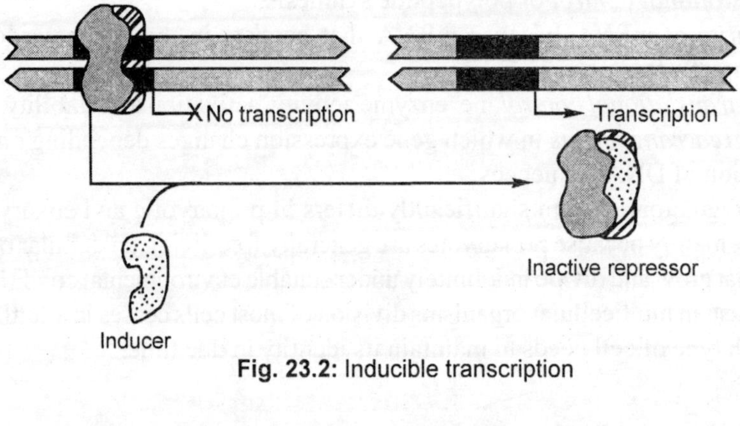

Fig. 23.2: Inducible transcription

ii. *Repressible transcription:* In repressible transcription, the default state is "on" until active repressor is formed to turn it "off". It is mediated by the regulatory protein called an *aporepressor*, it has no DNA-binding activity of its own. The active repressor is formed by the combination of aporepressor and a small molecule called *co-repressor.* Presence of the co-repressor thereby results in the cessation of transcription (Fig. 23.3). Repressible regulation is often found in the control of the synthesis of enzymes that participate in *anabolic pathways.*

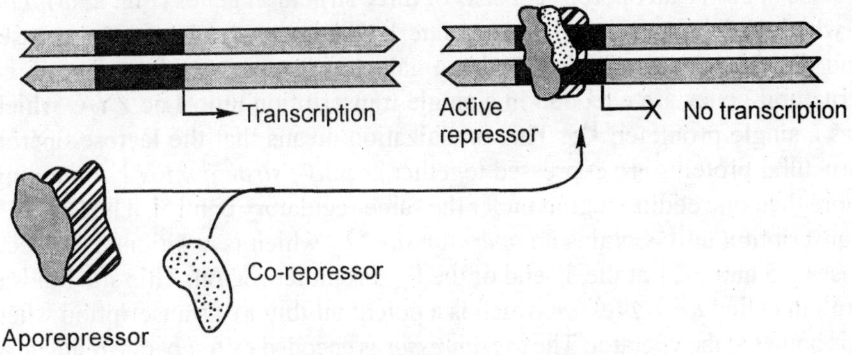

Fig. 23.3: Repressible transcription

b. Positive Regulation

In contrast to the negatively regulated system, in positively regulated system, the default state is "off" and binding of a regulatory protein is required to turn it "on". Such a regulatory protein is called a *transcriptional activator protein* (Fig. 23.4). Negative and positive regulations are not mutually exclusive. Some systems are both positively and negatively regulated. Negative regulation is more common in prokaryotes while positive regulation is more common in eukaryotes. Some genes exhibit *autoregulation*, which means that the protein product of a gene regulates its own transcription.

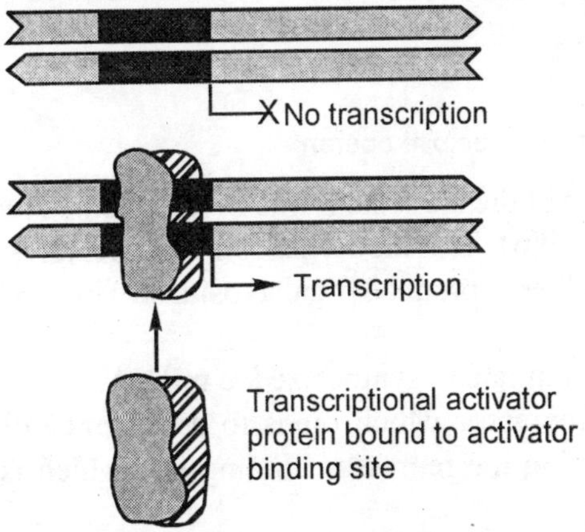

Fig. 23.4: Positive regulation of gene expression

B. Lactose Metabolism and Operon System of Gene Regulation

a. Lac Operon

Gene regulation was first studied in *E. coli* which degrades lactose. It is a classical example of negative regulation. In *E. coli,* two enzymes are necessary for lactose metabolism, *β-galactosidase,* which cleaves lactose to yield galactose and a transporter protein, *lactose permease* which is required for entry of lactose into the cell. Through mutant cell lines it was shown that lactose operon (*lac* operon) consists of three structural genes (Fig. 23.5): *Lac Z* which codes for β-galactosidase, *Lac Y* which encodes lactose permease; and the third *lac A* which encodes a galactoside transacetylase. The three structural genes are encoded in a single transcription unit, Lac ZYA, which has a single promoter, $\mathbf{P}_{lac}$. This organization means that the lactose operon structural proteins are expressed together as *polycistronic mRNA* containing more than one coding region under the same regulatory control. The *lac* ZYA transcription unit contains an *operator site* $\mathbf{O}_{lac}$ which is positioned between bases – 5 and + 21 at the 5'-end of the P_{lac} promoter region. This site binds a protein called *lac repressor* which is a potent inhibitor of transcription when it is bound to the operator. The *lac repressor* is encoded by a separate regulatory gene *lac I* which is also a part of the lactose operon; lac I is situated just upstream of P_{lac}.

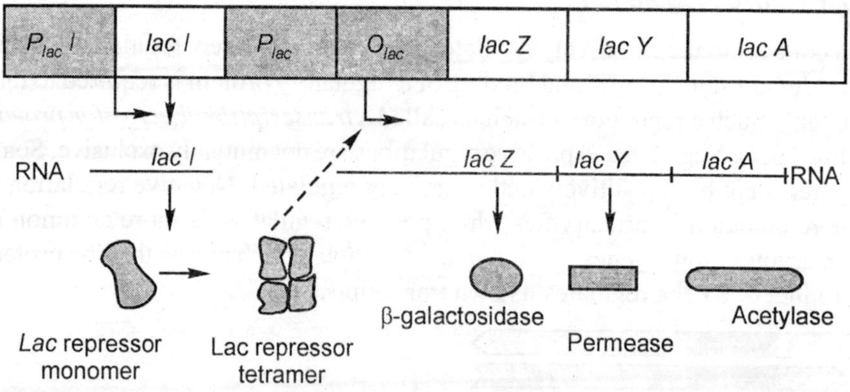

Fig. 23.5: Structure of lactose operon

The genetic regulatory mechanism of the lac system was first explained by Francois Jacob and Jaçques Monod in 1961. This model has following features:

1. Mutation in *lac Z* (lac Z^-) leads failure to produce β-galactosidase while *lac Y* mutant do not produce permease.
2. The promoter mutation (lac P^-) eliminate to synthesize *lac* mRNA.
3. The product of *lac* I gene is a repressor which binds to a sequence of nucleotides in the DNA constituting the repressor binding site which is called operator.

4. When the repressor is bound to the operator, initiation of transcription of *lac* mRNA by RNA polymerase is prevented.
5. Inducers stimulate mRNA synthesis by binding to and activating the repressor. In the presence of an inducer, the operator is not bound by repressor, and the promoter is available for the initiation of mRNA synthesis.

b. Induction

In the absence of an inducer, the *lac* repressor blocks all but a very low level of transcription of *lac* ZYA. When lactose is added to the cells, the low basal level of the permease allows its uptake, and β-galactosidase catalyses the conversion of some lactose to allolactase.

Allolactase acts as an inducer and binds to the lac repressor. This causes a change in the conformation of the repressor tetramer, reducing its affinity for *lac* operator (Fig. 23.6). The removal of the *lac* repressor from the operator site allows the polymerase to rapidly begin transcription of the *lac* ZYA genes. Thus addition of lactose, or a synthetic inducer such as *isopropyl-β-D-thiogalactopyranoside (IPTG)* very rapidly stimulates transcription of the lactose operon structural genes. The subsequent removal of the inducer leads to an almost immediate inhibition of this induced transcription, since the free *lac* repressor rapidly re-occupies the operator site and the *lac* ZYA transcript is extremely unstable.

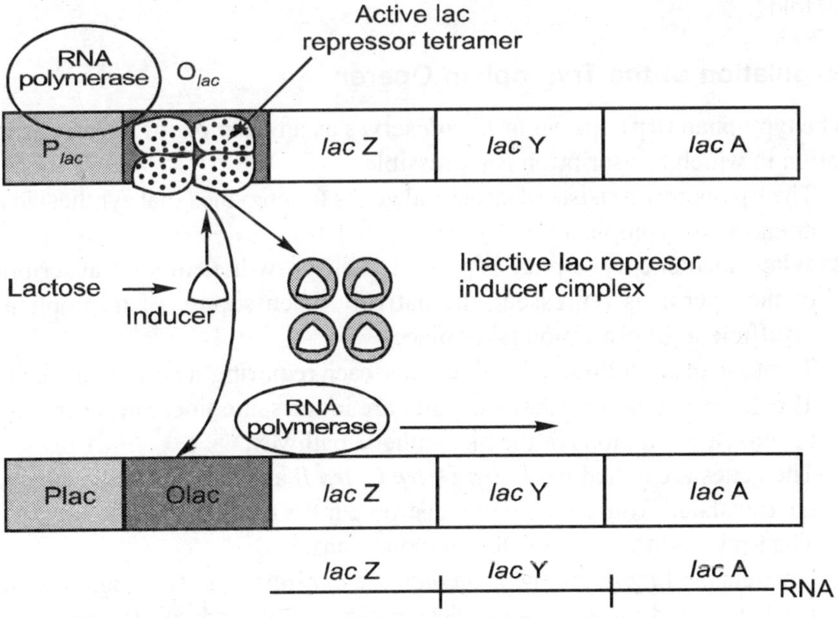

Fig. 23.6: Binding of inducer inactivates the *lac* repressor

c. Positive Regulation of Lac Operon (cAMP Receptor)

- Glucose is a preferred source of carbon and energy for *E coli,* and when both glucose and lactose are present in the growth medium, transcription of the *lac* operon does not take place.
- The inhibitory effect of glucose on expression of *lac* operon is indirect, mediated by a small molecule cAMP (cyclic adenosine monophosphate).
- cAMP is synthesized by an enzyme, *adenyl cyclase*, and the concentration of cAMP is indirectly regulated by glucose metabolism.

E coli and many other bacteria contain a specific activator protein called *cAMP receptor protein (CRP)*, encoded by the gene *crp.* Mutations in either *crp* or *adenylate cyclase gene* prevent the synthesis of *lac* mRNA. CRP and cAMP bind to one another, forming a *cAMP-CRP* complex that serves as a transcriptional activator that must bind DNA in the region of *lac* promoter in order for transcription to occur (Fig. 23.7). Unlike the repressor, which negatively regulates the *lac* operon. The cAMP-CRP complex positively regulates the operon. This regulation is mediated by CRP which exists as a dimer, cannot bind to DNA on its own nor regulate transcription. Glucose reduce the level of cAMP in the cell. When glucose is absent, the level of cAMP in *E. coli* increases and CRP binds to cAMP. The cAMP-CRP complex binds to the lactose operon promoter P_{lac} just upstream from the site of RNA polymerase. CRP binding induces a 90° bend in DNA and this is believed to enhance RNA polymerase binding to the promoter enhancing transcription by 50 fold.

Regulation of the Tryptophan Operon

The typtophan (trp) operon in *E. coli* serves as an example of negative regulation in which transcription is repressible:

1. The trp operon consists of structural genes for enzymes that synthesize the amino acid tryptophan.
2. When adequate tryptophan is present in the growth medium, transcription of the operon is repressed, alternatively when supply of tryptophan is insufficient, transcription takes place.
3. Tryptophan is synthesized in five steps, each requiring a particular enzyme.
4. The genes coding for these enzymes are in the same linear order in which the enzyme functions in the biosynthetic pathway.
5. The genes are called *trp E, trp D, trp C, trp B* and *trp A,* and the enzymes are translated from a single polycistronic mRNA molecule.
6. The trp E coding region is the first one translated.
7. Upstreams of *trp E* are the *promoter,* the *operator,* and two regions called the *leader* and the *attenuator,* designated as trpL and *trp A* respectively (Fig. 23.8).
8. The repressor gene *trpR* is located quite far from this operon.

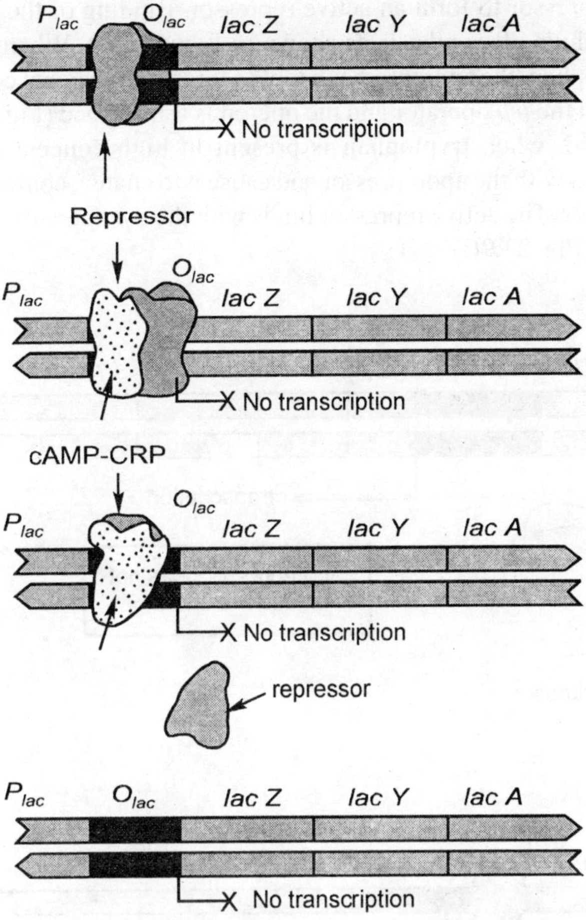

Fig. 23.7: Four regulatory states of the lac operon

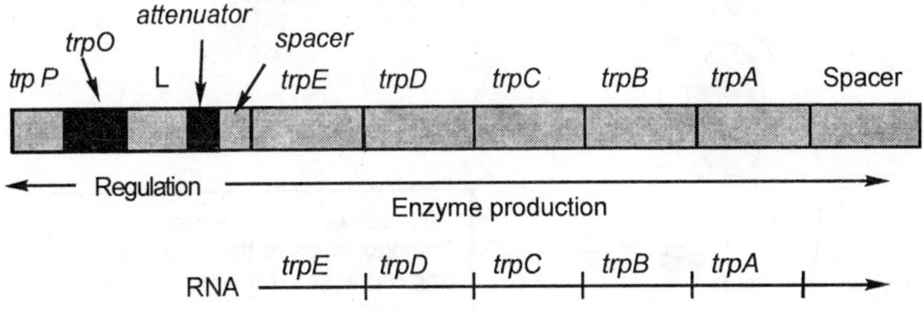

Fig. 23.8: The *E. coli* trp operon

The product of the *trpR* gene is the aporepressor protein. Mutations in either this gene or the operator cause constitutive initiation of transcription of trp mRNA. The role of tryptophan is that of a corepressor, which binds with the trp aporepressor to form an active repressor. Binding of the repressor to trp operator shuts off synthesis as shown in Figure 23.9. When there is not enough tryptophan, the aporepressor adopts a three-dimensional conformation unable to bind the *trp* operator and the operon is transcribed (Fig. 23.9A). On the other hand, when tryptophan is present in high concentration, some molecules bind with the aporepressor and cause it to change conformation into active repressor. The active repressor binds with the *trp* operator and prevents transcription (Fig. 23.9B).

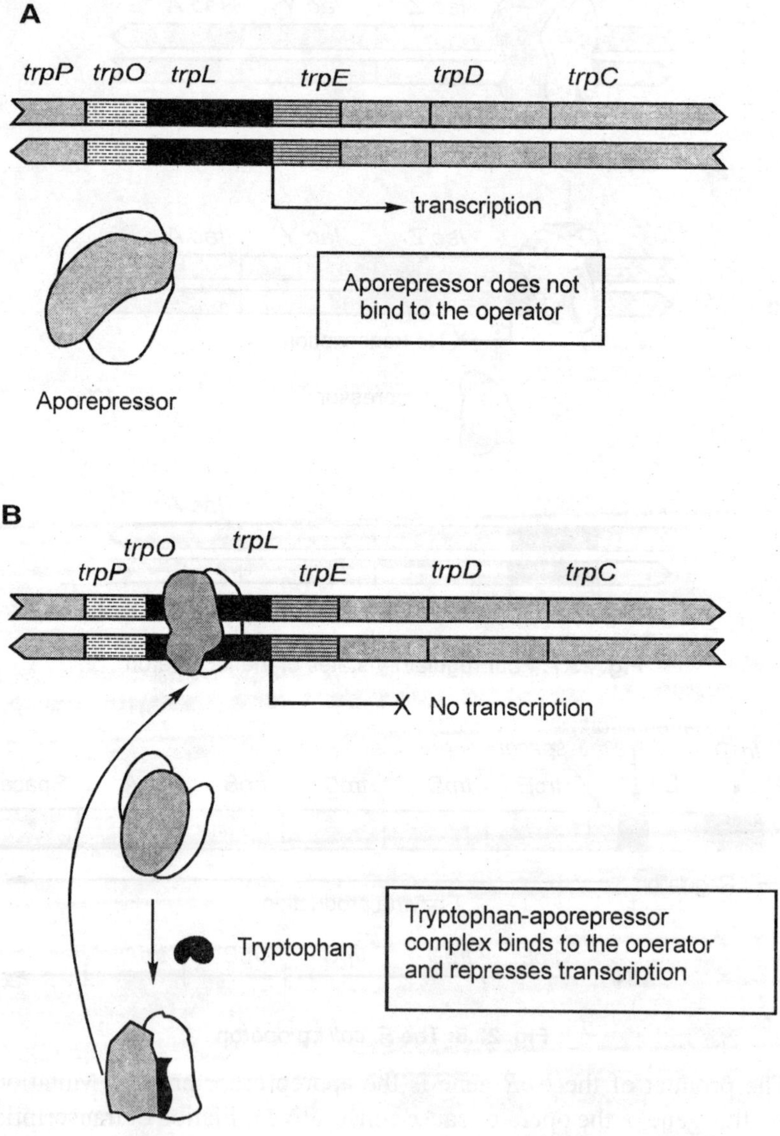

Figs 23.9A and B: Regulation of *E. coli trp* Operon

Attenuation

There is another level of regulation of *trp* operon that provides for fine tuning, which is employed when some tryptophan is present in the growth medium, but not enough to sustain the optimal growth. The *trp* operon responds to this situation by means of a regulatory mechanism in which the amount of transcription in the depressed state is controlled by the concentration of tryptophan in the cell. This mechanism is found in many operons responsible for amino acid biosynthesis and is called *attenuation.*

The attenuation mechanism controls whether transcription, once started, will continue through operon or be terminated prematurely. However, in the presence of even small amount of tryptophan, most of the mRNA molecules terminate in a specific 28-base regions within the leader sequence (Fig. 23.8) called the *attenuator.*

In *E. coli,* termination of transcription is determined by whether a small peptide encoded in the leader sequence can be translated. This coding sequence specifies a *leader polypeptide* 14 amino acids in length, and it includes two adjacent tryptophan codons at position 10 and 11. When there is sufficient changed tryptophan tRNA allow translation of these codons, the nascent transcript adopts a conformation in which the attenuator is hidden, and transcription continues through the entire operon. On the other hand, when there is insufficient charged tryptophan tRNA to allow translation of the leader polypeptide, the ribosome stalls at the tryptophan codons, which creates a situation in which the attenuator is exposed, terminating transcription. The attenuation forms the only mechanism of negative feedback control where there is no repressor-operator regulation.

C. Regulation by Alternative σ Factors

1. The *sigma (σ) factor* is responsible for recognition of consensus promoter sequences and is required for transcription initiation. In *E coli,* σ^{70} is responsible for recognition of – 10 and – 35 consensus sequences.
2. Alternative sigma factor σ32 regulates expression of around 17 proteins in *E. coli* σ32 has its own specific promoter consensus sequence.
3. A set of σ factors are responsible for regulation of sporulation in **Bacillus subtilis**.
4. Many bacteriophages synthesize their own σ factor in order to take over the hosts cells own transcription machinery by substituting the normal cellular σ factor and altering the promoter specificity of the RNA polymerase.
5. **B. subtilis** SPO-l phage expresses a cascade of σ factors which allow a defined sequence of expression of early, middle and late phage genes.

D. Regulation of Transcription in Bacteriophage (λ)

In bacteriophage λ lytic and lysogenic phase is regulated by CI and cro proteins that function as genetic switch: CI turns on lysogenic and cro turns on the lytic cycle.

The molecular basis on which the decision between the lysogenic and the lytic cycle is determined can be explained by Figure 23.10. Upon infection of *E. coli* by λ, the λ molecule circularizes and RNA polymerase binds at P_L and P_R and initiates transcription of N and Cro genes. The N protein acts to prevent termination of the transcription from P_L and P_R[9] thus allowing the production of CII protein. CII protein activates transcription at P_E and P_I, thus allowing the production of CI and *int* (integration) proteins. The CI protein shuts down further transcription from P_L and P_R and stimulates transcription at P_M, increasing its own synthesis in the positive autoregulation loop. Lysogeny is achieved if the concentration of CI protein reaches a level high enough to prevent transcription from P_L and P_R and to allow the **int** protein to catalyze site-specific recombination between the circular λ molecule and the *E. coli* chromosome at their respective attachment sites (**att**). The result is formation of a lysogen.

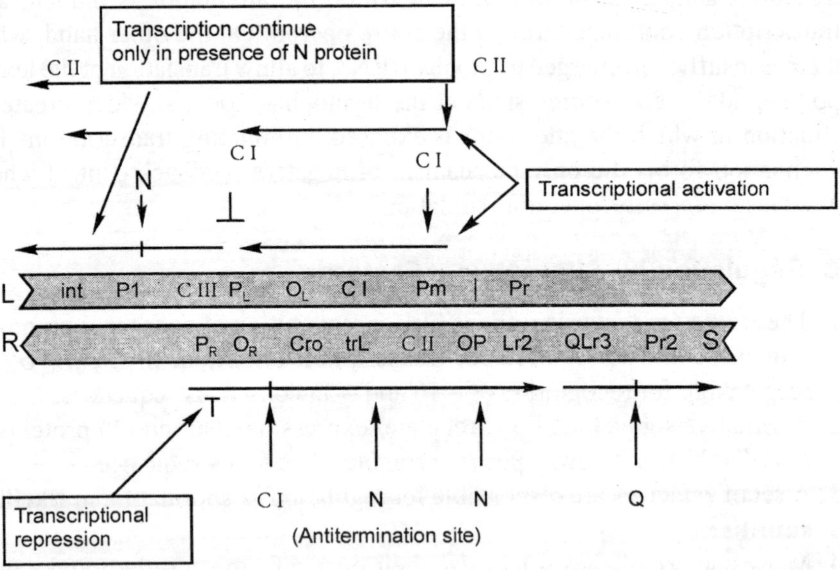

Fig. 23.10: Genetic and transcriptional map of the control region of bactriophage λ

The alternative pathway is lytic development, which takes place when the cro protein dominates. The cro proteins also bind to O_R and in doing so, blocks transcription from P_M.

If this occurs, then the concentration of repressor cannot rise to the level required to block transcription from P_L to P_R. Transcription continues from

P_R to P_{R2}, and N and Q proteins prevent termination in the rightward transcripts. Since the λ DNA is in a circular configuration, rightward transcription moves through genes S and R and thence through the head and tail genes. The production of proteins needed for cellular lysis and formation of phage particles ensues, followed by phage assembly and cellular lysis to release phage.

III. REGULATION OF TRANSCRIPTION IN EUKARYOTES

Many eukaryotic genes are house keeping genes that encode essentially metabolic enzymes. Other genes, which differ in their expression depending upon the cell types or at different stages of the cell cycle are often regulated at the level of transcription. Typically, level of expression of eukaryotic genes may differ 2-10 fold between uninduced and induced levels, in contrast to the more drammatic differences seen in prokaryotes, in which the ratio between uninduced and induced levels may be as great as 1000 fold.

A. Transcriptional Activator Proteins

Transcription activator proteins or transcription factors work by direct interaction with one or more components of the transcription complex, and in this way they recruit the transcription complex to the promoter of the gene to be activated. Many transcriptional activator proteins can be grouped into categories based on their motif (amino acid sequences) as below:

a. Helix-turn-helix Motif

Many transcriptional activator proteins can be grouped into categories on the basis of characteristics shared by their amino acid sequences. For example, one category has a helix-turn-helix which consists of a sequence of amino acid forming a pair of α-helices separated by a bend; the helices are so situated that they can fit neatly into the grooves of a doubled stranded DNA molecule. The helix-turn-helix motif is the basis of the DNA binding ability, although the sequence specificity of the binding results from other parts of the protein. This motif characteristic of DNA binding proteins containing 60-amino acid **homeodomain** which is encoded by a sequence called **homeobox**. The characteristic helix-turn-helix motif is also found in prokaryotes like bacteriophage DNA-binding proteins (Fig. 23.11) as the phage λ, cro δ repressor, **lac**, **trp** repressors and cAMP receptor proteins (CRP). In the Antennapedia transcription factor of ***Drosophila***, this domain consists of four α-helices in which helices II and III are at right angles to each other and are separated by a characteristic β-turn.

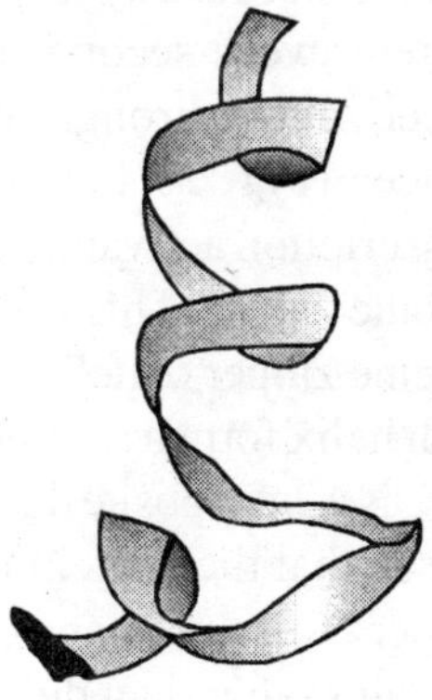

Fig. 23.11: Helix-helix core structure

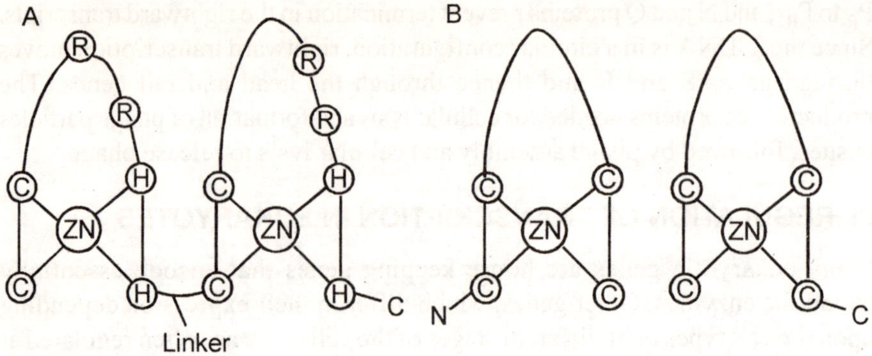

Figs 23.12A and B: (A) C_2H_2-zinc finger motif (B) Zn-finger folded structure

b. Zinc Finger Motif

A second large category of transcriptional activator proteins include a DNA-binding motif called a **zinc-finger motif** because the folded structure a zinc-ion. An already familiar example is the GAL4 transcriptional activator protein in yeast. The protein functions as a dimer composed of two identical GAL-4 polypeptides oriented with their zinc-binding domains at the extreme ends. The DNA sequence recognized by the protein is a symmetrical sequence, 17 base pair in length including a CCG triplet at each end that makes direct contact with a zinc containing domains (Fig. 23.12).

b. Leucine Zipper Motif

Leucine zipper proteins contain a hydrophobic leucine residue at every seventh position in a region that is often at the C-terminal part of the DNA-binding domain. These leucine lie in an α-helical region and the regular repeat of these residues forms a hydrophobic surface on one side of the α-helix with a leucine protein every second turn of the helix. These leucines are responsible for dimerization through interactions between the hydrophobic faces of the α-helices (Fig. 23.13). This interaction forms **coiled coil** structure. b Zip transcription activators contain a basic DNA binding domain N-terminal to the leucine zipper. This is present on an α-helix which is a continuation from the leucine zipper α-helical c-terminal domain. The N-terminal basic domains of each helix form a symmetrical structure in which each basic domain lies along the DNA in opposite directions interacting with a symmetrical DNA recognition site so that the protein in effect forms a clamp around the DNA. The leucine zipper is also used as a dimerization domain in proteins that use DNA binding domains other than the basic domain, including some homeodomain proteins.

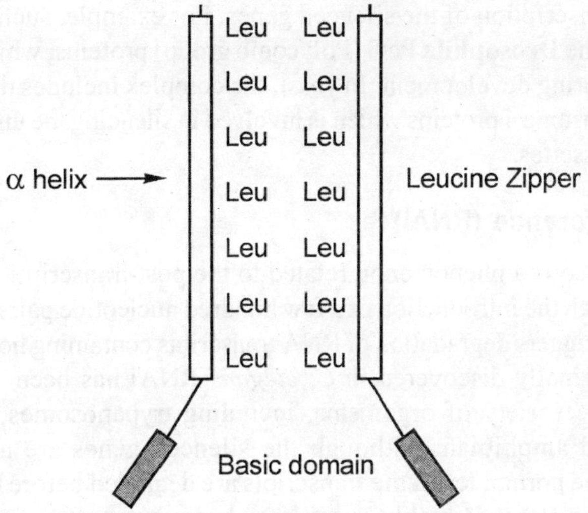

Fig. 23.13: The leucine zipper motif

B. Transcription Enhancers and Silencers

Some transcriptional activator proteins bind with particular DNA sequences known as **enhancers**. Enhancer sequences are typically rather short and are found at a variety of locations around the gene they regulate. Most enhancers are upstream of the transcriptional start site, others are in introns within the coding region, and a few are even located at the 3' end of the gene. One of the most thoroughly studied enhancer is in the mouse mammary tumour virus and determines the transcriptional activation by the glucocorticosteroid hormone. The consensus sequence of the enhancer in AGAQCAGQ in which Q stands for either A or T.

Enhancers are essential components of gene organization in eukaryotes because they enable genes to be transcribed only when the proper transcriptional activators are present. Some enhancers respond to molecules outside the cell, e.g., steroid hormones that form receptor-hormone complexes. Other enhancers respond to molecules that are produced inside the cell (e.g. during development), and these enhancers enable the genes under their control to participate in cellular differentiation, thought to be expressed in a tissue specific manner. Many genes are under the control of several different enhancers, so they can respond to a variety of different molecular signals, both external and internal.

Some genes are also subjected to regulation by transcriptional silencers, which are short nucleotide sequences that are targets for DNA-binding proteins, that once recruited to the site, promote the assembly of large protein complexes that prevent transcription of the silenced genes. For example, such silencing complexes are the **Drosophila** PcG (Polycomb group) proteins, which silence certain genes during development. In yeast, the complex includes the Rap 1p, SIR, ORC and histone 4 proteins which is involved in silencing the unexpressed mating-type cassettes.

C. RNA Interference (RNAi)

RNA interference is a phenomenon related to the post-transcriptional genes silencing in which the introduction of a few hundred nucleotide pairs of double stranded RNA triggers degradation of RNA transcripts containing homologous sequences. Originally discovered in *C. elegans* RNAi has been shown to occur in a wide variety of organisms, including trypanosomes, planaria, *Drosophila* and amphibian. Although the silenced genes are apparently transcribed at the normal level, the transcripts are degraded before they leave the nucleus. The effect is highly specific and requires only a few double-stranded RNA molecule per cell to trigger the silencing. The RNAi effect can be transmitted from cell to cell.

D. Examples of Transcriptional Regulation in Eukaryotes

Following are the examples of transcriptional regulation in eukaryotes.

1. SPl is a ubiquitous transcription factor which contains 3-zinc finger motifs and two glutamin-rich transactivation domain.
2. Steroid hormones enter the cell and bind to a steroid hormone receptor. The receptor dissociates from the bound inhibitor protein, dimerises and translocates to the nucleus. The DNA binding domain of the steroid hormone receptor binds to response elements, giving rise to activation of target genes. Thyroid hormone receptors act as transcription repressors until they are converted to transcriptional activators by thyroid hormone binding.
3. Interferon-γ activates JAK kinase which phosphorylates STAT 1-α. STAT 1α dimerises and translocates to the nucleus where it activates expression of target genes.
4. Tat proteins bind to an RNA sequence present at the 5'end of all human immunodeficiency virus (HIV) RNAs called TAR. In presence of Tat, HIV transcription terminates prematurely. The Tat-TAR complex activates TF II H in the transcription initiation complex at the promoter, leading to phosphorylation of the RNA Pol II carboxyl terminal domain (CTD). This permits full length transcription by the polymerase.
5. The expression of myo D and related genes (my 75, mr 74 and myogenin) can convert non-muscle cells into muscle cells. Their expression activates muscle specific gene expression and blocks cell division. Each gene

encodes a helix-loop-helix (HLH) transcriptional factor. HLH heterodimer formation and the known DNA binding inhibit the Id, give rise to diversity and regulation of these transcription factors.

6. The homeobox, which encodes a DNA-binding domain, was originally found in *Drosophila melanogaster.* Homoeotic genes that encode transcriptional factor which specify the development of body parts. The conservation of both function and organization of the homeotic gene clusters between *Drosophila* and mammals suggest that these proteins have important common roles in development.

STUDY QUESTIONS

1. Describe in brief the mechanism of regulation of gene expression in parokaryotes.
2. Describe various eukaryotic transcription factors and their motifs.
3. Write short notes on:
   a. *Lac* operon
   b. *Trp* operon
   c. Repressor
   d. Enhancer
   e. RNA i

MCQs

Select one most appropriate answer given for each questions below:

1. **Which of the following structural genes codes for the enzyme β-galactosidase?**
   A. Lac A
   B. Lac Y
   C. Lac Z
   D. Lac P
2. **Which of the following genes codes for the enzyme lactose permease?**
   A. Lac A
   B. Lac Y
   C. Lac Z
   D. Lac P
3. **The lac operon model was first described by:**
   A. Khorana
   B. Jacob and Monod
   C. Lehninger

4. **The difference between uninduced and induced levels of genes in prokaryotes may be:**
   A. 10 fold
   B. 100 fold
   C. 1000 fold
   D. 10,000 fold
5. **The difference between uninduced and induced levels of eukaryotic genes may be mostly:**
   A. 10 fold
   B. 100 fold
   C. 1000 fold
   D. 10,000 fold
6. **Helix-turn-helix motif is found in:**
   A. Prokaryotes only
   B. Eukaryotes only
   C. Both prokaryotes and eukaryotes
   D. Only in yeast.
7. **Zinc finger motif is found in:**
   A. Bacteriophage only
   B. Prokaryotes only
   C. Eukaryotes only
   D. All the above
8. **RNAi functions as:**
   A. Transcriptional enhancer
   B. Transcriptional silencers
   C. Transcriptional regulators (by both enhancing and silencing)
   D. None of the above.

1 C 2 B 3 B 4 C 5 A 6 C 7 D 8 B

24

Recombinant DNA Technology

I. INTRODUCTION

The deoxyribose nucleic acid (DNA) is the hereditary carrier of all the genetic information that is essential for defining the way in which each cell of an organism functions. This information in DNA is encoded in the form of a stretch of nucleotides. Such a sequence of nucleotides present in a particular position on a chromosome that encodes for a final functional product is defined as a gene. The technology that has brought about a revolution in the understanding and revealing of this genetic information at the molecular level is the modern day recombinant DNA technology. The technique of cloning a gene lies at the heart of this technology. Multiple copies of the desired gene can be made using this technique. In brief, it involves cutting out the gene of interest using restriction enzymes. This gene is then inserted into small circular vectors (plasmid) that have been cut using the same enzymes. The gene is then ligated into the vector using another enzyme called DNA ligase. Once constructed in this way, this recombinant plasmid vector carrying the gene of interest can be inserted into bacterial, yeast or cultured animal cells. The cells that have taken up the plasmid are separated from others in a selection procedure that takes advantage of drug-resistance genes found on the plasmid. A single colony of such cells is then picked up and it gives rise to a whole population of identical cells, or clones, by normal cell division. In this process, all of the resultant cells are expected to contain a copy of the plasmid carrying the inserted gene. Thus any gene of interest can be specifically selected and amplified using this technique. The various steps involved in gene cloning have been depicted in Figure 24.1.

II. STRATEGY OF GENE CLONING

Cloning a particular DNA fragment, for either amplification or expression of a particular gene, can be divided into the following subheadings:

1. Choosing a suitable vector as a vehicle for gene cloning.

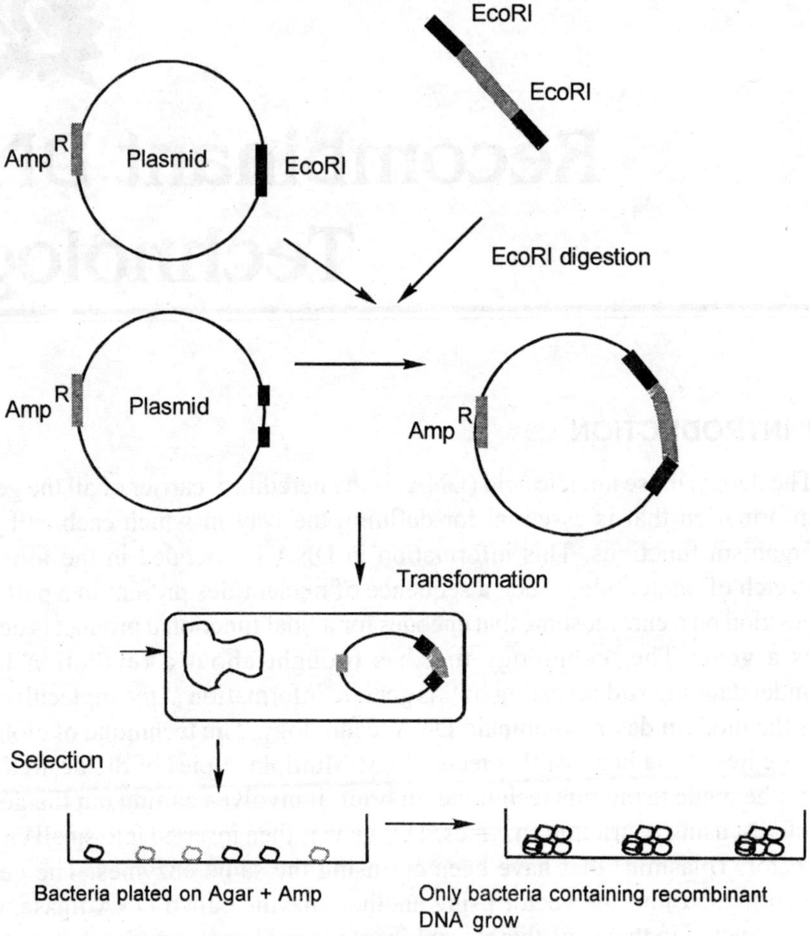

Fig. 24.1: The various steps involved in gene cloning

2. Manipulation of DNA (Vector and Insert).
3. Introduction of the recombinant DNA molecule into living cells.
4. Identification of recombinant clones.

All these steps are explained in detail one by one below.

A. Choosing a Suitable Vector as a Vehicle for Gene Cloning

A vector can be defined as an external DNA molecule, having its own origin of replication, capable of carrying a DNA fragment along with itself into a host cell. If a vector is used for just replicating the inserted DNA fragment, then it is called as a "cloning vector". If it is used for expressing the inserted gene inside the recipient host cell, then it is called as an "expression vector". All cloning vectors share the following common features:

1. They contain certain sequences that allow the vector to propagate in the host cell (bacteria or yeast).
2. They contain multiple cloning sites, which is a short continuous stretch of recognition sequences that are recognized by different restriction enzymes. This gives the user a choice to clone the DNA of interest using any of these enzymes.
3. A method of selecting for bacteria (or yeast) containing a vector with foreign DNA. This is usually accomplished by selectable markers for drug resistance present in the vector.

Examples of cloning vectors are plasmids, cosmids, and artificial chromosomes:

A. *Plasmid:* They can be defined as autonomously (self) replicating, extrachromosomal circular DNA molecules, distinct from the normal bacterial genome and usually not essential for cell survival. A number of artificially constructed plasmids are used as cloning vectors. Their cloning limit is up to 10,000 bp.

pUC19 is a typical example of cloning plasmid vector. It contains a multiple cloning site or polylinker that contains recognition sequences of several different restriction enzymes. It also contains its own origin of replication called as "ori" which allows its propagation in the bacterial host. Also ampicillin resistance gene present on this vector confers resistance to the bacteria which harbors this vector DNA.

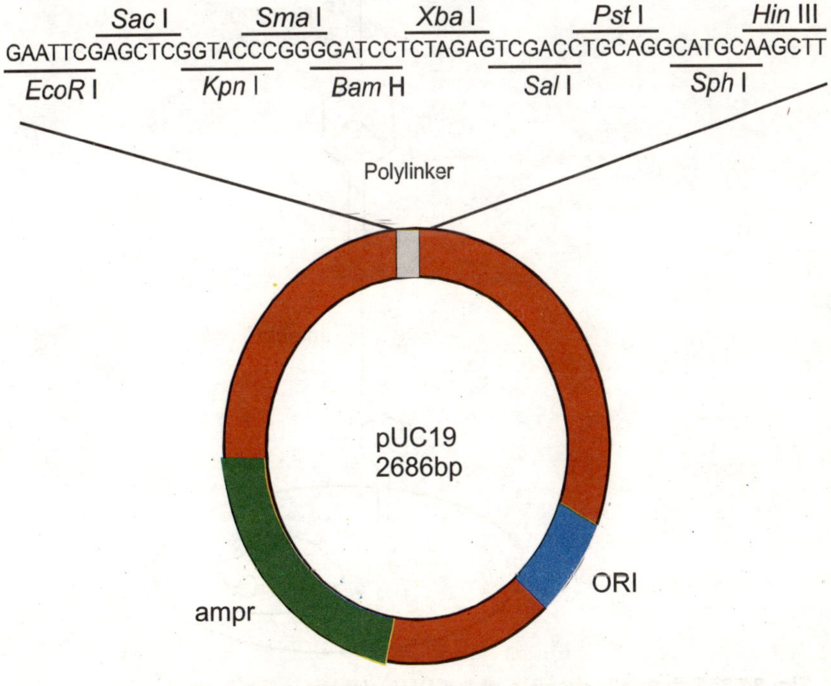

Fig. 24.2: Structure of pUC 19 vector

B. *Lambda phage:* λ phages are viruses that infect bacteria. The major advantage of the λ phage vector is its high transformation efficiency. They are about 1000 times more efficient than the plasmid vector.
The DNA to be cloned is first inserted into the λ DNA, replacing a nonessential region. Then, by an **in vitro packaging and assembly system**, the λ virion carrying the recombinant DNA can be formed. The λ genome is 49 kb in length and can carry up to 25 kb foreign DNA.

C. *Cosmid:* It is an artificially constructed cloning vector containing a gene (the cos gene) of the virus called phage lambda. The cosmid vector is a combination of the plasmid vector and the Cos site. This site is required for inserting the target DNA into the λ head. Cosmids can be packaged into lambda phage particles for infection into the bacteria *E. coli*. This permits the cloning of larger DNA fragments than can be introduced into bacterial hosts using plasmid vectors. Cosmid vectors have the following advantages:
- High efficiency of transformation.

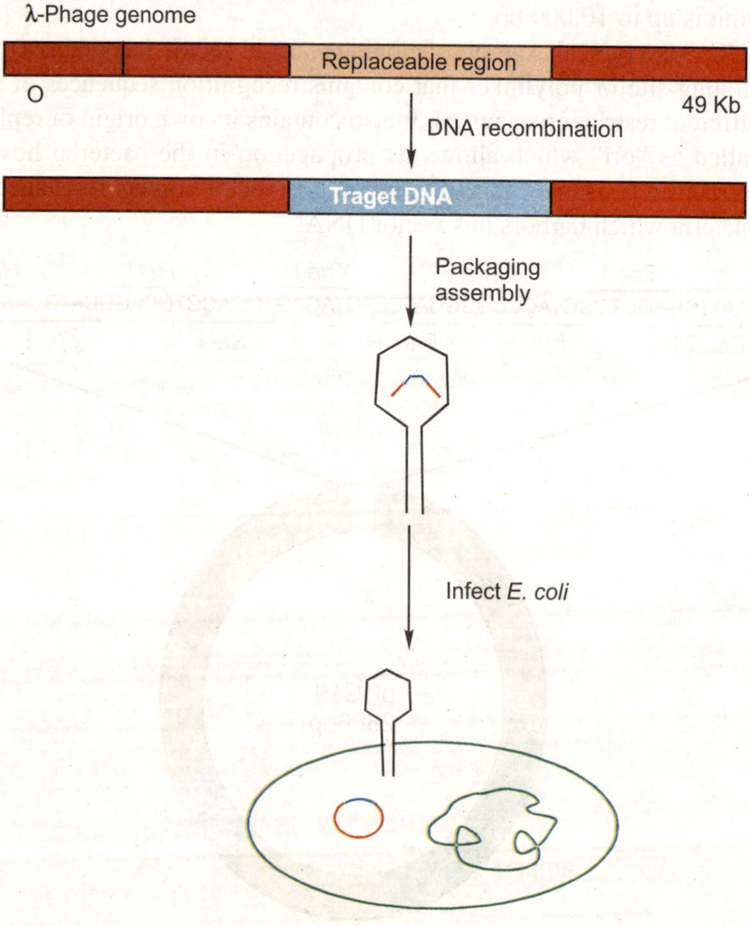

Fig. 24.3: Schematic drawing of the DNA cloning using λ phages as vectors (Adapted from http://www.web-books.com/MoBio/Free/Ch9A4.htm)

- The cosmid vector can carry up to 50 kb of insert DNA whereas the maximum insert size for cloning in plasmid and phage vectors is limited to 25 kb.

D. **Artificial chromosomes** contain all the elements that a DNA needs to function as a chromosome in the host organism. They can be subdivided into two categories:

YACs are yeast artificial chromosomes. A YAC is constructed from the telomeric (end), centromeric, and replication origin sequences which are required for the replication of the cloned DNA within yeast cells. The YAC vector is capable of carrying a large DNA fragment, but its transformation efficiency is very low. Its cloning limit is upto 3000 kb.

BACs (bacterial artificial chromosomes) are DNA constructs based on a fertility plasmid useful for cloning very large inserts in bacteria usually *E.coli.*

B. Manipulation of DNA (Vector and Insert)

The technique of cloning DNA necessitates the involvement of various enzymes which are valuable tools in cutting, modifying and joining together different DNA sequences. Bacterial enzymes called as restriction endonucleases are used as DNA cutters. These enzymes cleave the sugar-phosphate backbone of DNA. In most practical settings, a given enzyme cuts both strands of duplex DNA within a stretch of just a few bases. A bacterial cell uses these enzymes as a self-defense mechanism against intruding DNA of bacteriophages. They work by cutting up the foreign DNA in a process that is called as restriction. The bacterial cell, however, protects its own DNA from restriction by synthesizing methyl transferase enzyme. This enzyme adds methyl groups to those adenine or cytosine bases which are present in the sites recognized by the restriction endonuclease. This methylation prevents the bacteria's own DNA from being cut by the restriction enzyme which it harbors. This combination of restriction endonuclease and methylase activities is referred to as a *restriction-modification system*. The substrates for restriction enzymes are more-or-less specific sequences of double-stranded DNA called *recognition sequences*. The recognition site normally varies in length from 4, 6 or 8 bp. Most recognition sequences are pallindromes, i.e. they are tandem repeat sequences which read the same forward (5' to 3' on the top strand) and backward (5' to 3' on the bottom strand). Different restriction enzymes that cleave within the same target sequences are termed as isoschizomers. They are of importance because a DNA fragment cut by one enzyme can be reannealed with a DNA fragment cut by a different enzyme which is its isoschizomer. The restriction enzymes generate one of these three different types of ends after digestion of a DNA fragment:

- **5' overhangs:** These are generated when an enzyme cuts asymmetrically within the recognition site such that a short single-stranded unpaired segment extends from the 5' ends. e.g. *BamHI* cuts in this manner.
- 5'- G-G-A-T-C-C-3' *BamHI* → -G ⁵'G-A-T-C-

  3'-C-C-T-A-G-G-5' -C-C-T-A ⁵' C

3' overhangs: These are generated when a single stranded unpaired segment extends from the from the two 3' ends. e.g. *KpnI* cuts in this manner.

5'- G-G-T-A-C-C- 3' *KpnI* → -G-G-T-A-C ³' C-

3'-C-C-A-T-G-G- 5' -C- ³'-C-A-T-G-G-

Blunts: These are generated when enzymes cut at precisely opposite sites in the two strands of DNA and generate blunt ends without any unpaired overhangs. *SmaI* is an example of an enzyme that generates blunt ends.

5'-C-C-C-G-G-G-3' *SmaI* → 5'-C-C-C G-G-G-3

3- G-G-G-C-C-C-5' 3'-G-G-G C-C-C-5'

The 5' or 3' overhangs generated by enzymes that cut asymmetrically are called ***sticky ends*** or ***cohesive ends***, because they can be annealed with their partner by base pairing. The key enzyme that joins the cut ends of these DNA molecules is called as **DNA ligase**. This enzyme seals the nicks created by the above enzymes in a reaction requiring ATP and dNTPs. Single stranded nicks are thus religated to form a duplex double stranded DNA. The enzyme mediates the formation of phosphodiester bonds between adjacent 3'-OH and 5'-phosphate termini, thereby sealing the nicks formed by restriction enzyme digestion. This enzyme can join cohesive as well as blunt ends. Another enzyme called as alkaline phosphatase is used to remove phosphate groups from the linearized vector DNA so as to prevent recircularization of the vector alone. The phosphatase untreated insert DNA can thus easily ligate with the vector DNA. This ligation reaction is generally carried out at low temperature so as to increase the chances of vector and insert DNA to anneal together.

C. Introduction of the Recombinant DNA Molecule into Living Cells

After the desired fragment (gene) is ligated into a suitable vector, it is important to introduce it in to a host cell (mostly bacteria) for its amplification. Under normal conditions bacteria will not take up these plasmids. They have to be made 'competent' to do so by treating them with a solution of ice cold calcium chloride. These competent cells containing the ligation mix are then given a brief heat shock treatment (90-120 seconds) at 42°C followed by incubation at ice for 5-10 minutes. This enables them to take up the recircularized plasmid DNA. This process of introducing a foreign DNA into a bacterial cell is known as 'transformation'. The bacteria which take up the plasmid DNA are said to be transformed. These bacteria are plated out at low concentration on agar containing the same antibiotic whose resistance gene is present on the plasmid. In this way, only the cells containing the plasmid will survive. Each bacterium that survives will then divide to form a colony of identical cells, each of which will contain the plasmid.

D. Identification of the Recombinant Clone

Obtaining a clone of the specified gene is a critical test of any gene cloning experiment. Success or failure of this experiment often depending upon whether or not a strategy can be devised by which recombinant clones can be selected. To identify whether or not a recombinant clone is obtained, a direct selection procedure is applied whereby plating onto a selective medium (like an antibiotic) enables transformants to be distinguished from non-transformants. The next step involves the differentiation of tansformed colonies containing recombinant DNA from the ones containing self-ligated vector molecules. With most cloning vectors insertion of a DNA fragment into the plasmid destroys the integrity of one of the genes present on it. Therefore the recombinants can be identified, since the inactivated gene is no longer displayed by the host cells. This principle is referred to as insertional inactivation. One example of a gene which is usually inactivated for selection is β galactosidase gene. Beta galactosidase (normally coded by the *lac Z*, which resides on the *E. coli* chromosome) is one of a series of enzymes involved in the breakdown of lactose to glucose plus galactose. Some strains of *E. coli* have a modified *lac Z* gene, one that lacks the segment referred to as *lac Z'* and coding for the alpha-peptide portion of beta-galactosidase (beta-gal). These mutant strains can synthesize the enzyme only when they harbour a plasmid, such as pUC8, that carries the missing *lacZ'* segment of the gene. A cloning experiment with pUC8 thus involves selection of transformants by screening for beta-gal activity in addition to antibiotic resistance. Cells with a normal pUC8 are amp resistant and able to make the beta-galactosidase; recombinants are also amp resistant, but unable to synthesize beta-galactosidase. Screening for presence or absence of beta-galactosidase involves a lactose analogue called X-gal (5-bromo-4-chloro-3-indoyl-beta-D-galactopyranoside) which is broken down by beta-gal to a product that is deep blue in colour. If X-gal (plus and inducer of the enzyme such as isopropylthiogalactoside, IPTG) is added to the agar, along with amp, then non-recombinant colonies containing the cells which synthesize beta-gal, will be coloured blue, whereas recombinants with a disrupted *lacZ'* gene and unable to make beta-gal, will be white in colour. Thus recombinant clones can be directly distinguished from nonrecombinant vectors by this method. Further confirmation of clones is performed by picking up colonies, growing them and then performing restriction digestion after isolating plasmid with restriction enzyme will exhibit a pattern of DNA fragment which will be different from the parent vector.

III. IDENTIFICATION OF A CLONE FROM A GENE LIBRARY

A library is a collection of clones which contain all the genetic material of an organism. Two kinds of libraries are constructed nowadays in order to study the diverse genetic material:

1. *Genomic library:* A genomic library is a collection of clones sufficient in number to be likely to contain every single gene present in a particular organism. These are prepared by purifying total cellular DNA and then making a partial restriction digest, resulting in a collection of fragments that can be cloned into a suitable vector.
2. *cDNA library:* The *cDNA* library contains only complementary DNA molecules synthesized from mRNA molecules. It is a representative of all the mRNA molecules inside an organism. cDNA libraries still rank as a primary method by which researchers study gene expression. The advantage of cDNA library is that it represents the collection of expressed genes only. To prepare a cDNA library, the first step is to isolate the total mRNA from the cell type of interest. Because eukaryotic mRNAs consist of a *poly-A tail,* they can easily be separated. Then the enzyme reverse transcriptase is used to synthesize a DNA strand complementary to each mRNA molecule. After that the single-stranded DNA molecules are converted

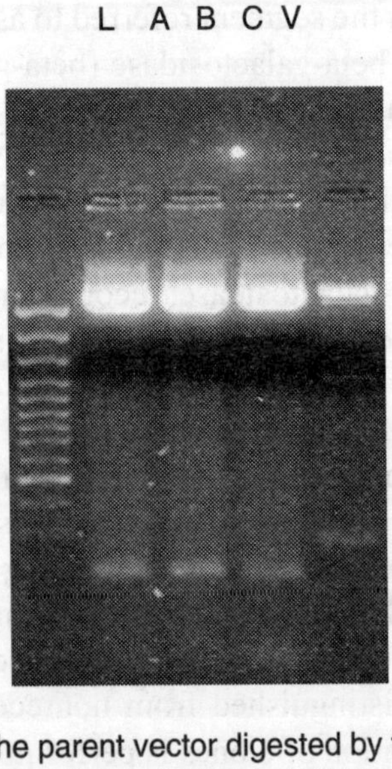

Fig. 24.4: The clones and the parent vector digested by the same restriction enzyme are shown in the figure. The pattern of digested DNA of clones A,B,C are different from the pattern of the vector labeled as V. A DNA ladder (L) is run along with the digested samples to know their sizes.

molecule. After that the single-stranded DNA molecules are converted into double-stranded DNA molecules by DNA polymerase and the resulting double stranded DNA is ligated into a suitable vector.

The desired clone from a library can be identified utilizing a number of procedures based either on the detection of the translation product of the cloned gene, or an easier method of identifying directly the correct recombinant DNA molecule. This can be achieved by an important technique of **hybridization probing.** Complementarity of bases between two single stranded nucleic acid molecules leads to the formation of a stable double stranded molecule. This nucleic acid hybridization can be used to identify a particular recombinant clone if a DNA or RNA probe, complementary to the desired gene, is available. It is not even necessary to purify each recombinant molecule. Instead an *in situ* probing method can be used. For this the colonies are transferred to a nitrocellulose or nylon membrane, then using various treatments, the DNA is denatured and fixed on the nitrocellulose or nylon membrane by a short period of high temperature or UV radiation. The membrane is then probed with a complementary DNA or RNA sequence, having a radioactive tag, which helps in its detection. These days non-radioactively labelled probes are more popular. Here the probe DNA is complexed with the enzyme horseradish peroxidase and is detected through the enzyme's ability to degrade luminol with the emission of chemiluminescence. The signal can be recorded on normal photographic film similar to autoradiography.

IV. APPLICATIONS OF GENE CLONING IN BIOTECHNOLOGY

Biotechnology can be unerstood as the use of biological processes in industry and technology. One of the reasons why this field had received so much attention during the last decade is because of our ability to clone genes. Although many useful products can be obtained from microbial culture, the list in the past was limited to those compounds naturally synthesized by microorganisms. Many important pharmaceuticals that are produced not by microbes but by higher organisms could not be obtained in this way. Gene cloning has changed this. Gene for an important animal or plant protein can now be taken from its normal host, inserted into an expression vector and introduced into a bacterium. If the manipulations are performed correctly the gene will be expressed and the protein synthesized by the bacterial cell. It may then be possible to obtain large amounts of the protein. Despite the development of sophisticated expression vectors, there are still numerous difficulties associated with the production of protein from foreign genes cloned in *E. coli*. These problems can be due to the sequence of the foreign gene or due to the limitations of *E. coli* as a host for recombinant protein (e.g. protein folding and post-translation modifications).

1. **Production of recombinant proteins** by eukaryotic cells: The problems associated with obtaining high yields of active recombinant proteins from genes cloned in *E. coli* have led to the development of expression systems for other organisms, such as yeast and filamentous fungus. For proteins with complex and essential glycosylation structures, an animal cell is the only type of host within which the active protein can be synthesized. A recent innovation in the recombinant protein production is to use a **transgenic animal**, an animal that contains a cloned gene in some or all of its cells, usually introduced by microinjection into a fertilized egg cell. The most successful approach to this has been to use farm animals such as sheep or pigs, with the cloned gene attached to the promoter for the animal's beta-lactoglobulin gene. Since this promoter is active in the mammary tissue, the recombinant protein is secreted in milk. Plants also offer a means of mass production of recombinant proteins. One very promising future possibility is that plants could be used to synthesize vaccines, providing the basis to a cheap and efficient vaccination programme.
2. **Production of recombinant pharmaceuticals:** A number of human disorders can be traced to the absence or malfunction of a protein normally synthesized in the body. Most of these disorders can be treated by supplying the patient with the correct version of the protein, which must be abundantly available. Gene cloning can be used to obtain large amounts of recombinant human proteins. A few examples of these are recombinant insulin, human growth hormone, recombinant factor VIII and IX, erythropoietin.
3. **Gene cloning** is also being used to identify and isolate genes responsible for certain human diseases. This provides an indication of the biochemical basis to the disease, enabling therapies for it. This technology can be used in **gene therapy**, which includes various methods to cure any disease by introduction of a cloned gene into the patient.
4. **In agriculture**, gene cloning provides a new dimension to crop breeding by enabling changes to be made to the genotype of a plant, circumventing the random processes inherent in conventional breeding. A good example of this is the development of plants that resist insect attack by synthesizing insecticides coded by cloned genes.
5. **It also provides a means** of developing methods for diagnosis of various disease conditions and is also used in forensic medicine for confirming a culprit using specialized techniques like DNA fingerprinting.

STUDY QUESTIONS

1. What strategy will you use to construct a cDNA library from rat brain and how will you select the clone of your interest from this library?
2. Describe the different kinds of vectors which are used in cloning?
3. List the applications of recombinant DNA technology?

MCQs

Select one most appropriate answer given for each questions below:

1. **If you want to check whether the expression of the Calpain gene is elevated in cancer of the liver or not, then which of these libraries from liver will you screen?**
   A. Genomic library
   B. cDNA library
   C. Either genomic library or cDNA library
2. **Which vector will you use for cloning and expression of a 2 kb cDNA in *E. coli?***
   A. Plasmid
   B. Cosmid
   C. Phage
   D. YAC
3. **In a cloning experiment, a cDNA fragment was cloned into the β galactosidase gene of a β lactamase carrying plasmid. After transformation in *E. coli,* the recombinant clones can be selected by plating them on LB-agar containing:**
   A. Ampicillin
   B. X gal
   C. Ampicillin + X gal
   D. Tetracycline.

---

1 B 2 A 3 C

PART IV:

APPLIED BIOTECHNOLOGY

•

Good Laboratory Practice *393*

•

Applications of Biotechnology *408*

•

Basic Tools in Bioinformatics *424*

•

Intellectual Property Rights *436*

25

Good Laboratory Practice

I. INTRODUCTION

A good laboratory practice (GLP) is essential for starting up and continuing research work in the area of cell biology, immunology, molecular biology and recombinant DNA technology. It requires planning, development of space, equipments and then handling, trained personnel, animal facility, working area and finally designing and conducting experiments, preparation of report, etc. In this chapter we will discuss about the basic components of GLP.

II. PLANNING

The major requirements of good biotechnology laboratory are the maintaining sterility and cleanliness. The introduction of laminar flow has greatly facilitated this requirement. Before a new laboratory is planned, several considerations need to be taken into account as below:

1. Consider for air inlet and extracts from the laminar flow hood.
2. Where the facilities for washing up and for sterilization of equipments, etc. be located? Will there be difference in the ceiling height between washing and sterilization area?
3. During planning the building, consider for ramp or elevator for carrying up the heavy equipments and machineries such as freezer, CO_2 incubator, ultra centrifuges liquid nitrogen cylinders, autoclaves, etc.
4. The location should be chosen carefully so as to avoid space constraints and awkward angles and projections in the rooms which limits flexibility.
5. Make sure that the door ways are wide and high enough to install equipments without much constraints.
6. Consider a separate animal room, sterilization and wash up room, storage room, quarantine and decontamination facility in addition to the proper experimental laboratory.
7. The preplanning about how many people will work in the laboratory, how long they will work and what kind of experiments they will do will determine

how much working area is required, how many equipments, pipettes and accessories they will share. It will give a fair idea of requirements of many minor equipments, space, working table, laminar flow hood, pipettes, etc. It will also give you an insight into how much area/size of the facility you require, what type of accommodation you require.

8. A standard biotechnology laboratory requires following accommodations.
   i. One or more aseptic area/room
   ii. An incubation room.
   iii. A dark room for fluorescent microscopy, photomicrography and gel documentation system.
   iv. A refrigeration room.
   v. A radiolabelling/radioisotope handling room.
   vi. A storage room.
   vii. Clean water and ice facility.
   viii. Decontamination and containment facility. A small laboratory plan has been shown in Figure 25.1.

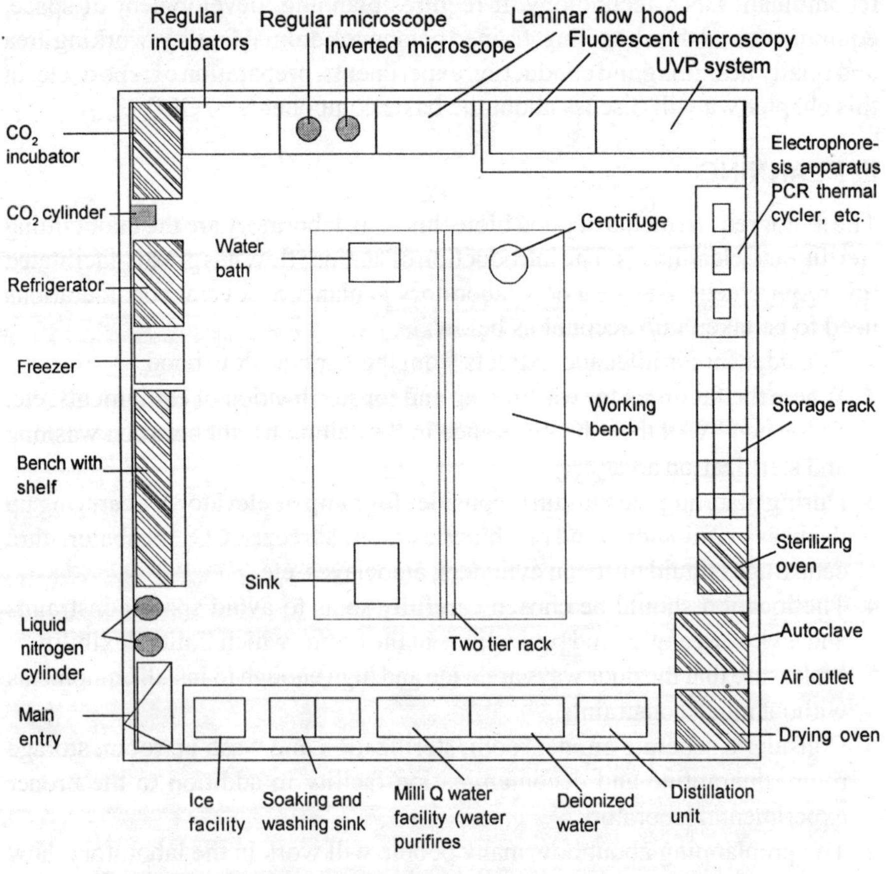

Fig. 25.1: Showing layout of a small biotechnology laboratory

A. Sterile Handling Area

Sterile work should be located in a quiet part of the laboratory, should be restricted to cell and tissue culture and PCR and there should be no traffic or other disturbance causing dust, etc. The work area should be in its simplest form. Nothing should be stored on the bench.

a. Laminar Flow

The use of laminar flow hoods affords a greater control of sterility at a lower cost and provides a separate sterile room. Individual free standing hood is preferred. The hood should be selected as per the available accommodation. Free standing or bench top, however, free standing hood should be selected in such a way that it provides sufficient leg space and accommodates pumps, aspirators and a small storage cabinet for pipettes, dispenser, syringes, tubes and disposables underneath. Chair should be of adjustable height, angled back and movable legs with wheels. In addition a small room or cubicle for use as a containment area will not be a bad idea. It will also require biohazard cabinet or pathogen hood.

b. Incubation

Incubation should be carried out in a separate incubator or thermostatically controlled room. Incubators lose more heat when opened frequently and recover slowly. Therefore, it is useful to kees them in a hot room or corner with other heating appliances. You may require humid incubator (water jacketed) or CO_2 incubator with regulated CO_2 supply.

c. Hot Room

The area for hot room need not be constructed rather it can be made using proper sublation. Avoid keeping wooden furnishing in the hot room as it absorbs heat and harbours microorganisms. A small bench with stainless steel or plastic laminate will be suitable for keeping flasks, minor equipments and note book. The temperature should be controlled at 37°C within the variation of ± 5°C. Heat should be supplied in the hot room via a fan heater. A second fan, positioned on the apposite side of the room and with the air flow opposing that of the fan heater, will ensure maximum hot air circulation. A thermostat is required for the maintenance of temperature. A temperature recorder should be installed to monitor temperature and should have a chart for recording temperature.

d. Cold Room

A cold room with proper insulation and maintaining low temperature will greatly reduce cost on other items such as refrigerators, refrigerated centrifuge and other minor equipments used in works at low temperature. It should have metallic (stainless steel) or plastic shelves to store chemicals, culture media and samples. The door of cold room should have automatic door closure and with proper gasket in order to minimize heating when it is closed.

e. Preparation

In a small laboratory, extensive preparation of media and solution should be avoided. It is useful if commercially prepared media, solutions, etc. should be used as far as possible. However, proper sterilization, pH and physiological conditions of the media and solutions should be checked. The reliable commercial media, solution and kits are difficult to obtain, in such cases a designated area should be used for preparations. If needed an extra horizontal laminar flow hood should be installed for such purposes. A sterile area should be used for sterilization/autoclaving media and accessories.

f. Wash-up

Wash up and storage facilities should be ideally located in a separate room with sufficient soaking and washing space. This room can be equipped with oven and drying facility, as well. The wash room should contain sufficiently larger sinks with hot and cold water tap, trolley for transfer of equipments and pipette washer, and a parking place for dirty and clean equipments/glasswares.

g. Storage

Sufficient storage facility should be provided for:

1. Sterile liquids
2. Sterile glasswares
3. Sterile disposable, plastic and gloves.
4. Disposable bags, containers
5. Filters, apparatus, screw cap, etc.
6. Liquid nitrogen tank or vessels.
7. Chemicals and reagents.

h. Containment Facility

The biotechnology laboratory should have a separate containment facility for handling infectious organisms. Depending upon the nature of the organisms the containment facility should be upgraded from P_1-P_4 facility. At least it should have double door fitted with air shower and a burn unit from which the burned air is exhausted out. The workers in this unit should be thoroughly protected by mask, gloves and disposable aprons. All the organisms should be handled within this confined space and nothing should come out of this facility without proper disinfection.

III. EQUIPMENTS MAINTENANCE AND HANDLING

Like any other laboratory, the equipments used in biotechnology laboratory can be divided into three categories: (i) Essential—the absolutely needed equipments without which laboratory cannot function; (ii) benificial—the work may be done quickly with less labour with these equipments, (iii) useful—these equipments make life easier, improve working and makes working environment more sophisticated and attractive.

A. Essential Equipments

Followings are the essential equipments to start up a biotechnology laboratory.

1. Laminar flow hood.
2. Shaker incubator
3. CO_2 incubators
4. Sterilization (Autoclave) and drying equipments
5. Refrigerators and freezers
6. Electronic balance
7. PCR thermal cycler
8. Centrifuges
9. Electrophoresis apparatus with power pack
10. Micropipettes
11. UV transilluminator
12. pH meter

1. Laminar-Flow Hood

Laminar-flow hood provides asceptic and clean environment for preparation of media, handling culture materials and even for PCR. Usually one hood is sufficient for 3-4 persons. Laminar-flow hood provides air curtain between user/environment of the lab and the materials to be used inside the hood. Two types of flow hoods are available, vertical and horizontal. A horizontal flow hood is cheaper and provides best sterile conditions for culture. It is basically suitable only for preparing medium, sterile re-agent, and for culturing non-primate cells. If you have to use hazardous materials such as radioisotope, carcinogen, human cell lines or replicating virus-containing cultures then you should use class I or class II biohazard cabinets.

Following points should be considered for procuring laminar-flow hood.

i. It should be large enough 4 ft wide × 2 ft deep.
ii. The front screen should be able to be raised, lowered or completely removed to accommodate materials.
iii. It should not make noise.
iv. It should be easily cleaned.
v. Comfortable to sit at.

In horizontal laminar-flow hood the air flow blows from the sides facing you, parallel to the work surface and is not recirculated. In the vertical model, the air flows down to the top on to the work surface and is either recirculated or vented. Horizontal flow hoods give the most stable air flow and best sterile protection to the culture and re-agents. For maintenance of the hood following points should be considered.

i. In this equipments a routine maintenance checks of the primary filters are required at 3-6 months intervals.
ii. With horizontal-flow hoods filters may be removed or discarded or washed with soap water.
iii. In every 6 months, the main 'High Efficiency Particulate Air (HEPA) filters should be monitored for air flow and holes.
iv. Weakly checks should be made below the work surface for spillage.
v. The work area should be swabed or wiped with 70% alcohol or 5% phenolic disinfectant.
vi. Continuous running of the flow-hood keeps the working area clean.
vii. While using switch off the ultraviolet light.
viii. Do not switch on ultraviolet light while your culture/plastic wares are inside the hood.

2. Incubators

If a hot room is not available, it is essential to buy an incubator. The incubator should have at least two shelves with a perforated and removable separator. Perforated separator allows even distribution of heat in both chambers while removable separator will allow you to keep larger flasks when required. The incubator should be larger enough, i.e. 1.5-6 ft 3 and it should have forced air circulation, temperature control within $\pm$ 0.2°C and a safety thermostat. Some incubators have a heated water jacket to distribute heat evenly around the cabinet. It is useful as it avoids colder spots as well as holds temperature for longer time in case with power failure and frequent opening of the door. It will be a good idea to have motorized incubator with a rotator for carrying out hybridization experiments.

If you require CO_2 environment for your culture, you can purchase a modular chamber in which a number of culture plates and flasks can be kept and 5% CO_2 gas with air can be introduced in the chamber after flushing out the initial air. It is fairly useful for short-term and even long-term cultures thus, purchase of costly CO_2 incubator can be avoided.

3. CO_2 Incubator

Although CO_2 incubators are costly, however, ease of use and superior CO_2 environment and temperature control justify their cost. Several brands of CO_2 incubators are available, however, before buying make sure that the supplier have required skill and workshop to repair it, since you may face frequent breakdown of different components in the Indian conditions.

It is advisable to clean and disinfect the interior of water jacketed and CO_2 incubators as there are chances of contamination if not handled properly.

4. Sterilization and Drying Equipments

The simplest and cheapest way of sterilization is by using *pressure cooker* that generates 100 kPa above ambient pressure. However, to comply for a good laboratory practice (GLP) one should have atleast a bench top autoclave with suitable programming and safety device. It will be ideal, however, to have at least two standing standard smaller model of autoclaves in each laboratory so that you have a back up in case with breakdown of one autoclave. Alternatively, one should have one larger pressure cooker and one autoclave per laboratory. Some precautions will be helpful for maintenance of the autoclave and for getting better service.

i. The autoclave should be frequently cleaned, gasket should be checked and replaced immediately in case with wear and tear.
ii. The chamber should be evacuated before sterilization (remove equipments, screw caps, swabs, etc) to allow efficient steaming for items to be sterilized.
iii. After sterilization, evacuate the chamber so that moisture, steam or air is removed.
iv. Use deionized or reverse-osmosis water in the autoclave.
v. Leave sufficient space around the autoclave for adequate ventilation, air extraction and removal of heat.

It will be useful to dry the autoclaved glasswares containers in the oven-in order to remove moisture. Table top ovens will be useful for a smaller or medium size laboratory.

5. Refrigerators and Freezers

Normally, a cold room is more useful and economical interms of space and if you have more people working in the laboratory. However, it is difficult to maintain cold room particularly in Indian conditions and in other temperate countries. In the absence of cold room, domestic refrigerators and freezers are sufficient for a medium biotechnology laboratory. It is advisable to have domestic refrigerator without freezer so that you have more space for your chemicals and reagents, enzymes or other items. In addition, it does not require defrostation. Another requirement is a freezer. Normally a –20°C freezer is sufficient for keeping blood and tissue samples and most of the re-agents. It is important to defrost freezers at least at 3 months' intervals. It is important to check freezers for unwanted or expired re-agents and samples so that you do not overburden your freezers. Refrigerators and freezers require less maintenance than the cold rooms.

6. Electronic Balance

One medium capacity electronic balance is a must for biotechnology laboratory. Different make of balances are available in the market. The balance should be kept in a corner where there is minimum disturbance. It should be properly calibrated and placed on an even surface. New balances are coming with levelling device. It should be cleaned regularly.

7. PCR Thermal Cycler

PCR thermal cyclers are of three types, (i) Regular thermal cyclers (ii) Gradient thermal cyclers and (iii) Real time PCR thermal cycler. However, regular thermal cycler is sufficient to carry out most of the experiments. A standard PCR thermal cycler has either 24/96 well block, it is programmable and have autorecovery of cycles in case with power failure (Fig. 25.2). Select a thermal cycler which can adjust eppendort vials of any make. The thermal cycler should have sufficient memory to store 4-5 programmes. Check the temperature to ensure that all wells have similar temperature. Secure the lid in case you run an experiment. The PCR machine should have interchangeable blocks, e.g. it should accommodate 24 well, 96 well and even slides in case of such requirement. It is better to keep PCR thermal cyclers under laminar-flow hood in order to avoid contamination.

Fig. 25.2: PCR-thermal cycler

8. Centrifuges

You may require to make cell suspension from cultures, isolation of DNA and RNA. In such situation, it is better to have one large size 4 × 1 litre or 6 × 1 litre capacity of medium speed (20,000 rpm) centrifuge and one microfuge for smaller size samples (DNA/RNA isolation). A bench type centrifuge will be most suitable.

9. Electrophoresis Apparatus and Power Pack

Electrophoresis apparatus (both vertical and horizontal) along with a suitable power pack is must for any biotechnology laboratory. Power packs are commercially available with several provisions. You must select a power pack which has minimum variations in power and voltage over extended time period. It is advisable to get the electrophoresis chamber of required dimension fabricated if commercially available electrophoresis chamber do not suit your requirement. Clean the electrophoresis chamber before using it and after use remove buffer from the tank and clean and dry it again. Frequent cleaning and drying will prevent deposition of salts in the tank and increases its efficiency. Figure 25.3 shows electrophoresis apparatus with power pack.

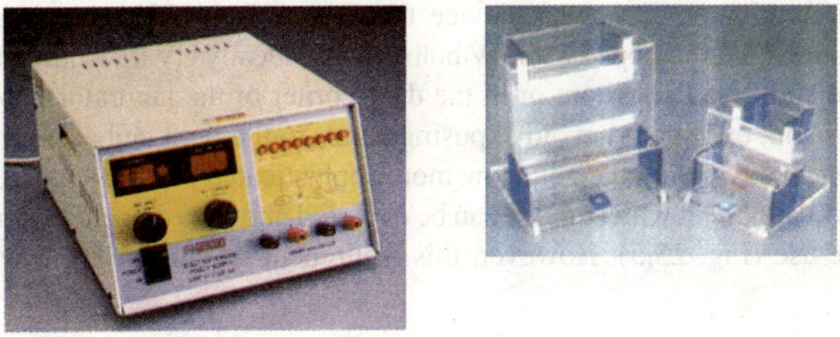

Fig. 25.3: Electrophoresis apparatus and power pack

10. Pipettes Micropipettes

Simple pipetting is one of the most frequently used and cheap device in the routine biochemistry immunology and microbiology laboratory but it is good only to dispense larger quantity of samples. In DNA and RNA work one requires smaller quantities in the fractions of 1 ml volume. Several commercially available fixed volume or variable volume micropipettes are available (Fig. 25.4). It is useful to have at least one 1-10 μl, one 10-20 μl and one 5-200 μl pipette in any biotechnology laboratory. The pipettes should be properly calibrated. In such pipettes, only tip needs to be sterile. While pipetting care should be taken to touch the liquid only with removable tip, the non-sterile shank must be touch the liquid. Put only required pressure to withdraw liquid and one step further while ejecting, so that total liquid is ejected out. It is advisable to carefully read manufacturers instructions before you start using the micropipette. Multidispensor and motorized pipettes are also available for repeated dispensing of larger volumes.

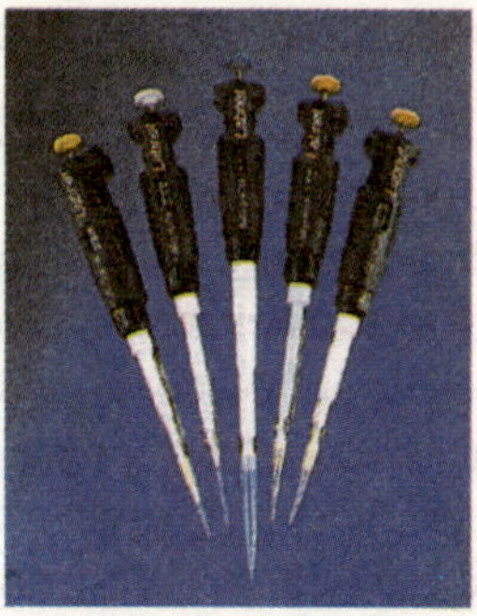

Fig. 25.4: Micropipettes

11. UV Transilluminator

UV transilluminator is a cheap device to observe DNA/RNA bands after electrophoresis. It is equipped with UV bulb or tube. Ideally UV transilluminator should be kept in a dark room or in the dark corner of the laboratory. When UV light is on, one should avoid exposing hands to the light and also protect your eyes when viewing the gel. Now more sophisticated computerized UVP system is available in which image can be captured and stored in the computer for future use (Fig. 25.5). However, this equipment is expensive but worth buying.

Fig. 25.5: UV transilluminator

12. pH Meter

pH meter is another basic equipment useful for any laboratory. It is less expensive. It also requires frequent calibration. The bulb should be kept in distilled water and should not be allowed to dry out. You will have to adjust sample temperature while taking the reading in the pH meter. Read manufacturer's instructions before using the equipment. If the bulb has been kept dry for a longer time, reactivate the sensor as per manufacturer's instruction before using it. Keep it off when not in use.

B. Benificial Equipments

Followings are the benificial equipments:

1. Inverted microscope
2. Cryostat
3. Refrigerated centrifuge
4. Multidispenser/Multipipettes/Pipetting aids
5. Liquid nitrogen cylinders
6. Magnetic stirrer
7. Roller racks
8. ELISA reader
9. –80°C freezer.

C. Useful Equipments

Useful equipments for a biotechnology laboratory are as follows:

1. Liquid scintillation counters
2. Fluorescence microscope/flow cytometers
3. Real-time PCR thermal cyclers
4. Ultra centrifuge
5. Low temperature cell freezer
6. UVP system/Gel scan
7. Video camera and monitor
8. Double beam spectrophotometer.
9. Luminometer
10. Electrophorator
11. Phosphor imaging system.

I V. TRAINING

It is vital that when a new staff or students join the laboratory they should be trained properly for maintaining cleanliness and hygeine, handling maintenance of equipments and risk as well as safety aspect. While training the staff a brief introduction to principles and theory is important. Once trained, the personnel should be initially allowed to work under supervision of instructor or senior experienced staff before they are allowed to work independently. In case with sophisticated equipments, the users should be trained by manufacturer's or company agents.

V. GENERAL PROTOCOL

Following general protocols should be followed while doing experiments.

1. Swab clean the working table, flow-hood with 5% detergent followed by 70% alcohol.
2. Bring all the required items such as solutions, pipettes, needles, syringes, tips, etc. and stack them on the table within your reach.
3. If glass items, open the lid of the container so that it is easily available as and when required.

4. If plastic items, remove the wrap (outer packaging) and stack them on the table individually.
5. Keep petridish, multi well plate, etc. closed with lid as longer exposure may contaminate them.
6. Keep a discarding flask for liquid waste and separate discarding boxes/ bags for plastics, needles, papers and paper towels, cottons, etc.
7. Use needle destroyer once you have finished using it and then discard it.
8. Keep a separate log book for each major equipments and always enter date, time (duration), user name and comments if any.
9. Always maintain an experiment record book. Enter date, name of the experiments, observations, results and your comments. If data is recorded in terms of charts, graph, photograph, etc. pin it on the same page of your note book.
10. When you are finished with the experiment, do not forget to swab clean the working place so that it is available to you or other worker neat and tidy.
11. Do not forget to label your solutions, re-agents, sample, etc.
12. Wear apron, gloves, etc. while carrying out your experiment.
13. It is advisable to keep marking pen, pencil, a small scissor/blade, timer, adhesive tape etc. in your apron pocket so that you do not have to look around at the time when you need them.
14. All the bottle, vials and, re-agent bottles should be wiped clean and stored properly after use.

These little points will ease your tension and make your life easy while carrying out experiments.

VI. SAFETY

Inspite of the scientific background and proper training of the personnel, accidents do happen-mostly due to a casual approach in handling regular biological, toxic and radioactive materials. It is, therefore, important to identify potential hazards but **do not over emphasize** the risk. Unnecessary overemphasizing may lead to disregard to the laid down guidelines and its non-compliance. Safety aspects constitute the followings:

A. Risk Assessment

Risk assessment is an important principle that has become incorporated into most of the safety legislations. One must consider following points while assessing the risk.

i. The way in which a hazardous material/equipment is used.
ii. Extent and frequency of use.
iii. General environmental condition.
iv. Who uses it?

v. Is users trained?
vi. What are the procedure for handling hazardous material?
vii. What precautions, clothing, protection the users take?

It is important that the biohazards are categorized properly, i.e. toxic chemicals, corrosive chemicals, solvent, ionizing radiation, carcinogens and mutagens, fire, electrical shock and sharp equipments/broken glass, etc.

B. Standard Operating Procedures

Hazardous substances, equipments and condition should be taken into consideration in order to develop a standard operating procedure. There are four different stages of SOP: procurement, storage, handling and disposal. Presence of one or more hazardous material within the procedure will compound the risk and may complicate the necessary precautions. In a biotechnology laboratory, one must consider devicing operating procedure for **genetically manipulated** products as well, which could be different from the general hazardous materials.

C.Safety Regulations

General safety regulations may be available with the local and central authority or with health and safety commissions. Some of the safety regulations are available with US Department of Health and Human Services, and UK guidelines may be obtained from the web site www.open.gov.uk/hse/condocs. In addition, several books and manuals are available on this subject.

D. Biohazard in Biotechnology Laboratory

Following hazardous materials must be conisdered while defining the safety procedure.

1. Sharp and heavy equipments.
2. Glasswares and other sharp items.
3. Chemcial toxicants
4. Gases (CO_2, O_2 N_2)
5. Liquid nitrogen
6. Fire
7. Radiation
8. High energy source
9. Human biopsy materials, blood, sputum, etc.
10. Genetic manipulations.

E. Containment and Disposal

A seal pathogen cabinet, laminar flow hood and lead separation can be used to contain infections and radioactive materials respectively. Toxic chemicals, irritants, carcinogens and mutagens must be used after wearing protective clothing masks and gloves.

Use of fume hood is advisable in case with vapours or toxic gas, etc. However, the need for protection against biological hazards is also defined by the suppliers/source of the materials as well.

Potential hazardous materials should be sterilized and autoclaved before disposal. Different commercially available disinfectants such as hypochlorite is often used to disinfect the reusable equipments/materials. In case with radio-active materials, properly follow the guidelines of radiation protection authority/ organizations.

STUDY QUESTIONS

1. What are the essential requirements of a biotechnology laboratory? With suitable diagram illustrate layout of a small biotechnology laboratory.
2. What essential equipments are required for a biotechnology laboratory. Describe thelaminar flow hood, its uses and maintenance.
3. What are the essential components of safety evaluation in a biotechnology laboratory? Name the sources of biohazard in a biotechnology laboratory.

MCQs

Select one most appropriate answer given for each questions below:

1. Which one of the following is an essential equipment for a biotechnology laboratory:

A. Refrigerated centrifuge
B. Laminar flow hood
C. Multi dispenser
D. Double beam spectrophotometer.

2. Autoclave is used for:

A. Automatic washing glassware
B. Automatic dispensing solutions
C. Sterilization equipments and glasswares, etc.
D. Automatic disposal of laboratory waste.

3. Which one of the following equipment comes under useful equipment catagory:

A. PCR thermal cycler
B. Low temperature cell freezer
C. UV transilluminator
D. Multi-dispenser

4. The abbreviation HEPA filter stands for:

A. High Energy powered Air filter
B. High Efficiency purified Air filter

C. High Efficiency Particulate Air filter
D. High Energy Powered Air filter

5. **Which of the following statements is false:**
   A. Horizontal laminar flow is cheaper and provides best sterile conditions.
   B. In horizontal laminar flow the air flow from the sides parallel to the work area and the same air is recirculated.
   C. In vertical laminar flow hood, the air flows down to the top on the work surface
   D. Vertical laminar flow hoods also provides sterility to the culture.

1 B 2 C 3 B 4 C 5 B

26

Applications of Biotechnology

I. INTRODUCTION

Biotechnology is broadly defined as "using organisms or their products for the benefit of mankind". Traditional biotechnology has been in practice for a very long time, it being used in baking breads, brewing alcoholic beverages and breeding food crops and domestic animals. Recent developments in molecular biology have given biotechnology a new prominence. Genetic engineering is being used in the production of pharmaceuticals, gene therapy and the development of transgenic plants and animals. Certain drugs and pharmaceuticals as well as diagnostic products are already in use. Environmental biotechnology processes make it possible to more efficiently clean up hazardous waste without the use of caustic chemicals. Industrial biotechnology applications have led to cleaner processes with lower production of wastes and lower energy consumption.

II. HUMAN APPLICATIONS

Diagnosis, prevention and treatment of diseases are all being transformed by advances in gene technology. Some of the important areas of applications in humans are as follows:

- Diagnostics, therapeutics and biopharmaceuticals
- Gene therapy
- Stem cells
- DNA fingerprinting.

1. Diagnostics, Therapeutics and Biopharmaceuticals

Diagnosis of infectious diseases is a profound application of new DNA technology. Tuberculosis, AIDS, papillomavirus and many other infectious diseases, in addition to inherited disorders like cystic fibrosis or sickle cell anaemia, are diagnosed within hours by the PCR technique. Diagnosing a disease early is often the key to successfully controlling it. Gene technologies means that diagnosis can be far quicker and more sensitive; the greatly increased sensitivity and speed of the PCR technique, as compared with traditional methods, allows earlier intervention and treatment. The power of PCR to amplify specific DNA sequences has made it a valuable tool in diagnosis of genetic defects and mutations. Diagnosis of diseases at DNA level permits (a) identification of carriers of inherited disorders (b) prenatal diagnosis and (c) early diagnosis before the onset of symptoms. For example, donated blood is screened to protect the blood supply from HIV and hepatitis. Monoclonal antibody based diagnostic tests are also available

For example home pregnancy tests are based on the detection of the pregnancy hormone hCG using monoclonal antibodies labeled with a color-producing enzyme that will cause the test window of the kit to turn color if the hormone is detected in a woman's urine. This basic kit design is used in many diagnostic kits; however, each kit is set up a little bit differently depending on the type of samples tested and the target consumer environment in which the kit will be used.

Biotechnology contributes to human health through the development of new vaccines. Vaccines help body to recognize and fight diseases. Subunit vaccines produced by biotechnological methods consist only of the antigen and not the whole organism. These vaccines are safer because they prevent the spreading of the disease causing organism itself. Currently, a vaccine for recombinant hepatitis B vaccine – Engerix-B (Smith Kline Beecham) is approved for human use. The vaccine is produced by inserting the gene responsible for producing the hepatitis surface antigen into yeast cells. During the fermentation process a large quantity of the antigen is produced which is then purified. When injected into the body, the antigen stimulates the production of antibodies that combats the hepatitis virus. Researchers are also working on vaccines to combat influenza, AIDS and herpes viruses.

Biotechnical methods are now used to produce many proteins for pharmaceutical and other specialized purposes. Hundreds of these drug products and vaccines are in clinical trials. Some of the US-FDA approved biotechnology drugs are shown in the Table 27.1.

2. Gene Therapy

Gene therapy is an experimental medical intervention that involves modifying the genetic material of living cells to fight disease. It is being studied in clinical trials (research studies with humans) for many different types of cancer and for other diseases. One of the goals of gene therapy is to supply cells with healthy copies of missing or altered genes.

The first disease approved for treatment with gene therapy was adenosine deaminase (ADA) deficiency. The enzyme adenosine deaminase is essential

Table 27.1

Product name	*Company*	*Application*	*Approved year*
1. Alphanate (human antihaemophilic factor)	Alpha therapeutics Corp.	Treatment of haemophilia A or acquired factor VII deficiency	1997
2. Alphanine (virus filtered human coagulation factor IX)	Alpha therapeutics Corp.	To prevent and control bleeding in patients with Factor IX deficiency due to haemophilia B	1996
3. Acctimmune (gamma interferon)	Genentech, Inc.	Approved for use in the management of chronic granulomatous disease	1990
4. Humulin (recombinant human insulin)	Eli Lilly & Co	Approved for treatment of diabetes	1982
5. Humatrope (Somatropin)	Eli Lilly & Co	Approved for treating human growth hormone deficiency in children	1987

for effective immune system function and children lacking it have a severe combined immune deficiency (SCID) and are vulnerable to every passing germ. Although ADA deficiency can be treated with a drug called PEG-ADA the drug is expensive and must be taken for life through the intravenous route. ADA deficiency was selected for several reasons:

- The disease is caused by a defect in a single gene, which increased the likelihood that gene therapy would succeed.
- The gene is always 'on' and its regulation is simple
- The amount of ADA present was not needed to be precisely regulated. Even small amounts of the enzyme were known to be beneficial while larger amounts were tolerated well.

The treatment consisted of removal of white blood cells from the child's body, letting the cells grow *in vitro*, inserting the missing genes into the cells and then infusing the genetically modified cells back into the patient's blood-stream. This procedure was not a cure since the treated white blood cells work only for a few months and the process must be repeated.

The biology of gene therapy is very complex and there are many techniques that still need to be developed and diseases that need to be understood more fully before gene therapy can be used appropriately.

One of several approaches may be used for correcting faulty genes:

- A normal gene may be inserted into a nonspecific location within the genome to replace a nonfunctional gene. This approach is most common.

- An abnormal gene could be swapped for a normal gene through homologous recombination.
- The abnormal gene could be repaired through selective reverse mutation, which returns the gene to its normal function.
- The regulation (the degree to which a gene is turned on or off) of a particular gene could be altered.

In most gene therapy studies, a "normal" gene is inserted into the genome to replace an "abnormal," disease-causing gene. A carrier molecule called a vector must be used to deliver the therapeutic gene to the patient's target cells. Currently, the most common vector is a virus that has been genetically altered to carry normal human DNA. Target cells such as the patient's liver or lung cells are infected with the viral vector. The vector then unloads its genetic material containing the therapeutic human gene into the target cell. The generation of a functional protein product from the therapeutic gene restores the target cell to a normal state. Some of the different types of viruses used as gene therapy vectors include retroviruses, adenoviruses, adeno-associated viruses, herpes simplex viruses, vaccinia virus.

The several nonviral options for gene delivery include:

- Direct introduction of therapeutic DNA into target cells. This is the simplest method but the approach is limited in its application because it can be used only with certain tissues and requires large amounts of DNA.
- Liposome mediated transfer of therapeutic DNA; the liposome is capable of passing the DNA through the target cell's membrane.
- Therapeutic DNA also can get inside target cells by chemically linking the DNA to a molecule that will bind to special cell receptors. Once bound to these receptors, the therapeutic DNA constructs are engulfed by the cell membrane and passed into the interior of the target cell. This delivery system tends to be less effective than other options.

Research is on at introducing a 47th (artificial human) chromosome into target cells. This chromosome would exist autonomously alongside the standard 46—not affecting their workings or causing any mutations. It would be a large vector capable of carrying substantial amounts of genetic code, and it is anticipated that, because of its construction and autonomy, the body's immune systems would not attack it. A problem with this potential method is the difficulty in delivering such a large molecule to the nucleus of a target cell.

Current gene therapy is experimental and has not proven very successful in clinical trials. Factors that have kept gene therapy from becoming an effective treatment for genetic disease include

- **Short-lived nature of gene therapy:** For a permanent cure it is essential that the therapeutic DNA introduced into target cells must remain functional and the cells containing the therapeutic DNA must be long-lived and stable. Problems with integrating therapeutic DNA into the genome and the rapidly dividing nature of many cells prevent gene therapy from achieving any long-term benefits.

- **Immune response:** The risk of stimulating the immune system in a way that reduces gene therapy effectiveness is always a potential risk. Furthermore, the immune system's enhanced response to invaders it has seen before makes it difficult for gene therapy to be repeated in patients.
- **Problems with viral vectors:** Viruses while serving as efficient vectors present a variety of potential problems to the patient like toxicity, immune and inflammatory responses, and gene control and targeting issues. In addition, there is always the fear that the viral vector, once inside the patient, may recover its ability to cause disease.
- **Multigene disorders:** Disorders that arise from mutations in a single gene are the best candidates for gene therapy. Unfortunately, some of the most commonly occurring disorders, such as heart disease, high blood pressure, Alzheimer's disease, arthritis, and diabetes, are caused by the combined effects of variations in many genes. Multigene disorders such as these would be especially difficult to treat effectively using gene therapy

3. Stem Cells

Stem cells have three important characteristics that distinguish them from other types of cells.

i. Stem cells are ***unspecialized.***
ii. Stem cells are ***capable of dividing and renewing themselves for long periods.***
iii. Stem cells can ***be induced to give rise to specialized cells*** under certain physiologic or experimental conditions

There are two kinds of stem cells in animals and humans:

A. Embryonic stem cells
B. Adult stem cells

A. Embryonic Stem Cells

Embryonic stem cells, as their name suggests, are derived from embryos. Specifically, embryonic stem cells are derived from embryos that develop from eggs that have been fertilized *in vitro.* The embryos from which human embryonic stem cells are derived are typically four or five days old and are a hollow microscopic ball of cells called the blastocyst. The blastocyst includes three structures: the *trophoblast-* the layer of cells surrounding the blastocyst; the *blastocoel*—the hollow cavity inside the blastocyst; and *the inner cell mass*—a group of approximately 30 cells at one end of the blastocoele.

Human embryonic stem cells are isolated by transferring the inner cell mass into plastic culture dishes that contain a nutrient culture medium. The cells divide and spread over the surface of the dish, the inner surface of which is typically coated with a feeder layer of non-dividing cells. The feeder layer gives the inner cell mass an adherent surface to which they can attach. Also, the feeder cells release nutrients into the culture medium. Efforts are being made to devise ways of growing embryonic stem cells without the mouse feeder cells that carry the risk of transmitting viruses or other macromolecules present in them to the human cells.

Over the course of several days, the cells of the inner cell mass proliferate and they are subcultured repeatedly for many months.The cells that have proliferated in cell culture for six months or more without differentiating, are peluripotent, and appear genetically normal, and are referred to as an embryonic stem cell line. Once cell lines are established, or even before that stage, batches of them can be frozen for future use.

A. *Adult/ somatic stem cells:* An adult stem cell is an undifferentiated cell found among differentiated cells in a tissue or organ, can renew itself, and can differentiate to yield the major specialized cell types of the tissue or organ. The primary roles of adult stem cells in a living organism are to maintain and repair the tissue in which they are found. Unlike embryonic stem cells, which are defined by their origin (the inner cell mass of the blastocyst), the origin of adult stem cells in mature tissues is unknown.

   Adult stem cells have been identified in many organs and tissues including brain, bone marrow, peripheral blood, blood vessels, skeletal muscle, skin and liver. There are a very small number of stem cells in each tissue. Stem cells are thought to reside in a specific area of each tissue where they may remain quiescent for many years until they are activated by disease or tissue injury.

B. *Embryonic vs.adult stem cells:* Adult and embryonic stem cells differ in the number and type of differentiated cell types they can become. Embryonic stem cells can become all cell types of the body because they are peluripotent. Adult stem cells are generally limited to differentiating into different cell types of their tissue of origin. However, some evidence suggests that adult stem cell plasticity may exist, increasing the number of cell types a given adult stem cell can become.

   Large numbers of embryonic stem cells can be relatively easily grown in culture, while adult stem cells are rare in mature tissues and methods for expanding their numbers in cell culture have not yet been worked out

   A potential advantage of using stem cells from an adult is that the patient's own cells could be expanded in culture and then reintroduced into the patient. The use of the patient's own adult stem cells would mean that the cells would not be rejected by the immune system. This represents a significant advantage as immune rejection is a difficult problem that can only be circumvented with immunosuppressive drugs.

   Embryonic stem cells from a donor introduced into a patient could cause transplant rejection. However, whether the recipient would reject donor embryonic stem cells has not been determined in human experiments.

C. *Potential uses of human stem cells*: Human stem cells can be used in basic research and in clinical research in many ways.

Studies of human embryonic stem cells: may yield information about the complex events that occur during human development and differentiation. It is known that turning genes on and off is central to this process. Conditions, such as cancer and birth defects, are due to abnormal cell division and differentiation. A better understanding of the genetic and molecular controls of these processes may yield information about how such diseases arise and suggest new strategies for therapy.

4. DNA Fingerprinting

DNA fingerprinting is the process of crossmatching two strands of DNA. Each person has a unique DNA fingerprint. A DNA fingerprint is the same for every cell, tissue, and organ of a person. It cannot be altered by any known treatment. Consequently, DNA fingerprinting is rapidly becoming the primary method for identifying and distinguishing among individual human beings.

A. Making DNA fingerprints: DNA fingerprinting requires following steps:

a. Isolation of DNA: DNA is recovered from the cells or tissues of the body. Only a small amount of tissue - like blood, hair, or skin - is needed. For example, the amount of DNA found at the root of one hair is usually sufficient.

b. Cutting, sizing, and sorting: Restriction enzymes are used to cut the DNA at specific places and the DNA pieces are sorted according to size by agarose gel electrophoresis.

c. Transfer of DNA to nylon: The separated DNA fragments are transferred to a nylon sheet by placing the sheet on the gel and soaking them overnight.

d. Probing: Adding radioactive or colored probes to the nylon sheet produces a pattern called the DNA fingerprint. Each probe typically sticks in only one or two specific places on the nylon sheet. The final DNA fingerprint is built by using several probes (5-10 or more) simultaneously.

B. Uses of DNA fingerprints: DNA fingerprints are useful in several applications of human health care research, as well as in the justice system.

i. Biological evidence: In criminal investigations, DNA from samples of hair, body fluids or skin at a crime scene is compared with those obtained from the suspects.

Paternity determination is possible because a child's DNA pattern is inherited, half from the mother and half from the father. To establish paternity, DNA fingerprints of the mother, child and the alleged father are compared. The matching sequences of the mother and the child are eliminated from the child's DNA fingerprint; what remains comes from the biological father. These segments are then compared for a match with the DNA fingerprint of the alleged father.

ii. **Personal identification**: Because every organ or tissue of an individual contains the same DNA fingerprint, the US armed services have just begun a program to collect DNA fingerprints from all personnel for use later, in case they are needed to identify casualties or persons missing in action. The DNA method will be far superior to the dogtags, dental records, and blood typing strategies currently in use.

iii. **Diagnosis of inherited disorders:** DNA fingerprinting is used to diagnose inherited disorders in both prenatal and newborn babies in hospitals around the world. These disorders may include cystic fibrosis, hemophilia, Huntington's disease, familial Alzheimer's, sickle cell anemia, thalassemia, and many others.

Early detection of such disorders enables the medical staff to prepare themselves and the parents for proper treatment of the child. In some programs, genetic counselors use DNA fingerprint information to help prospective parents understand the risk of having an affected child. In other programs, prospective parents use DNA fingerprint information in their decisions concerning affected pregnancies.

iv. **Developing cures for inherited disorders:** Research aimed at locating inherited disorders on the chromosomes depend on the information contained in DNA fingerprints. By studying the DNA fingerprints of relatives who have a history of some particular disorder, or by comparing large groups of people with and without the disorder, it is possible to identify DNA patterns associated with the disease in question. This work is a necessary first step in designing an eventual genetic cure for these disorders.

III. AGRICULTURAL BIOTECHNOLOGY

The goals of agricultural biotechnology are:

- Increase crop yield
- Improve crop quality
- Reduce production costs through decreased use of chemical pesticides and herbicides.

Biotechnology is proving its value in meeting these goals. The plants are genetically and physiologically more complex than single cell organisms and hence the necessary technologies are developing more slowly.

1. Improvements in Crop Yield and Quality

Modern biotechnology has offered opportunities to produce more nutritious and better tasting foods, higher crop yields and plants that are naturally protected from disease and insect pests. Similarly, research is in progress to engineer plants that would grow under adverse conditions such as saline soil, drought prone areas anaerobic and poor soil conditions.

Genetic engineering methods to improve fruit and vegetable crop characteristics such as taste, texture, size, colour acidity or sweetness and ripening process are being explored as potentially superior to the traditional method of cross-breeding that produces numerous non-selective changes. Genetic modifications have produced fruits that can ripen on vine for better taste, yet have longer shelf lives through delayed pectin degradation. By modifying the genes that control ripening, agronomists can provide supplies of seasonal fruits and vegetables for extended periods of time. Tomatoes and other produce containing increased levels of certain nutrients such as vitamin C, vitamin E and beta carotene help protect against the risk of chronic diseases such as some cancers and heart disease. Methods for improving field crops such as wheat corn and soybeans are also being developed. **Golden rice** has attracted much attention because it contains building blocks for vitamin A. The grain got its name because it glows with the golden colour of beta carotene, the yellow orange compound that gives carrots their colour and the world's most common source of vitamin A. Similarly, introducing genes that increase available iron levels in rice may supplement iron deficiency that causes anaemia in a large number of people.

Research in agricultural biotechnology is complicated by the fact that many of a crop's trait are encoded not by one gene but by many genes working together. Therefore, it is necessary to first identify all the genes that function as a set to express a particular trait. This knowledge can then be applied to altering the germlines of commercially important food crops.

2. Biopesticides

Biotechnology is central to the search for effective environmentally safe and economically sound alternatives to chemical pesticides. It is possible to insert genes in plants that give them properties that provide defense against insects. To transform a plant into a genetically modified plant, the gene that produces the trait of interest is identified and separated from the rest of the genetic material from a donor organism. A donor organism may be a bacterium, fungus or even another plant. For example in the case of **Bt-corn** the donor organism is a naturally occurring soil bacterium *Bacillus thuringiensis* (Bt) and the gene of interest produces a protein **Bt-delta endotoxin** that kills Lepidopteran larvae, in particular the European corn borer. The Bt-delta endotoxin is very selective generally not harming insects in other orders (such as beetles, flies, bees and wasps). For this reason genetically modified plants that have the Bt gene are compatible with biological control programmes because they harm the insect predators and parasitoids much less than broad spectrum insecticides.

The Bt endotoxin is considered safe for humans, other mammals, fish, birds and the environment because of this selectivity. Besides, since the bacterial pesticide is a protein that is fragile it degrades rapidly in the environment thereby prohibiting spread to ground water. Also, by reducing the need for pestcontrol, impacts and resources spent on the land are less, thereby preserving the topsoil. Bt-gene has been inserted in several crops like potatoes, corn, soybeans and cotton giving them built - in resistance to insects.

Dormant spores of **Bt** may also be dusted directly on the crops. The spores then become active and multiply, covering plants with bacteria poisonous to the target insects that feed on them. The Bt gene may also be put into endophytes that multiply within the host plant and move throughout the plant's vascular system forming a microscopic defense against the feeding insects.

3. Biofertilizers

Modern high yield agriculture entails consumption of vast amounts of chemicals for use as fertilizers and any means that will permit the plat to do this work itself would result in significant savings for the farmer. Leguminous plants produce their own source of usable nitrogen fertilizer by a process known as **nitrogen-fixation**. This process is made possible by a bacterium – *Rhizobium* that grows symbiotically on plant's roots. The nitrogen fixing bacteria capture the atmospheric nitrogen and biochemically convert them into water soluble nitrogen. This form of nitrogen is an essential nutrient for increasing the quantity and quality of plant yield. Research on *Rhizobium* and other nitrogen fixing bacteria and legumes may one day provide the basis for the modification of either the bacteria or non-leguminous plants, thereby making many crops more nearly self-sufficient in obtaining nitrogen.

IV. ENVIRONMENTAL BIOTECHNOLOGY

Environmental biotechnology is used in waste treatment and pollution prevention. Many microbes can more efficiently clean up many wastes than conventional methods and greatly reduce our dependence on traditional methods of land-based disposal.

Every organism ingests nutrients to live and produces by products as a result. Different organisms need different types of nutrients. Some bacteria naturally feed on chemicals and other wastes, including some hazardous materials. They consume those materials, digest them and excrete harmless substances in their place. This property of microorganisms is exploited in cleaning up of wastes in two ways:

- Bioremediation and
- Biotreatment

1. **Bioremediation** It uses natural as well as recombinant microorganisms to breakdown toxic and hazardous substances already present in the environment. This can be illustrated with an example. In 1975, a massive leak from a military fuel storage facility in South Carolina, USA released 80,000 gallons of kerosene-based jet fuel. The spill could not be contained inspite of extensive remedial measures and it soaked into the permeable sandy soil, reaching the underlying water-table. Soon ground water was leaching such toxic chemicals as benzene from fuel-saturated soils and carrying them toward nearby residential area. It became a serious environmental problem by 1985. Removing the contaminated soil was technically impossible and removing contaminated ground water did not address the source of contaminants.

   Bioremediation offered a solution to the problem because it was shown that microorganisms naturally present in the soils were actively consuming fuel-derived toxic compounds and transforming them into harmless carbon dioxide. Furthermore, it was shown that the rate of transformations could be greatly increased by the addition of nutrients. By stimulating the natural microbial community through nutrient addition, it was theoretically possible to increase rates of biodegradation and thereby shield the residential area from further contamination. This theory was put into practice and bioremediation worked. Similarly, oil-eating bacteria help to clean up oil-spills.

   Most of the cost associated with traditional cleanup technologies is associated with physically removing and disposing contaminated soils. Because engineered bioremediation can be carried in place by delivering nutrients to contaminated soil, it does not incur removal-disposal cost. The bacteria digest the waste at the site and turn it into harmless byproducts. After the bacteria consume the waste materials , they die off or return to their normal population levels in the environment.

2. **Biotreatment** It is a much broader term, which refers to all biological treatment processes, including bioremediation. Biotreatment can be used to detoxify process waste streams at the source—before they contaminate the environment—rather than at the point of disposal. For example, enzyme bioreactors are being developed that will pretreat some industrial waste and food waste components and allow their removal through the sewage system rather than through solid waste disposal mechanisms.

   Waste can also be converted to **biofuel** to run generators where, the byproducts of pollution fighting microorganisms are themselves useful. For example, methane can be derived from a form of bacteria that degrades sulfur liquor, a waste product of paper manufacturing. This methane can be used as a fuel or in other industrial processes.

V. INDUSTRIAL BIOTECHNOLOGY

Industrial biotechnology focuses on developing environmentally benign processes for the utilisation of renewable raw materials by using enzyme catalysis, by engineering metabolic pathways and by creating novel proteins. The biotechnological methods have led to cleaner processes with lower production of wastes and lower energy consumption.

The goals of industrial sustainability are:

1. Decrease material and energy input
2. Maximize renewable resources and biodegradable substances as input
3. Minimize generation of harmful waste products
4. Produce recyclable or biodegradable products

In addition to working toward sustainability by using biomass-based material and energy inputs, biotechnology offers us many options for minimizing the environmental impact of manufacturing processes by decreasing energy use and replacing harsh chemicals with biodegradable molecules produced by living things.

Living systems manage their chemistry more efficiently than man-made chemical plants, and the wastes that are generated are recyclable or biodegradable. Biocatalysts, and particularly enzyme-based processes, operate at lower temperatures and produce less toxic waste, fewer byproducts and less emissions than conventional chemical processes. They may also use less purified raw materials. Use of biotechnology can also reduce energy required for industrial processes.

Many industrial biotechnology products, such as beer, antibiotics or enzymes, are produced by growing microorganisms, under clean, highly controlled conditions. Some biotechnological processes use whole organisms (e.g. yeasts to make wine), while others use enzymes that have been purified from the organism that makes them (e.g. enzymes in detergents to dissolve grease and protein)

1. Biocatalysts

The biological products that offer us the greatest potential for decreasing the environmental impact of industrial manufacturing processes are the enzymes or biocatalysts. Enzymes are proteins produced by all living organisms. Some of the most common enzymes are proteases, which breakdown protein; cellulases, which breakdown cellulose; lipases, which act on fatty acids and oils; and amylases, which break starch down into simple sugars. In today's laundry detergents, enzymes such as proteases and amylases are some of the active ingredients and help to breakdown stains that are otherwise hard to remove with conventional surfactants alone.

About 60% of enzymes are the product of modern biotechnology, and new uses are being developed for environmental, food, agricultural and textile industry uses.

- Cellulases are complexes of enzymes which, together, breakdown the cellulose polymers of plant tissue. Cellulases produced by a fungus called *Trichoderma* are used for 'stonewashing' jeans – but their activity has to be carefully controlled to get the desired effect without destroying the cotton material. Changing the relative proportions of the enzymes in the cellulase complex creates different effects: true 'designer jeans'.
- Many enzymes have been genetically modified to improve their useful properties. For example, subtilisin (the bacterial component that is the key enzyme ingredient in 'biological washing powders) has been modified to make it more stable at temperatures used for washing, and less likely to be inactivated by the calcium-chelating 'builders' that are required for good performance of detergents.

Through bioprospecting, scientists are discovering novel biocatalysts that will function optimally at the relatively extreme levels of acidity, salinity, temperature or pressure found in some industrial manufacturing processes. Information from genomic studies of microbes is helping researchers capitalize on the wealth of genetic diversity in microbial populations. Researchers use DNA probes to fish, on a molecular level, for genes that express enzymes with specific biocatalytic capabilities. Once snared, such enzymes can be identified and characterized for their ability to function in industrial processes, and if necessary, they can be improved with biotechnology techniques.

2. "Green" Plastics

Green plastics are biodegradable, non-petroleum-based commercial wrappings that look, feel and act like traditional plastics, but breakdown later into organic components.These use plant material to replace petroleum based chemicals to make plastic and synthetic fabrics. One example is starch-based packaging generally made from agricultural commodities such as corn or potatoes. These dissolve in prolonged contact with water and heat. It takes weeks often months to break them down.

3. Biomining

In the past few years, the mining industry has been turning to the more efficient microorganisms to leach out minerals in an environmentally safe manner. For example, using a bacterium *Thiobacillus ferooxidans* to leach copper from mine tailings has improved recovery rates and reduced operating costs. Moreover, it permits extraction from low grade ores—an important consideration in the face of the depletion of high grade ores. Currently, 25% of all copper worldwide is produced through bioprocessing. The method is also being used to economically extract gold from very low grade sulfide gold ores, once thought to be worthless. Future development of the biomining process appears to be promising.

VI. OTHER APPLICATIONS

1. Human Genome Project

The structure of DNA was deciphered by James Watson, a geneticist, and Francis Crick, a physicist, thus marking the beginning of molecular biology in the 20th century. Their determination of the physical structure of the DNA molecule became the foundation for modern biotechnology, enabling scientists to develop new tools to improve the future of mankind. The Human Genome Project is a major biotechnological endeavor, the aim of which is to makc a detailed map of human DNA. The hereditary instructions inscribed in DNA guide the development of the human being from fertilized egg cell to death.

2. Adult Cloning

One area that has aroused considerable interest has been in the field of adult cloning. Until recently it was not possible to clone an adult, however, all that changed with Dolly the lamb that was cloned from a cell of an adult sheep. With this form of cloning (which is also known as *somatic cell cloning)*, a cell from a developed animal has to be fused with an egg cell that has been 'emptied' of its own genes. This is done by removing the nucleus from the egg cell.

For the livestock industry, adult cloning opens many possibilities. If a breeder has a prize animal he can replicate it without diluting its desirable features with those of the mate. Cloning can compress the time frame required to translate the fundamental discoveries into applications. It also offers a means of genetic 'rescue' where the genetics of flocks endangered by disease can be saved. Likewise, animals with special qualities nearing the end of their lives can be duplicated so that stud producers can continue to access their genetic value. There are also potential human applications in which cloned transgenic animals such as pigs could be used as a supply of organs for human transplantation, a process known as *xenotransplantation.*

3. Biosensors

Some of the more exciting frontiers of biotechnology include protein-based "biochips" which may replace silicon chips. It is believed that biochips would be faster and more energy efficient. Biochip implants in the body could deliver precise amounts of drugs to affect heart rate and hormone secretion or to control artificial limbs. Biosensors are monitors that use enzymes, monoclonal antibodies, or other proteins to test air and water quality, to detect hazardous substances, and to monitor blood components *in vivo.*

VII. CONCLUSION

With the increase in our understanding of genomics and genome manipulation, we now have the capability to make permanent genetic changes to many organisms that were previously not possible. The tools of genomics can alter the vigor and productivity of organisms, allow production of high value products in nontraditional production organism, raise or lower the susceptibility of a host to pathogens and pests, and modify the virulence of a pathogen. As we learn more about gene introduction and control of the expression of native genes, we will be able to better fight genetic diseases and make corrections in those individuals who are predisposed to other health and medical problems. The impact of this knowledge on our environment and our future is unknown.

It is widely accepted by the scientific community that biotechnology has great potential for contributing to the solution of many existing problems and to greater productivity in a wide range of industries. However, the power of modern biotechnology and the pace, scale and range of its application has raised justifiable concerns within the general public. These concerns include the safety of GM foods; the impact of GM crop production on biodiversity; potential health risks of using antibiotic resistance genes as markers in GM plants; ethical and moral issues, such as the acceptability of transgenic species, patenting of germplasm, human reproduction applications, animal cloning, threats to third world farmers, etc. Whatever the future of these particular ventures, it seems molecular biology and biotechnology will be important sciences of the coming century.

STUDY QUESTIONS

1. Name various applications of biotechnology. Describe one of them in detail.
2. Write short notes on
   a. Biopesticide
   b. Environmental biotechnology
   c. DNA fingerprinting
   d. Stem cell transplantation
   e. Biosensors
   f. Gene therapy
3. Describe in brief uses of biotechnology in Forensic Science and Biomedicine.

MCQs

Select one most appropriate answer given for each questions below:

1. ADA deficiency can be corrected by:

A. stem cell transplantation
B. Gene therapy
C. Biopharmaceuticals
D. DNA fingerprinting

2. **Stem cells are:**
   - **A.** Specialized cells
   - **B.** Differentiated cells
   - **C.** Unspecialized cells
   - **D.** Growth factor secreting cells
3. **DNA fingerprinting can be used to identify:**
   - **A.** Persons
   - **B.** Organs
   - **C.** Cells
   - **D.** Age of the person
4. **Which of the following statement is false:**
   - **A.** Every cell, organ, tissue of a person have same DNA fingerprint
   - **B.** DNA fingerprint cannot be altered by acid treatment
   - **C.** In DNA fingerprinting DNA, fragments are sorted according to electrical charge.
   - **D.** In DNA fingerprinting, DNA fragments are sorted according to size of the fragment
5. **Golden rice is important because it contains:**
   - **A.** Building blocks of protein fibres
   - **B.** Building blocks of vitamin A
   - **C.** Building blocks of nucleic acids
   - **D.** Building blocks of vitamin D
6. **Mouse feeder cells are required for growing human embryonic stem cells. However, feeder cells may provide one of the following complications:**
   - **A.** Contamination of mouse cells in human stem cells.
   - **B.** Transmission of viral gene into human stem cells
   - **C.** Apoptosis of human stem cells
   - **D.** Immunological reaction in human stem cells.

---

1 B **2** C **3** A **4** C **5** B **6** B

27

Basic Tools in Bioinformatics

I. INTRODUCTION

Computer and Internet has almost revolutionized the biological research. They help in managing enormous amount of biological data, discovery of newer relationship between biological molecules, genes and genome. Such informations have contributed to the development of a new branch of science called 'Bioinformatics'.

Now research papers, journals, nucleotide sequence, molecular structures are digitalized and distributed through centralized data base or journal websites. Many journals are published online. Nearly every research group have their own website offering various services such as software downloads to automated data processing services. The Internet has become an useful source of information for biological research. However, one must know where to find useful data, how to judge its usefulness and how to use softwares to perform required functions? In this chapter we will understand the art of finding suitable basic informations on biological research, DNA, RNA and protein sequences, genomic data, gene expression data by searching biological data base.

II. USING SEARCH ENGINES

A simple web search for *bioinformatics* will give you tens of thousands of results. Thus you will be searching a needle in a haystack. Dozens of search engines such as *Googles, Lycos, Alta Vista, Hotblot,* etc. exist that will help you to find your way around billions of pages that make up the web. Therefore, understanding how to structure your queries to weed out unwanted junk is very useful. Formulating Boolean querie limits the search space.

A. Boolean Searching

In order to search effectively, a simple way is to use Boolean logic. Boolean queries restrict the result that are returned from a data base by joining a series of search terms with the operators, AND, OR, and NOT. The meaning of these operators are:

i. Joining two key words with **AND** finds the documents that contain only *keyword 1,* and *keyword 2.*

ii. Using **OR** finds documents that contain either *keyword 1* or *keyword 2* (or both).

iii. Using **NOT** finds documents that contain *keyword 1* but not *keyword 2.*

iv. Some search engines consider or *space* an OR, so when you type for protein *space* structure or 'Excite' search engine you are asking for 'protein or structure' and you come up getting thousands of unwanted data (junk).

v. However, another search engine 'Google' defaults to AND, so you will find only references that contains 'Protein and Structure' and that is what you want to know.

Thus it is important to know that how the search engine you are using, interprets your queries.

vi. Boolean queries are read from left to right, like text.

vii. Parenthesin can structure more complex Boolean queries.

viii. Many search engines allow you to use quotation marks to specify phrases, e.g."*protein structure*".

ix. Suppose you want to know information on BRCA gene mutation in breast cancer by two authors, Satya N Das and TM Valarmathi. You might structure a Boolean query statement as follows:
[(BRCA gene AND "Breast Cancer") AND (Das OR Valarmathi)]

x. There are many web tutorials available on Boolean searching.

B. Search Engine Algorithms

The key features to look at in a web search engine is its data building and indexing, link to the connected pages and a sensible ranking strategy for results.

a. Hot Blot

It is probably the best search engine for new users. It is relatively comprehensive and regularly updated. It requires simple query statement.

b. Google

Google is extremely comprehensive and a favourite search engine for frequent users. It has indexing of over 1 billion VRLS. Pages are ranked and linked to well-connected pages/topics. It allows specialized services such as locating all the pages linked to the pages of interest. It copies the pages of your interest so that the content is accessible even if your server is offline. I found it most comprehensive and most useful. The home page of Google has been shown in (Fig. 27.1).

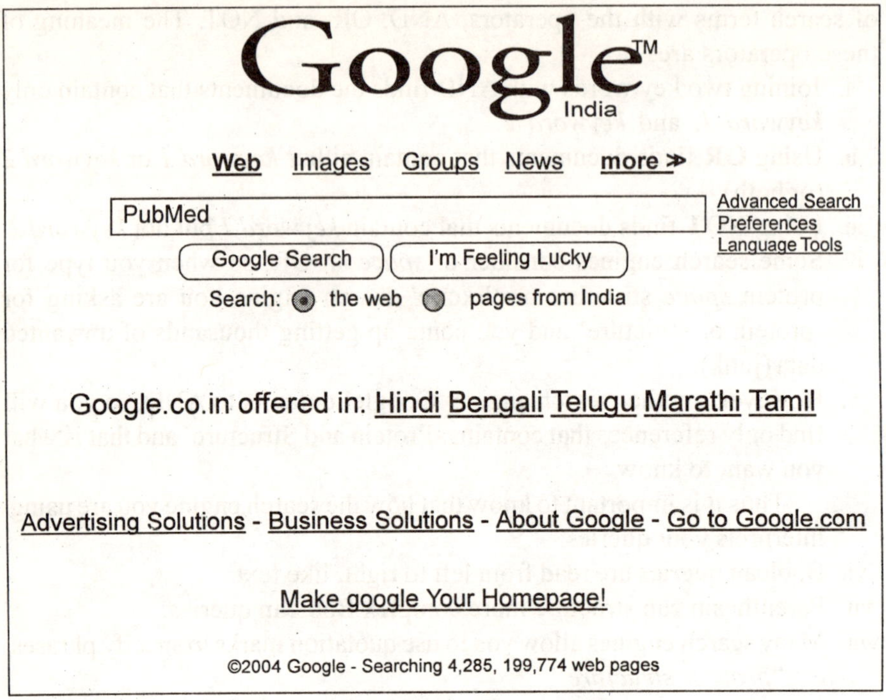

Fig. 27.1: Google search engine

III. SEARCHING SCIENTIFIC JOURNALS AND ARTICLES

The traditional of publishing an article is to conduct experiments, write results with background informations, interpretations, methods and appropriate references. The manuscript is critically reviewel by the editorial board and published in the journals. The abstract of the article is mostly freely available on the internet however, full articles are mostly available on the websites of the respective journals on payment or on subscription.

Another recent trend is online publication of the journals having no hardcopy counterpart. For example, *Human Mutation* an electronic publication contains full article which is available on the website of Wiley interscience. However, its hard copy counterpart contains only abstracts of the articles. Abstracts of the scientific papers published in this journal is also available in other resources such as PubMed.

An excellent resource for searching scientific literature is the free server sponsored by *National Centre for Biotechnology Information (NCBI)* at the National Library of Medicine, USA. This server can be accessed by anyone with a web brouser to search the Medline data base. Many government organizations and academic research groups have reference materials on their websites. NCBI has a good practical tutorial on use of the BLAST sequence alignment program. For biologists and biomedical scientists PubMed is another valuable resource available on the web containing almost 4000 listed journals related to biology, biochemistry, genetics and clinical publications for medical professionals. Quality of the journals are based on their *impact factor*. High impact factor journals are **Cell, Nature, Journal of Biological Chemistry, Science, etc.**

PubMed

PubMed (http://www.ncbi.nlm.nib.gov/entrez/query.fcgi) uses keyword based search allowing Boolean operators AND, OR and NOT in query statements. Users can specify field name, search term, key words, authors name, subject and year of publication for search. In addition, users can search Medical Subject Heading (MeSH). The MeSH browser (http://www.nlm.nih.gov/mesh/meshhome.html) allows users to enter a key word or word fragments to find key words in the MeSH library. PubMed automatically finds MeSH term related to query terms.

In PubMed, you can limit your search to Author Name or Subject, Journal and field (Figs 27.2a and b). Clicking AND, OR or NOT buttons joins the new term to your previous query using the specified Boolean operator. The preview/index menu allows you to build a detailed query. You can select the articles, see abstracts, save it in a file for offline preview or take print of the titles, abstract or summary of the articles. You can also find links to the specified article or related subjects. The URL of your search will appear on the location field at the top of the web browser, so that you can bookmark it.

IV. THE PUBLIC BIOLOGICAL DATABASE

In the field of biology at the molecular level, the nomenclature problem is immense so much so that only a scientist who works with a particular gene or gene product can recognize what the commen name refers to. There are many things to be known about a given gene such as: its source organism, chromosomal location, activator sequences, regulatory proteins, etc. There are various ways of its categorization, such as based on their function, product, RNA, metabolic pathway, structure, etc. Correct annotation of genomic data, i.e. putting sufficient information into the database is an active area of research. The solution has been to give each new entry into the database, a new serial number and then to store it in a relational database that knows proper linkage between the serial numbers, any number of genes, gene product and other informations about the genes.

A. Data Annotation and Formats

i. Data representation and distribution still remains a challenge in bioinformatics.
ii. Mostly data is represented in the form of nucleotide sequence of DNA, RNA and amino acid sequence of proteins.

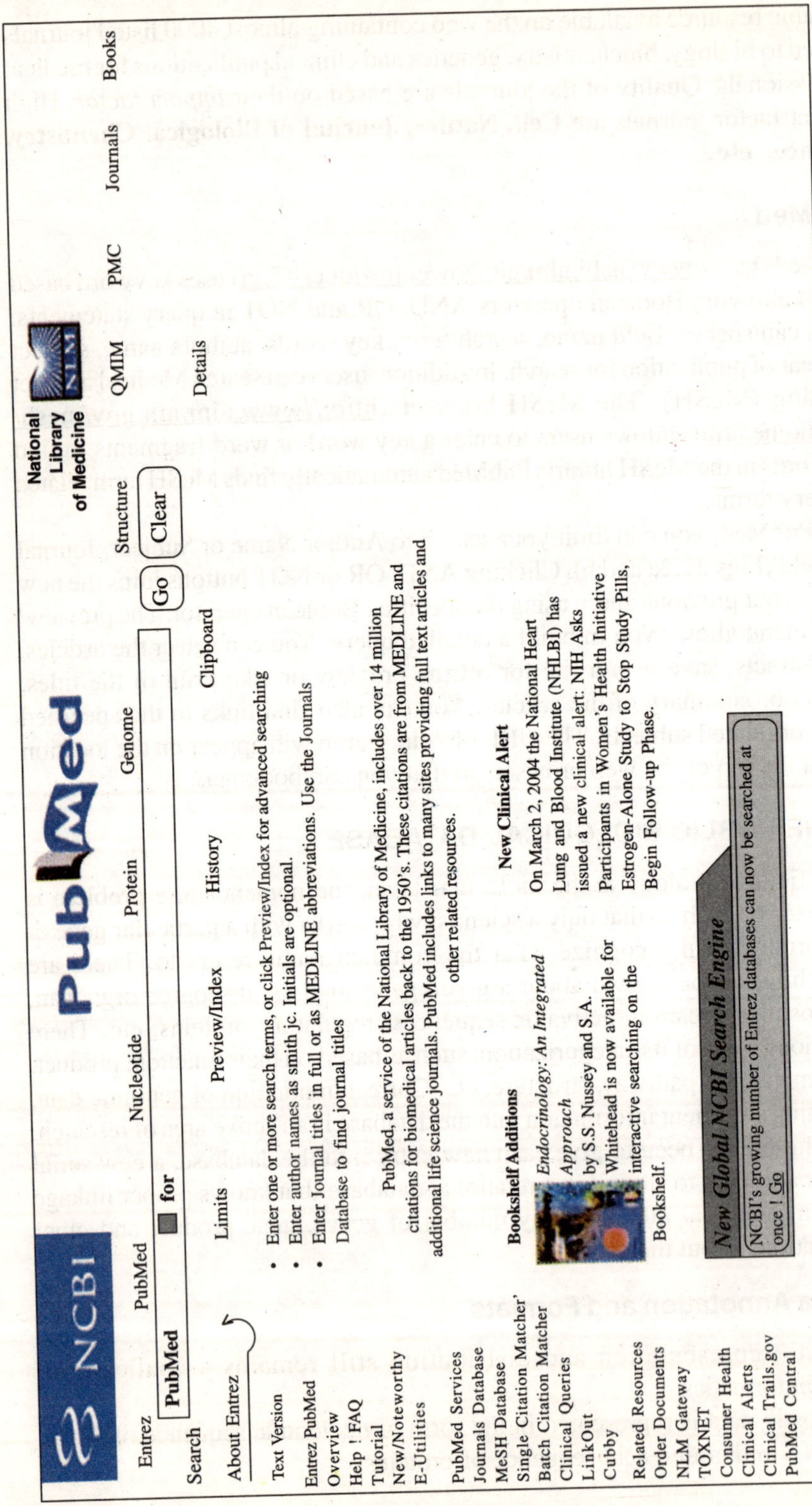

Fig. 27.2a: PubMed main menu

NCBI **PubMed**

Entrez PubMed Nucleotide Protein Genome Structure OMIM PMC Journals Books

PubMed Help

Last updated: March 26, 2004

- PubMed Searches
  - Author
  - Journal Titles
  - Automatic Term Mapping
  - Truncation
  - Phrase Searches
  - Additional Search Rules
    - Boolean and Syntax & Search Field Qualification
  - Subsets
  - Cookies
- Creating a Web Link
- Features Bar - Limits Preview/Index History Clipboard Details
- Documents
  - Display
  - Sort
  - Send to: Text File Clipboard E-mail Order
  - Import Citations into a Reference Management Program
  - Print
- Links
  - Related Articles - Computation Algorithm
  - Books
  - LinkOut (full-text)
  - NCBI Databases
- Cubby - Registration Stored Searches LinkOut Preferences Document Delivery Services
- Other Services
  - Text Version
  - E-Utilities
  - Journals Database
  - MeSH Database
  - Single Citation Matcher
  - Batch Citation Matcher
  - Clinical Queries
- Customer Support and Training Publications
- References
  - Search Field Descriptions and Tags
  - MeSH Subheadings & Families of MeSH Subheading Explosions
  - NLM Author Indexing Policy
  - Publication Types
  - Stopwords
  - MEDLINE Display Format
  - Grant Abbreviations and Institute Acronyms

http://www.ncbi.nlm.nih.gov/entrez/query/statoc/help/pmhelp.html

Fig. 27.2b PubMed help menu

iii. Biological sequence and structure databases have developed in parallel in USA and Europe.
iv. The use of proprietary software for data analysis has contributed a number of proprietary data formats—as given below:

a. *3D Molecular Structure Data*

i. The first public molecular biology data base was Protein Data Bank (PDB), the central repository for X-ray crystal structures for protein molecules.
ii. PDB contains data in the format of its crystal structure.
iii. However, protein structure shows redundancy.
iv. Many statistical studies of protein structure is based on sets of protein chains that have not more than 25% of their sequences in common.

b. *DNA, RNA and Protein Sequence Data*

i. Data can be acquired and stored for DNA, RNA and protein sequences.
ii. The first DNA sequence database was the Gene Sequence Data Base (GSDB) established in 1979 at Los Almos National Lab.
iii. Uptodat gene sequence information is still available from GSDB through the National Center for Genome Resources.
iv. Other data bases are European Molecular Biology Laboratory (EMBL) DNA Database of Japan and National Institute of Health (NIH) which is available through GenBank.

c. *Genomic Data*

i. In addition to Human Genome Project, there are now separate genome projects for other model organisms and plants.
ii. The major available genome data include viruses, bacteria, archea, yeast and commonly studied plant model system.
iii. Among animal kingdom, NCBI's Entrez Genome data base contains about 900 species including *C. elegans*, mice, rats, fruit flies, *Mycobacterium tuberculosis*, *Plasmodium*, puffer fish, etc.
iv. Among plante kingdom genomic data of *A. thaliana*, rice, maize, etc. are available.
v. These databases are evolving constantly.

d. *Biochemical Pathway Data*

i. A few research group have worked on developing database for functional studies of biochemical pathways of enzymes, hormones and other biologically active compounds.
ii. The important databases are "What Is There" (WIT) and Kyoto Encyclopedia of Genes and Genomes (KEGG).

iii. These databases contain reconstructed metabolic pathways of organisms where genomes has been fully sequenced.
iv. Other specialized databases focus microbial degradation pathways and intracellular signal transduction pathway.

e. Gene Expression Data

i. The recent development in the area of DNA-microarray technology has contributed significantly in the development of Gene expression data base.
ii. In this technique DNA are arranged on glass plate or DNA chips. Labelling the DNA or RNA with fluorescent molecules allows to measure expression levels of a particular gene.
iii. Such databases provide informations related to gene function.
iv. Such studies have laid the foundation of a new branch of genome biology called, *"Functional Genomics"*.
v. Two such important microarray data sites are, *National Human Genome Research Initiative's Microarray Project Site* and *Stanford Genome Resources Site.*

V. SEARCHING BIOLOGICAL DATABASES

There a number of databases available on the web. One should be aware of what data type is required for the individuals interest and what are the central data repositories for various data types. The two most established databases are NCCBI's GenBank for DNA sequence data and Protein Data Base (PDB) for molecular structures of proteins.

A. GneBank Database

1. NCBI in collaboration of EMBL provides the most complete collection of DNA sequence data.
2. It can be accessed at
   http://www.ncbi.nlm.nih.gov/entrez/viewer.fcgi?db=nucleotide
3. NCBI maintains data from every organism, every source, every type of DNA-from mRNA to cDNA clones to expressed sequence tag (EST) to high thoroughput genome sequencing data.
4. The main sequence types in a full GenBank search include:
   i. mRNA sequence—may be partial or complete sequence of transcription product of genomic DNA.
   ii. cDNA sequence—A DNA sequence generated by reverse transcription of mRNA.
   iii. Genomic DNA sequence—may contain both coding and non-coding sequence of DNA.

iv. EST sequence-short DNA sequence generated from mRNA extracted from a cell.
v. GSS sequence (Genome Survey sequence) single-pass sequence direct from the genome projects.

Each GenBank entry contains annotation-informations about gene's identity, source, gene, mRNA, CDS, translation product and exous as well as labeled with a specific accession number allotted to the sequence as shown below (Fig. 27.3).

b. Data Submission and Revision

Nucleotide sequence data can be submitted to GenBank database using local program called sequin and BankIt. The PC based program, sequin allows the user to annotate a sequence, perform analysis, and submit it to GenBank. Using sequin the outful files for direct submission can be sent to GenBank by email to *gb-sb@ncbi.nlm.nih.gov* or by mailing the submission file copied to floppy disk by post.

The Web based submission firm is called BankIt, available at *http://www.ncbi.nlm.niti.gov/BankIt/Index.html*. You have a choice to enter your sequence following the steps provided, or revise your submission.

PDB (Protein Data Bank)

1. PDB provides only one type of data, i.e. molecular structure of proteins.
2. It is accessible at: *http://www.rcsb.org/pdb/*.
3. A four letter PDB identifier can be entered directly for the data search.
4. The PDB makes data available in two formats: The legency PDB flat format and mmCIF data format.
5. Protein structure is automatically displayed using the installed and configured plug ins as shown in Figure 27.4.

c. *The Cardiff Human Gene Mutation Database*

1. This database includes known gene mutations responsible for a variety of human inherited diseases
2. It is available at *http://www.uwcm.ac.uk/uwcm/mg/hgmdO.htm/*
3. The data comprises various types of mutations in the coding regions of human gene.
4. The database does not include polymorphism data.

d. *Gene Cards: Encyclopedia of Human Genes Proteins and Diseases*

1. The Gene cords encyclopedia integrates information about human genes and their products
2. The data can be extracted from the following databases:
   i. GDB : Contains informations about genes and their features
   ii. MGD : Contains information about genes of laboratory mouse, markers, clones.

NCBI Entrez Nucleotide

Entrez PubMed Nucleotide Protein Genome Structure PMC Taxonomy Books

Search Nucleotide for Go Clear

Limits Preview/Index History Clipboard Details

Display default Show: 20 Send to File Get Subsequence Feature

1: AF507078.Homo sapiens IRCH...[gi:24431188] Links

LOCUS AF507078 328 bp DNA linear PRI 12-AUG-2003
DEFINITION Homo sapiens IRCHS6B breast and ovarian cancer susceptibility protein (BRCA1) gene, exon 20 and partial cds.
ACCESSION AF507078
VERSION AF507078.1 GI:24431188
KEYWORDS.
SOURCE Homo sapiens (human)
ORGANISM Homo sapiens
Eukaryota; Metazoa; Chordata; Craniata; Vertebrata; Euteleostomi; Mammalia; Eutheria; Primates; Catarrhini; Hominidae; Homo.
REFERENCE 1 (bases 1 to 328)
AUTHORS Valarmathi,M.T., Agarwal,A., Deo,S.S.V., Shukla,N.K. and Das,S.N.
TITLE BRCA1 germline mutations in Indian familial breast cancer
JOURNAL Hum. Mutat. 21 (1), 98-99 (2003)
MEDLINE 22384107
PUBMED 12497638
REFERENCE 2 (bases 1 to 328)
AUTHORS Valarmathi,M.T., Agarwal,A. and Das,S.N. TITLE Direct Submission
JOURNAL Submitted (30-APR-2002) Department of Biotechnology, All India Institute of Medical Sciences, Ansari Nagar, New Delhi, Delhi 110 029, India
FEATURES Location/Qualifiers
source 1..328
/organism="Homo sapiens"
/mol_type="genomic DNA"
/specimen_voucher="IRCHS6B"
/db_xref="taxon:9606"
/chromosome="17"
/map="17q21"
/cell_type="lymphocytes"
/country="India"
gene <1..>328
/gene="BRCA1"
mRNA <82..>165
/gene="BRCA1"
/product="breast and ovarian cancer susceptibility protein"
CDS <82..>165
/gene="BRCA1"
/codon_start=1
/product="breast and ovarian cancer susceptibility protein"
/protein_id="AAN61425.1" /db_xref="GI:24431189"
/translation="HDFEVRGDVVNGRNHQGPKRARESQDRK"
exon 82..165
/gene="BRCA1"
/number=20
variation 200
/gene="BRCA1"
/replace="a"
ORIGIN
1 tatatgacgt gtctgctcca cttccattga aggaagcttc tctttctctt atcctgatgg
61 gttgtgtttg gtttctttca gcatgatttt gaagtcagag gagatgtggt caatggaaga
121 aaccaccaag gtccaaagcg agcaagagaa tcccaggaca gaaaggtaaa gctccctccc
181 tcaagttgac aaaaatctcg ccccaccact ctgtattcca ctcccctttg cagagatggg
241 ccgcttcatt ttgtaagact tattacatac atacacagtg ctagatactt tcacacaggt
301 tctttttca ctcttccatc ccggccac

Fig. 27.3: Showing NCBIs DNA sequence data

DEPOSIT data
DOWNLOAD files
browse LINKS
BETA TEST new features
BETA XML files

RCSB PDB
PROTEIN DATA BANK

RCSB Home | wwPDB Home | Contact Us | Help

Did you find what you wanted?

Welcome to the PDB, the single worldwide repository for the processing and distribution of 3-D biological macromolecular structure data.

ABOUT PDB | NEW FEATURES | USER GUIDES | FILE FORMATS | DATA UNIFORMITY | STRUCTURAL GENOMICS | SOFTWARE | PUBLICATIONS | EDUCATION

Current Holdings

25004 Structures
Last Update: 06-Apr-2004
PDB Statistics

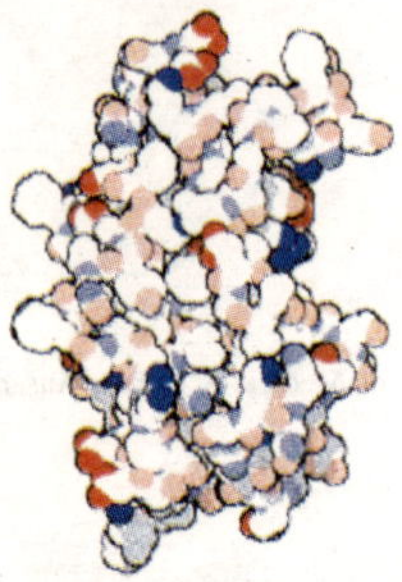

Molecule of the Month: Growth Hormone

The Protein Data Bank (PDB) is operated by Rutgers, The State University of New Jersey; the San Diego Supercomputer Center at the University of California, San Diego; and the Center for Advanced Research in Biotechnology of the National Institute of Standards and Technology — three members of the Research Collaboratory for Structural Bioinformatics (RCSB).

The RCSB PDB is supported by funds from the National Science Foundation (NSF), the National Institute of General Medical Sciences (NIGMS), the Department of Energy (DOE), the National Library of Medicine (NLM), the National Cancer Institute (NCI), the National Center for Research Resources (NCRR), the National Institute of Biomedical Imaging and Bioengineering (NIBIB), and the National Institute of Neurological Disorders and Stroke (NINDS).

Search the Archive

Enter a PDB ID or keyword [Search] Query Tutorial

○ PDB ID ○ Authors ◉ Full Text Search
☒ match exact word ☐ remove similar sequences

QuickSearch! search Web pages and structures
SearchLite keyword search form with examples
SearchFields customizable search form
Status Search find entries awaiting release

News

Complete News | pdb-l Archive
Newsletter | Subscribe

06-Apr-2004

A new 'QuickSearch' feature is available

After a period of beta testing, a new "QuickSearch" feature is now available on all production sites. This is a new keyword search feature which searches across the PDB archive and/or the RCSB PDB website static pages. [MORE...]

PDB Mirrors

Please bookmark a mirror site

San Diego Supercomputer Center, UCSD*
Rutgers University*
Center for Advanced Research in Biotechnology, NIST*
Cambridge Crystallographic Data Centre, UK
National University of Singapore
Osaka University, Japan
Max Delbrück Center for Molecular Medicine, Germany

OCA / PDB Lite MORE...

**RCSB partner*

In citing the PDB please refer to:

HM Berman, J. Westbrook, Z. Feng, G. Gilliland, T.N. Bhat, H. Weissig, I.N. Shindyalov, P.E. Bourne: The Protein Data Bank. *Nucleic Acids Research,* **28** pp. 235-242 (2000)

ABOUT PDB | NEW FEATURES | USER GUIDES | FILE FORMATS | DATA UNIFORMITY | STRUCTURAL GENOMICS | SOFTWARE | PUBLICATIONS | EDUCATION

© RCSB

Fig. 27.4: Home page of PDB

iii. OMIM : Contains a catalog of human genes and genetic disorders.
iv. SwissProt : Contains informations about proteins, their sequences cellular functions.
v. HGMD : A source of informations about disease-causing mutations in genes.
vi. Doctor's Guide to the Internet : A web service presenting news about biomedical research and its application
vii. Genatlas : A catalog of gene, markers and phenotypes with many links to major data sources.

3. Gen Cards is a project of the Weizmann Institute Genome Center and the Weizmann Institute Bioinformatics Unit.
4. It can be accessed at *http://bioinformatics.weizman.ac.il/cards*

STUDY QUESTIONS

1. Explain different data formats used in database. Describe in brief about GenBank database.
2. What do you understand by Search Engines. Describe any one Search Engine for searching scientific literature.

MCQs

Select one most appropriate answer for the questions given below:

1. PubMed is widely used for:
- **A.** Searching scientific literature
- **B.** Listing novels and textbooks
- **C.** Searching information about medical professionals
- **D.** Searching list of medicines for public use

2. GenBank:
- **A.** Is a repository for clones
- **B.** Is a repository for vectors
- **C.** Contains data on gene sequence
- **D.** Is a repository for genes.

3. Swiss Prot contains informations about:
- **A.** Gene sequence, functions
- **B.** Protein sequence, functions
- **C.** mRNA sequence and function
- **D.** Repository for DNA sequence

4. Impact factor of a journal indicates:
- **A.** Scientific quality of the journal
- **B.** Subscription of a journal by libraries and organizations
- **C.** Performance of journal in terms of marketing
- **D.** Publicity of the journal

---

1 A 2 C 3 B 4 A

28

Intellectual Property Rights

I. INTRODUCTION

The world trade has undergone a significant change in the last decade. Many new issues have come to the forefront to influence the trade patterns. The concerns of environment, biosafety, food and nutritional security, assertions of human rights, the sympathy for child labour, biopiracy of folklore and traditional knowledge-based concepts and innovations to mention a few, have been exerting dominant influences. Transfer of polluting industries to developing countries, using poor economies for pollution control, exploitation of cheap labour, out-sourcing from developing countries to the developed ones are all together new issues defying the relevance of the traditional theories of trade and the international boundaries.

At the same time, the measures for protecting of the inventions and intellectual property rights and its trade have gained momentum. These rights were being dealt with in the international trade even earlier, but the scope and coverage of trade-linked intellectual property rights became enlarged with the coming in being of World Trade Organization (WTO), Trade Related Intellectual Property Rights Agreement (TRIPS) and the measures such as protection of Geographical Indications got place in the charter WTO at the behest of developing nations.

R and D generates products and processes. These are protected using IP Rights. The products combined with the know- how forms a technology, which is transferred to the industry for production of the product for consumption by the public and industry, in the process generating the profits for the company and royalties for the inventor as well as host institute and project supporting institutions. The whole process may be represented as shown in Figure 28.1.

Intellectual property is a major component in what is traded during the process of technology transfer. To understand the underpinning better let us have an overview of what constitute IP rights, type of intellectual property rights, how to protect them, patents, criteria of patentability patent specifications, procedure, importance and uses of patent information.

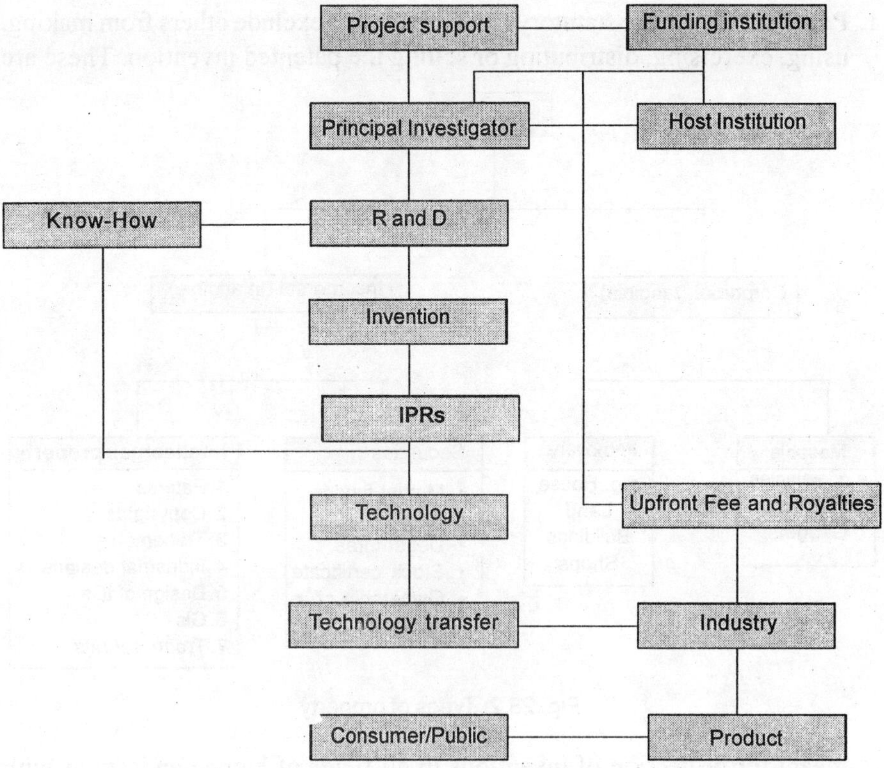

Fig. 28.1: The process of invention and its commercialization at R and D institutions

II. INTELLECTUAL PROPERTY

Property may be of two types, *i. Corporeal property* and *ii. incorporeal property*. This may be represented as in Figure 28.2. All the forms of property other than the intellectual property are well known and traded. Intellectual property is that property which results from the intellectual efforts of individuals and organizations. In other words, such property is the **"Creation of human mind and human Intellect"**. The ownership of property of any kind is always required to be protected in order to prevent such property from being misappropriated or misused. The rights associated with the protection of intellectual property are referred to as *intellectual property rights* (IPR). These are similar to any property consisting of movable or immovable thing wherein the owner may use his property as he wishes and no one else can lawfully use his property without the owner's permission.

A. Types of Intellectual Property

There are seven different forms of intellectual property, which are recognized by the Patent Offices all over the world. These are:

1. **Patents**: Patents are statutory rights granted to exclude others from making, using, exercising, distributing or selling the patented invention. These are

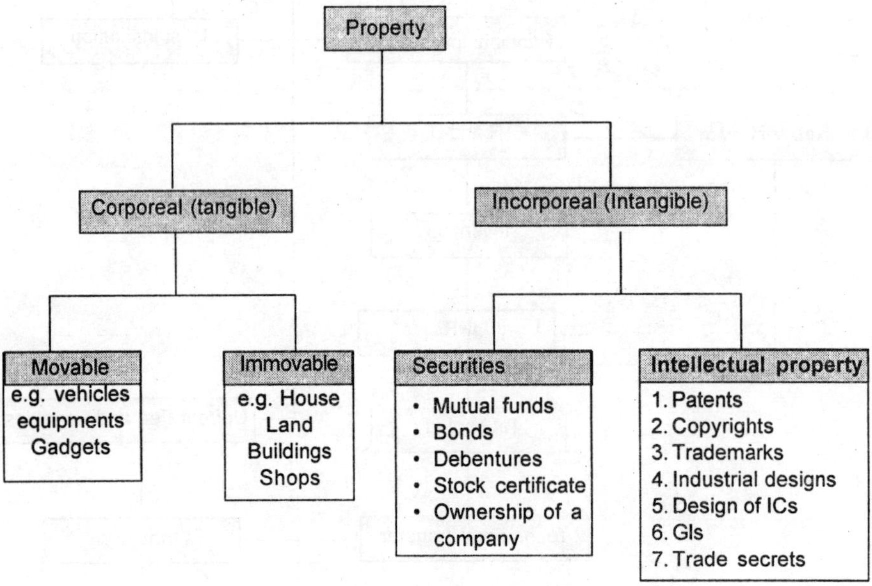

Fig. 28.2: Types of property

meant for protection of inventions in all fields of human endeavour with certain exceptions in certain countries. Patents result in a full public disclosure of a patented invention. They are limited by the territory of the country and are issued for a fixed period of time (20 years).

2. **Copyrights**: A right that protects original works of authorship fixed in a tangible medium of expression. Copyrights can include published and unpublished works—literary, dramatic, musical and dance compositions, films, photographs, audiovisual works, paintings, sculpture, and other visual works of art, as well as computer programs—from being copied. Copyright protects the expression of ideas (not the ideas themselves), and gives their authors exclusive rights to reproduce the copyrighted material. These rights are, therefore for protection of literary, artistic, dramatic, musical works, cinematographic films, recordings, broadcasts, softwares, research papers, journal articles, published books, magazine articles and other published materials.

3. **Trademarks**: A non-functional word, logo, slogan, symbol, design-or any combination of these—that distinguishes a product or service. Essentially, brand names and trademarks promote competition by giving products corporate identity and marketing leverage.

4. **Industrial designs:** Industrial designs refer to creative activity which result in the ornamental or formal appearance of a product and design right refers

to a novel or original design that is accorded to the proprietor of a validly registered design. The essential purpose of design law it to promote and protect the design element of industrial production. It is also intended to promote innovative activity in the field of industries. The existing legislation on industrial designs in India is contained in the New Designs Act, 2000 and this Act will serve its purpose well in the rapid changes in technology and international developments.

5. **Design of integrated circuits:** It provides protection for semiconductor IC layout designs. Layout design includes a layout of transistors and other circuitry elements and includes lead wires connecting such elements and expressed in any manner in a semiconductor IC. Semiconductor IC is a product having transistors and other circuitry elements, which are inseparably formed on a semiconductor material or an insulating material or inside the semiconductor material and designed to perform an electronic circuitary function.
6. **Geographical indications:** Geographical indications of goods refers to the place of origin of a product, which is characterized by certain specific qualities or reputation that the product has acquired due to the place of origin giving it a distinctiveness.
7. **Trade secrets:** Trade secrets are used for protection of undisclosed information. It may be a formula, pattern, a device, manufacturing process, method of doing business, or technical know-how that gives its holder a competitive advantage. Trade secrets cover a wide spectrum of information, including chemical compounds, machine patterns, customer lists and software. No Indian law protecting trade secrets exists. The subject matter of a trade secret must be secret. A trade secret can be an invention that is either patentable or non-patentable. The trade secrets involve no public disclosure of the information.

Of these patents are the most important form of protection for R and D organizations. In certain areas like computer softwares, databases, technical research papers, copyright is the most sought after protection. Trademark is used in combination with patent and or copyright to create brand name and brand equity. More recently patenting in the field of biotechnology has become important.

B. Protection of Intellectual Property

The owner of an invention can best protect his property if he keeps this out of the knowledge of rest of the world. But in this situation the progress of his research and that of the nation will be hampered and also the owner of the property will not be in a position to use his property at his will. Therefore, appropriate national legislations govern the intellectual property rights (IPRs).

The national legislation specifically describes the inventions, which are the subject matter of protection and those, which are excluded from protection.

For example, methods of treatment of humans and animals by surgery or therapy and inventions the use of which would be contrary to law or morality, or inventions injurious to public health are excluded from patentability in the Indian legislation.

IPRs are intangible rights, which grant an exclusive right to the inventor to impede others to freely exploit an invention. IPRs in general, and the patent system in particular, were established originally as a way to reward creativity and promote innovation. They allow the holder of the rights to recoup the investment in research, required for developing the new invention; in exchange the society receives the benefit of the disclosure of the new invention.

Scientists should always be cognizant to the legal ramifications of their research. Because of the nature of a scientific researcher's profession, IPRs could account for a large portion of his personal assets. This should prompt the researcher to be well versed in the procedures, which must be followed in order to protect intellectual property rights.

There are seven principal means by which IP may be legally protected: *patents, copyrights, trademarks, trade secrets, industrial designs, design of integrated circuits and geographical indications.* These forms of legal protection are primarily distinguishable on the basis of type of subject matter that each protects, as described in the previous section. Out of all these forms patents, trade secrets, copyrights and geographical indications are the IP rights, which may be associated with the biotechnological subject matter and only first two are used to protect IP rights of new inventions. The two forms of protection are mutually exclusive. Whereas patents result in a full public disclosure of the subject matter, the trade secrets require the invention to be kept confidential. If an invention is determined to be non-patentable, then trade secret is the only choice of protection of the invention. However, if an invention is considered to be patentable, then a choice has to be made between patent and trade secret. Figure 28.3 illustrates these options.

In biotechnology, we will be dealing with patentable invention all the time; therefore, we will deal only with patents in this chapter.

Governments grant patents to inventors on filing an application for an invention, which meet the statutory requirement of patenting. An inventor residing in one country can apply for a patent in other countries as well. Once a country grants a patent, the legal rights attendant this are enforceable, only within the territory of the granting country. Therefore, for protection of an invention in many countries, an inventor needs to obtain a patent in each country where he wishes to have protection.

Although the benefits bestowed on patentees are substantial and are immediately apparent, these are realized only when the patent is commercialized. The limited monopoly granted to the patentee is simply the small price that society pays in return for inventor's full disclosure of his new invention.

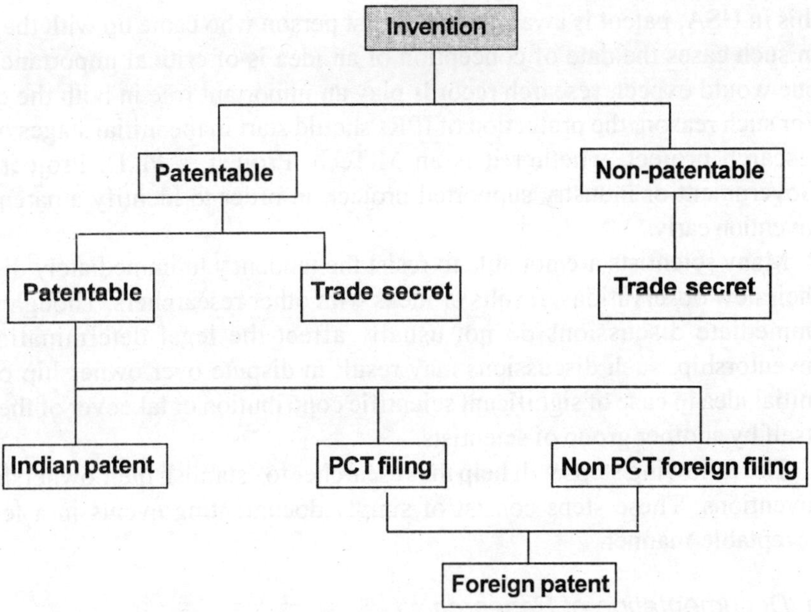

Fig. 28.3: Options for protection of new inventions

The basic statutory requirements for an invention to be patentable, the invention should be *i. Novel*, *ii. non-obvious* and *iii. useful*. For testing these conditions, the application has to pass through a lengthy and stringent procedure. Because of many barriers for obtaining a patent, a skilled patent practitioner, agent or attorney should carry out the patenting process. Every registered patent agent has a scientific background and has passed an examination testing the agents knowledge of patent laws. This examination is conducted by the Indian patent office every year in the month of November and the advertisement/ information for the same is available from the patent offices at Kolkata, Chennai, Mumbai and New Delhi six to eight months in advance. The information is also available through the website of the Controller General of Patents and Trademarks at http://www.ipindia.nic.in.

The protection of IPRs should be practicised right from the initial stages of any research project to its end. In addition, if during the course of working of an existing project, a researcher obtains a lead, an event or a result that satisfy the patentability criteria or leads to an idea for solving another problem, this new idea should also be handled in a manner which will preserve its legal identity as the researchers intellectual property.

One of the primary reasons for taking these early steps to protect IPRs is to keep the research records that will help to identify the invention at an early stage. In India, we have a first to file system like most other countries in the world. Patents are awarded to the first person who applies for a patent. Unlike this in USA, patent is awarded to the first person who came up with the idea, in such cases the date of conception of an idea is of critical importance. As one would expect, research records play an important role in both the cases. For such reason, the protection of IPRs should start in the initial stages of any research project, whether it is an M.Tech. Project, a Ph.D. Project or a Government or industry supported project, in order to identify a patentable invention early.

Many scientists are not able to resist the tendency to immediately discuss their new observations, results or ideas with other researchers. Though these immediate discussions do not usually affect the legal determination of inventorship, such discussions may result in dispute over ownership of the initial idea in case of significant scientific contribution or takeover of the idea itself by another group of scientists.

The following steps will help the researcher to establish their ownership of inventions. These steps consist of simply documenting events in a legally acceptable manner.

a. Documentation of Research

When a researcher first has an idea for solving a problem, he should record the idea in a bound laboratory notebook. A member of the laboratory can sign the notebook page as a witness and dates it. Any disclosure of the idea should be made only to those persons who need to know about it. The recording of ideas helps to clarify the honest confusion as to the inventor-ship, and it may discourage the recipients of the information from attempting to be identified as an originator of that idea. It will also help a patent attorney to prove the true identity of an inventor in subsequent legal proceedings.

Generally, research results are positive or negative based on the particular design of the work plan and the desired result. It is just as important to document work that gives a negative result as it is to documents work that gives a positive result. In case of negative results, it is possible that they may lead to another significant invention.

The primary purpose of recording research results is to prove that the particular specific research was done: (a) by a particular person, (b) at a particular time and (c) under particular conditions. In order for the recordation to effectively achieve this, the documentation must be:

i. Legible,
ii. Done in a pen, and
iii. Entered into a bound laboratory notebook.

Any deviation from these minimum requirements may cast doubts on the accuracy or authenticity of the record.

Data entered in a notebook should be accurate and legibly written. In-house memos as well as reports to supervisors and associates can also be used to document the details of the experiments. Therefore, such reports should also contain sufficient details to enable other researchers to understand the procedure discussed in the report.

b. Maintaining Confidentiality

The primary goal of any research programme is to produce result that will benefit the researcher, the employer, the funding agency and eventually the public. Protection of laboratory notebooks can be of great importance in the pursuit of this goal. Accurate and complete laboratory records facilitate efficient protection of the patent rights.

The researchers must take steps to protect the confidentiality of their work until the time comes when the legal protection in the form of filing of a patent is in place and a decision is made to publish or otherwise disseminate the work. From the beginning of the project, scientists should take precautions to prevent unwanted disclosure of their work. An inadvertent loss of confidentiality of research work/data can jeopardize efforts to obtain a patent protection. Similarly, an incomplete search of literature or patent database may also have the same effect.

A key part of this confidentiality is to limit access to research notebooks and research seminars. A researchers notebook should not be left unattended in an area that is accessible to persons outside the researchers immediate workgroup. If an outside person, even a researcher in the same institute or department, wishes to review or discuss research work done by another researcher or group, then such outside person should make the proper request to the researcher who is directly involved in the work.

c. Communication to the Colleagues and Associates

Communication of research work to associates can be in the form of discussions, seminars or written reports. Much can be gained from such discussions and, therefore, such communications are desirable. However, it is advisable to limit the number of persons receiving such confidential communications. Only persons who can contribute useful ideas or have a particular need to be aware of the subject matter should be invited to the seminars. Written reports should be sent to only these persons as well. There should be no recording or unnecessary copies made of the communication. Also care must be taken not to leave copies of a report in a place accessible to people outside the research group. Leaving a copy on a seminar table or in a library can lead not only to a loss of patent rights to an invention but also a premature disclosure of the confidential research work to the public.

Reporting or disclosure can be of two types viz., in-house disclosure and disclosure to people outside a research institute. Almost all scientists want to communicate the result of their work to their peers as soon as possible to gain some recognition and to add to their biodata another publication, which may be necessary to march forward in their career progression. But, if they do so prematurely they can lose all legal and patent rights to the invention inherent in it. A premature communication occurs when the communication precedes the establishment of a legal position on patent filing. However, if the communication is directed to a limited group of associates within the institute, then such communication does not normally constitute an impediment to obtaining patent rights. This kind of communication is least likely to cause a problem, when the group receiving the information is under some kind of obligation to the researcher to maintain confidentiality.

The disclosure to people outside a research institute may pose additional confidentiality problems. Therefore, a confidential relationship must be established before any communication is made. Thus limiting and controlling the communication of research is necessary for protecting the patent rights of inventions. There should be some communication in most instances, but it should be limited as far as possible and all communications should be done in a confidential manner. Figure 28.4 summarizes the steps to maintain confidentiality of work.

d. Establishing a Confidential Relationship

Before a researcher discloses confidential information to any person outside his research group, steps should be taken to insure that the recipient of the disclosure would keep the information in confidence. Impressing upon the disclosee that the information is, in fact, confidential, can do this or he must be informed in writing that the information must be kept in confidence. Also all documents associated with the disclosure should be clearly marked as

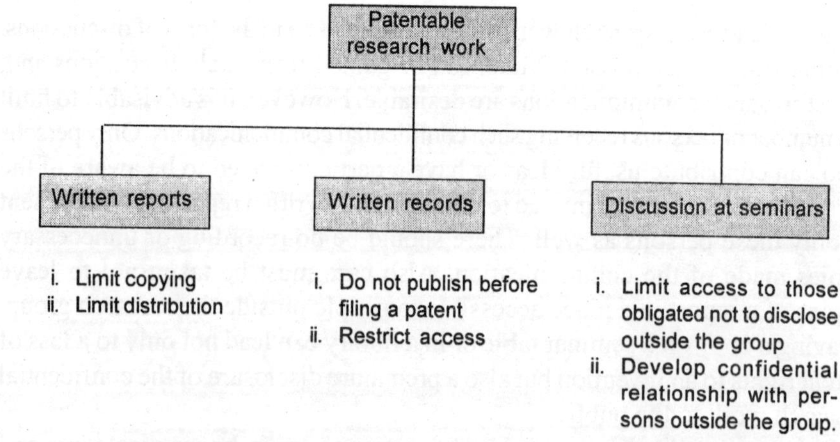

Fig.28.4: Steps to maintain confidentiality of research work

"Confidential". These precautions will remind the disclosee of his duty to keep the information in confidence.

When the information is to be transmitted to the people outside the institute, say to have test runs, or to have biological structures determined, the discloser should develop a confidential relationship with the disclosee, so as to transmit the information without any fear and to prevent unauthorized use of the confidential information. This involves the use of a **"Confidential disclosure agreement".**

A confidential disclosure agreement is a written agreement between the discloser and the disclosee of the information. Once both the parties sign the agreement, it is a legally binding contract. Any disclosure can be made after this to the disclosee and he is obliged to keep it confidential. If the agreement is drafted and executed properly, it can provide legal framework for protecting research work prior to obtaining formal patent protection. If there is a breach of agreement by any party receiving the information, then the usual legal remedies are available for the breach of contract. For example, a suit can be filed in a court of law seeking the damages and/or injunction to prevent further harm. These are procedures for maintaining confidentiality. However, when the researcher wishes to publish the research, then the confidentiality is short-lived, and the effects on legal protection can be significant. We will see this in the next section.

e. Publishing Research Work

A prime desire of most scientists is to publish their research work for peer review. However, premature publications can have certain adverse effects upon any subsequent attempt to obtain patent protection for the work disclosed in the publication. Therefore, it is very important for the scientist to realize the legal consequences of publishing research work.

The term publication has a specific legal meaning in the context of patents. A publication is a conveyance of information in a coherent way in a specified format that the information is available to the general public. It must have sufficient details to enable a person skilled in the art to practice the work. Such a publication is known as "an enabling disclosure". For example, if the publication pertains to making of a genetically modified (GM) product with enhanced protein content, and the details given are sufficient to enable a biotechnologist to make the GM product, then the publication is an enabling disclosure. Because it is an enabling disclosure, it destroys the confidentiality of the invention. Once the confidentiality is destroyed, it can no longer be recaptured. Therefore, if publication is being considered, the researcher should first make a decision regarding the preservation of patent rights. Figure 28.5 depicts the effect of publications on patents.

The date of publication of an article or paper is when the publication actually becomes available to the public, e.g. in a journal or on a website. The date an article is received for publication by a journal is not publication date. In India and in most foreign countries except USA, there exist a feature of absolute novelty, which means that any publication of an invention prior to the filing of a patent application bars the grant of a patent.

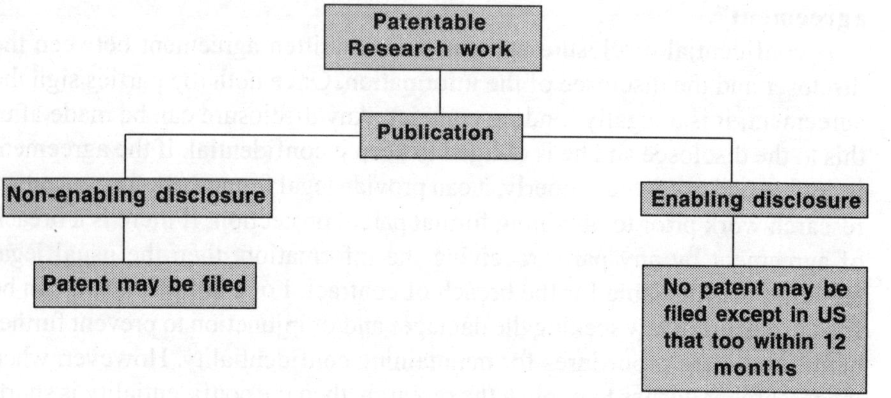

Fig. 28.5: Effect of publications on patent rights, with enabling and non-enabling disclosures

Thus, a premature publication can jeopardize both Indian and foreign patents and a full assessment of patentable invention for Indian and foreign patents must be done before publication. The relationship between publication and patent may be summarized as in Figure 28.6.

f. Prior Public Use

Like publication, prior public use is another way by which the confidentiality of the invention can be lost. The term "Public Use" like publication has a specific legal meaning. A public use occurs when an invention is used in such a way that the general public is not prevented from observing the use. The public use of an invention has important legal meaning. For example, under Indian law, if a complete invention is in public use, there arises a "statutory bar" to the subsequent patenting of the invention in India, i.e., a patent can never be issued for this invention. However, a public use, which is necessary to evaluate and perfect an invention, is referred as an "experimental use". An example of experimental use in biotechnology area is the testing of genetically modified (GM) crop in the field. Even when a GM crop is effective in green house tests, there may remain some uncertainty as to whether it will be effective in the field under the contemplated use conditions. Therefore, when a GM crop undergoes such field testing, it is advisable to clearly state

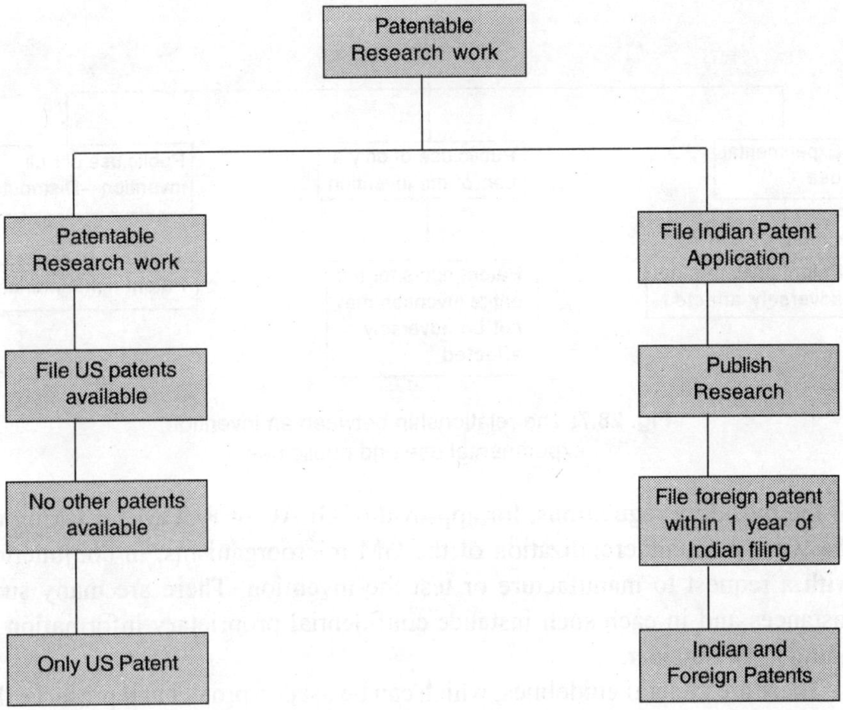

Fig. 28.6: The relationship between publication and patent

on the testing protocols that it is an experimental testing. It is only after such field tests are completed, and the GM crop has been shown to be effective, biologically safe and non-toxic for human consumption that public use may be made. The distribution of samples for consumer acceptance and efforts directed to marketing after effectiveness of the GM crop has been established, constitute public use and not experimental use.

The experimental use is also used in a completely different patent law perspective, i.e. in infringement determinations for granted patents. Use of a patented invention by an individual other then the patentee or licensee constitutes an infringement. However, a use, which is for experimental, non-commercial or research purpose, does not constitute an infringement. The relationship between an invention, its experimental use and public use is shown in Figure 28.7.

g. Disclosure to a Government Department

The scientists need to disclose the research work to a government department for a number of reasons including with a grant application, as progress report of the work for grant of next year, with a completion report or for taking care

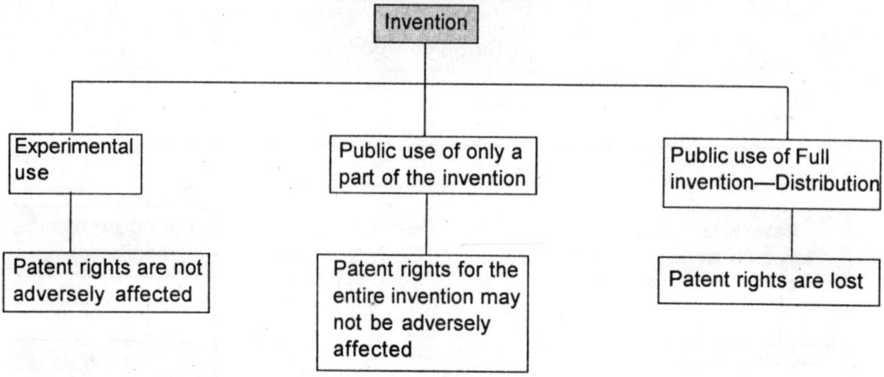

Fig. 28.7: The relationship between an invention, experimental use and public use

of the biosafety regulations, for approval of GEAC or RCGM for taking up the work or commercialization of the GM microorganisms, in conjunction with a request to manufacture or test the invention. There are many such instances and in each such instance confidential proprietary information is going to an outsider.

There are general guidelines, which can be used to protect and preserve the confidentiality when making such a disclosure to a government department. In most cases, the government department funding the research work also supports the filing of a patent application and is also one of the assignee of the patent. In such cases, there is obligation on the part of the department to keep the information confidential. Any disclosure to a government department should be sufficiently complete and accurate so that the department can make appropriate decision on the issues involved. The following guidelines should be followed when a submission is made:

i. The material, which is confidential, should be clearly marked.
ii. File a patent with the patent office or through the department, if the invention is patentable, before submission.
iii. Provide sufficient details to the government department for complying with the regulations, while retaining the confidential vital parts of the invention.
iv. Do not disclose the production yield data that is not relevant to the department's consideration.

The overall nature of the submissions to the government departments should reflect a total spirit of compliance. Any questions arising in a particular situation might best be addressed by communicating with the Officer concerned at the government department before making a submission. Finally, if a patent action is possible it should be taken before any submission.

III. PATENTING THE INVENTION

Once a decision is made to go ahead with a patent application, then participation by a patent attorney, whether through the institute, university, company or through the funding department, is virtually indispensable. The inventor should disclose all the invention details and closest prior art to the attorney. Patent attorney or the expert committee at the funding department will then ascertain the patentability of the invention. In addition, in biotechnology, there are often materials, such as microorganisms, which require special attention. Such living materials have to be deposited at one of the many culture depositories (International Depository Authorities–IDA) for a foreign patent (PCT) so that the culture is available to the public if the patent is granted. Institute of Microbial Technology, Chandigarh, has been designated as the IDA for the purpose of patent procedure after India's accession to the Budapest Treaty for the deposition of microorganisms in December 2001.

A. Depositing a Novel Microorganism

The filing of a patent application for a biotechnological invention involving the use of a novel microorganism necessitates making the microbe available to certain parties. In order to make a novel microorganism available to the public, the microbe has to be deposited with a culture depository before filing of the patent application. These depositories assign an accession number to the culture. By referring this accession number in the text of the patent application, a patentee makes that novel microorganism available to the public. The Budapest Treaty for deposition of microorganisms provides that a culture deposit in an acceptable culture depository recognized by all the member countries that require culture deposits in conjunction with patent application filing. This, then obviates the necessity of having to make deposits in every country where a patent application is filed. India became a member of the Budapest Treaty on 21st December 2001. No matter which depository is used, the advantage of depositing under Budapest Treaty is that only one deposit is required to fulfill the patent requirement of a number of countries. From 1977, when the Budapest Treaty came into force, there are more than 53 contracting states including India and 33 International Depository Authorities (IDAs) in various countries. Institute of Microbial Technology, Chandigarh has become 34th IDA, which has been recognized under the Budapest Treaty on 21st October 2002. IMTech, Chandigarh has state-of-the art facilities for deposition, storage and maintenance the cultures of bacteria, yeast, fungi and plasmids. There are several other laboratories in India, which have facilities for deposition of microorganisms in India. Few of them are as follows:

1. Indian Agricultural Research Institute, New Delhi (Plant Tissue Culture, Seeds, Germplasm, blue green algae).

2. National Centre for Cell Sciences (NCCS), Pune (Animal and human cell lines, tissues, hybrid cells, vectors)
3. National Dairy Research Institute, Karnal (dairy culture)
4. National Facility for Marine Cynobacteria, Tiruchanapalli
5. Forest Research Institute, Dehradun (pathgen collections, fungi).

B. Patenting Process in India

Once an application for patent is filed with the Indian Patent Office, a series of events ensues that can lead to the granting of a patent. This series of events is collectively known as the *prosecution* of the application. A knowledge of these events may help the inventor to aid the patent attorney during the prosecution of the patent application before the patent office and also help alleviate some of the anxiety of the inventor while the application is pending with the patent office.

a. Jurisdiction

In India, there are four Patent Offices at Kolkatta, New Delhi, Chennai and Mumbai with head office at Kolkata. All the head offices deal with the patent applications independently up to the stage of acceptance. However, head office issues a serial number, seals and grants the patent. The patents act 1970 does not allow any person to file application for patent in any of the offices .The applications are required to be filed according to their jurisdiction in accordance with the rule 4 of the Patent Rules 1972. If the application is filed in wrong jurisdiction, it will be rejected *ab-initio* and abandoned. The terrestrial limits for the purpose of jurisdiction has been shown in Table 28.1.

Table 28.1: Terrestrial limits for jurisdiction

Patent office	*States*
Patent Office Branch, Delhi W-5, West Patel Nagar, New Delhi	UP, Uttranchal, HP, J&K, Punjab, Haryana, Rajasthan, Delhi
Patent Office Chennai, Rajaji Bhawan, Chennai	Tamil Nadu, Karnataka, Kerala, Andhra Pradesh, Lakshsdweep, Pondicherry
Patent Office Mumbai Lower Parel (West), Mumbai	Maharastra, Gujrat, Madhya Pradesh, Goa, Daman, Diu, Nagar, Dadra Nagar Haveli, Chattisgarh
Patent Office, Kolkata Nizam Palace, Kolkata	Rest of India

b. The Persons who can Apply

Inventor himself can file an application for patent solely or in case of several inventors jointly. The application can also be filed by the assignee of the inventor claiming right to make application, solely or jointly or by legal representative of any deceased person who immediately before his death was entitled to make such application. In case, the application is made by the assignee, an assignment or endorsement duly signed by the inventor or inventors to be filed.

c. Kinds of Applications

An applicant can file an ordinary application under the Patents Act 1970 or an international application under Patent Cooperation Treaty (PCT). In case of International applications under PCT, the applicant is required to file application with receiving office at New Delhi, Chennai, Mumbai or Kolkata, as per jurisdiction or with Bureau of WIPO directly by paying the requisite fee. In case of ordinary application under the Patent Act 1970, the following types of applications can be filed:

I. Ordinary application with complete or provisional specifications.
II. Cognate application
III. Divisional application
IV. Patent of addition
V. Conventional application

d. The Documents Required

An applicant desiring to file an application for the grant of a patent in India is required to file the following documents in the Patent Office.

i. Application form
ii. Assignment or endorsement in case application is filed by assignee.
iii. Provisional or complete specification
iv. Declaration of inventor-ship if application was filed with provisional specification or complete specification
v. Statement and undertaking regarding corresponding foreign applications (optioned for Indian applicant).
vi. Power of attorney in case application is filed through registered patent agent/attorney
vii. Drawing, if any
viii. Abstract (Summary) of the invention
ix. Prescribed Fee (Table 28.2)

Table 28.2: The fee schedule

India

Item payable	*Form no.*	*Individual*	*Company*
Prov./Comp. Spec.	2	1500	5000
Request for Extention of Time	4	250	1000
Change of Applicant	6	500	2000
Notice of Opposition	7	1500	5000
Sealing of Patent	9	1500	5000
Certified Copy/Certificate		700	2500
On a petition		1000	3000

PCT International Fees

Basic Fee	-	US$ 382 (Upto 30 sheets)
Sheets in excess of 30	-	US$ 9 *
Designation Fee (Max 5)	-	US$ 82* (maximum 6, any designation over 6 being free of charge)

ISA Search Fee

Austria (At)	-	US$ 170
Australia (Au)	-	US$ 510
China (Cn)	-	US$ 100
European Patent	-	US$ 846
Sweden (Se)	-	US$ 866
USA	-	US$ 700
PCT Easy Reduction	-	US$ 125

* Fee reduced by 75% for natural persons from India.

e. Examination of Patent Application

The patent applications are taken up for examination by the Patent Office in the order of their filing. However, the divisional applications and patent of addition are taken up for examination on priority basis due to their shorter term. Also applications for inventions relating to food and drugs were taken up for examination on the priority basis when their term of patent was seven years from the date of filing of complete specification or 5 years from the date of sealing, whichever is shorter.

f. Opposition

When the application is accepted by the Patent Office, its acceptance is notified in the Gazette of India, Part III, Section 2. Any interested person can oppose the grant of patent within four months from the date of advertisement or notification of such acceptance in the Gazette or within such further period not exceeding one month in the aggregate (total 5 months). However, request for such extension must be filed before the expiry of such period of four months. The following grounds are available to any interested person for opposing the grant of patent.

i. The applicant for patent has wrongly obtained the invention
ii. The invention has been published before priority date or before filing date in any specification filed on or after 1st January 1912, or any documents published in India or elsewhere.
iii. The invention has been claimed in any of the complete specification published on or after the priority date of the applicant's claims for the invention.

iv. The invention as claimed was publicly known or used in India.
v. The invention is obvious and does not involve the inventive step.
vi. The invention is not an invention within the provisions of the Indian Patent Act 1970 or not a patentable invention.
vii. The invention is not sufficiently and clearly described in the patent specification.
viii. The application has failed to disclose the information u/s 8, regarding corresponding foreign applications or has furnished false information.
ix. The application for patent was not made within 12 months from the date of conventional priority date in case of conventional application.

g. The Patent Examiner

Once the patent application has been submitted to the patent office accompanied by certain formal documents and proper filing fees. The list of documents and the schedule of charges is shown in Table 28.2 above. Patent Office issues a patent application number (PAN) and a date of filing of patent, which is also known as priority date. The application is then routed to a particular examination group/specific examiner who is technically trained/ competent in the subject matter/area related to the patent application. For example, a patent application relating to biotechnology invention will be assigned to an examiner knowledgeable in the area of biotechnology. The examiner's primary role is to make sure that the invention in question fulfills all the statutory requirements of novelty, inventiveness and usefulness, under the patent law.

The patent examiner will first review the application to make sure that all formal documents/requirements have been met. For example, all patent applications must have specifications and one or more claims. The specifications generally consist of a title, an abstract, prior art or background of the invention, drawings and description of drawings and a detailed description of the invention. Once the examiner is satisfied that the requirements with respect to the format of the application have been satisfied, he goes to the next step. The next step is to evaluate the invention against the requirements of the Patent Act for a patentable invention. These statutory requirements are that the claimed invention be *i. novel*, *ii. non-obvious* and *iii. useful*.

For each patent application an examiner must:

i. Read and understand the invention.
ii. Search and read the prior art from all possible sources.
iii. Examine the publication of Indian patent specifications filed as or after 1st January 1912.
iv. Examine claimed invention in any claim of complete specifications filed in India, which is published on or after the date of filing of applicant's invention.

v. Compare the closest prior art with the invention claimed in the patent application.
vi. Prior public knowledge or public use of the invention, if any in India are also considered to ascertain novelty.
vii. The issue of obviousness is also considered along with the other grounds.
viii. Decide whether to allow claims or reject them under one or more sections of statute.
ix. Communicate the examination results via a written document called *first office action* to the patent attorney who filed the application.

The following exceptions for anticipation are also considered if the invention has been displayed publicly in exhibition or used publicly in India.

a. The application can be filed within six months from the date of such display in the exhibition.
b. The patentee or any person with the consent of patentee or applicant can file the application within one year from the date of working of invention, if such working was for the purpose of reasonable trial only.

The prior publication of the invention without the consent of the applicant or authorized person will also not be considered on anticipation. Further, the invention will not be deemed to have been anticipated if it was communicated to the Government or any person authorized by the government to investigate its merits.

h. Grant of Patent

When the complete specification in support of application for patent has been accepted and no opposition is filed or where the application was opposed and the opposition was finally decided in favour of the applicant or application has not been refused by the Controller under the Act, an applicant is required to file a request for sealing along with the prescribed fee at the Patent Office Kolkata, within six months from the date of notification of acceptance of such application in the Gazette of India Part III, section 2 or within extended period of 3 months in aggregate. If such fee for sealing is paid, the Controller shall cause the patent to be sealed with the seal of the Patent Office and the date on which the patent is sealed, shall also be entered in the register of grant. The patent, then, stands granted. The patent is granted for a period of 20 years from the date of grant for any invention.

i. Maintenance of Patent

The Patent Act 1970 not only provides for the rights of patentee u/s 48 of the Act but also provides certain duties to be performed by the patentee when the patent is granted. They are as follows:

1. The patentee is required to pay the renewal fee as prescribed within the prescribed period to maintain the patent to keep it in force failing which

the patent will cease to have effects on the expiry of the period .In other words, the patent will lapse. However, the period prescribed for the payment of renewal fee can be extended to such a period not being more than six months on making the request in the prescribed manner. In this connection, it may be noted that no renewal fee to be paid in respect of the 1st and 2nd years of the term of patent. However, the patentee is not entitled for damage if infringement occurs during the period when the patent was not in force. No renewal fee is required to be paid in respect of the patent of addition. The renewal fee schedule is shown in Table 28.3.

Table 28.3: The renewal fee schedule

Renewal Fee to be paid for the year	*Individual*	*Legal Entity*
3rd to 6th year	600	3200
7th to 10th year	1500	4500
11th to 14th year	3500	10000
15th to 19th year	5000	15000

2. The patentee or his licensee is required to furnish information periodically every year regarding the commercial working of patent in India. In case the patentee does not give such information, the Controller may publish such patents as "not commercially worked" in India, in the Gazette of India. In such situation, the Controller may grant compulsory license to any person if he has filled application for such compulsory license.

IV. NATIONAL LEGISLATION

The various rules and regulations related to patent are listed in this section. Patent Office has its own website at http://www.patentoffice.nic.in. The website provides information regarding the following matters:

a. Patent Act 1970
b. Patent Rules 1972, and as ammended in 1996 and 1999. These rules contain various forms required to be filled and various fees required to be paid from time to time.
c. A manual of Patent Office Practices and Procedures, which gives information regarding practice followed by the patent office regarding patentability, PCT applications, etc.
d. Frequently asked questions (FAQ) and their answers regarding Geographical Indications.
e. Current News and update.

V. INTERNATIONAL TREATIES

India is a member of the following international Treaties concerning with patent matters:

a. Convention to establish the World Intellectual Property Organization
b. Paris Convention for the Protection of industrial property with effect from 7th December 1998
c. Patent Cooperation Treaty with effect from 7th December 1998
d. WTO with effect from 01.01.1995
e. Budapest Treaty on international deposition of microorganisms for the purpose of patent procedure with effect from 17.12.2001.

VI. RELATED WEBSITES

The following websites provide information about the patents in the respective countries and the patent databases for a free patent search and full patent documents:

http://www.uspto.gov : USPTO

http://www.wipo.int : WIPO. This provides linkages to the websites of the various intellectual property offices of the world.

http://www.epo.org : EPO

STUDY QUESTIONS

1. Write in brief about the patenting process in India. What are the conditions for opposing a patent application?
2. Write short notes on the following:
   a. Roles of a Patent Examiner
   b. Maintenance of a Patent
   c. Intellectual Property
   d. Limitations of a patentable research work

MCQs

Select one most appropriate answer given for each questions below:

1. Intellectual Property Rights (IPR) protects:

A. Products only
B. Products, process and ideas
C. Products and process
D. Process and ideas only

2. Patents granted for an invention is valid for:

A. 10 years
B. 15 years
C. 20 years
D. 25 years

3. In India 'Patent Act' was inforced in the year:

A. 1950

B. 1960

C. 1970

D. 1980

4. Which one of the following statements is false:

A. A new treatment method of human disease can be patented under International Patent Law.

B. Inventions that are injurious to public health cannot be patented.

C. Intellectual property rights were established to reward creativity and to promote innovations.

D. IPR allows the researcher to recoup their investment.

1 B 2 C 3 C 4 A

Index

A

Acid-base properties of the amino acids 51
Active and passive immunization 263, 264
Acute rejection 230
Adenosine derivatives 70
Adhesion molecules 184
Adoptive immunotherapy 247
Adult cloning 421
Affinity chromatography 63
Affinity maturation and generation of memory cell 177
Agents that damage DNA 289
Agricultural biotechnology 415
Allograft 224
Allosteric enzymes 38
Alpha-helix 56
Amino acid classifications 51
Amino acids and charging of tRNAs 343
Aminoacyl tRNA synthetase 343
Amino-terminal sequence determination 60
Ampholytes, polyampholytes, pI and Zwitterion 19
Analysis of DNA structure 75
Anatomical barrier 93
Antagonism 196
Anti parasite immunity 259
Antibacterial immunity 256
Antibody 242
  diversity 135
  mediated effector functions 138
  protein fragments 135
  structure, function, antibody engineering 128
Antigen
  antibody interactions and its applications 151
  receptors (TCR) 183
Antigenic determinants (Epitopes) 107
Antigenic determinants of antibody molecules 134
Antigens 105
  immunogens and immunization 87
Antiviral immunity 253
Applications of
  antigen-antibody reactions 156
  biotechnology 408
  genetic markers 80
  immunology 92
Attenuation 371
Autografts 224
Autoimmune diseases 215
Autoimmunity 215

B

B cell activation, proliferation and differentiati 176
B cell subsets 175
Bacterial infections 256
Basic principles of biochemical reactions 17
Basic tools in bioinformatics 424
Benificial equipments 402
Beta -sheets 57
Binding of 43s pre-initiation complex to mRNA 355
Biocatalysts 419
Biochemical pathway data 430
Biofertilizers 417
Biohazard in biotechnogy laboratory 405
Biological consequences of the antigen—antibody interaction 155
Biology of T lymphocytes and generation of cell-mediated immune response 180
Biomining 420
Biopesticides 416
Biosensors 421
B-lymphocytes and generation of humoral immunity 172
Boolean searching 424
Brief history 114
Buffering 19

C

Calibration 165
Cancer immunotherapy 245
Carbohydrate nomenclature 43
Carbohydrates 43
Carboxy-terminal sequence determination 61
Cardiff human gene mutation database 433
Cascade induction 196
Catalytic antibodies (Abzymes) 142
Cathepsins 123
Cd_4^+ helper/inducer 191
Cd_8^+ suppressor/cytotoxic T cells 192
Cell
  migration and homing 205

shape 3
size and the reason for multicellularity 4
surface markers 174
walls provide rigid structure around cells 14
Cells
are joined by a variety of intracellular junction 14
don't end at their outer membrane; possess an extracellular matrix (ECM) 14
a brief history 3
Cellulose 47
Centrifugation 400
of proteins 65
Characteristics of tRNAs 341
Chemical
complexity 106
digestion of proteins 62
nature 106
reaction order 30
reactions and rates 29
synthesis of oligonucleotides 75
Chemokine
receptors 205, 208
super family 206
chemokine receptors 205
Chemotaxis 101
Chimaeric and humanized antibody 143
Choosing a suitable vector as a vehicle for gene cloning 380
Chromatography 75
Chronic rejection 231
Classification of enzymes 26
Clinical
manifestations of graft rejection 230
significances 59
Clip 124
Clonal selection, theory 89
CO_2 incubator 398
Cold room 395
Collaboration of proteins at the replication fork: the replisome 285
Communication to the colleagues and associates 443
Complement activation 155
Containment
and disposal 405
facility 396
Contour plot 167
Core enzyme 312
Co-stimulatory ligands 184
Cytokine
and cytokine receptor therapy 202
antagonists 202
receptors 199
general properties 195
Cytoplasm: site of protein synthesis 10
Cytoskeletal system provides internal fibrous structure 12
Cytosolic pathway 121
Cytotoxic cells 95

D

3d molecular structure data 430
Damage bypass 302
Data annotation and formats 427
submission and revision 432
Deciphering the genetic code 336
Depositing a novel micro-organism 447
Designing vaccines for active immunization 264
Destruction 102
Disclosure to a government department 447
Disease susceptibility 125
DNA damage, types 290
DNA
fingerprinting 414
markers for genetic mapping 78
polymerases 282
repair 289
replication 273
replication and repair 273
vaccines 266
DNA, RNA and protein sequence data 430
Documentation of research 442
Dot display 167
Double
immunodiffusion 156
strand break repair 302

E

Effector
cells 94
functions 100
functions of the host immune cells 227
mechanisms of the immune system 91
T cell and their functions 191
Electronic balance 400
Electrophoresis 75
apparatus and power pack 401
of proteins 63
Electrostatic forces 58
Elongation 320, 356
Endocytic pathway 122
Endogenous
and exogenous antigens 107
super antigens 190
Endomembrane system: moving materials into different compartments 11
Enthalpy 21
Environmental biotechnology 417
Enzymatic synthesis of RNA 308
Enzyme
linked immunosorbent assay 162
relative to substrate type 28

and mechanisms of enzyme action 26
as biological catalysts 31
substrate interactions 28
Epitopes recognized by B and T cells 108
Equipments maintenance and handling 397
Essential equipments 397
Establishing a confidential relationship 444
Etiopathogenesis of auto-immune diseases 217
Eukaryotic cell structure and cell components 9
Eukaryotic
promoters 317
RNA polymerases 314
Evasion mechanism of host defence by bacteria 258
Evasion of
anti tumor immunity 242
immune defences 254
Evidences of anti-tumour immune response 236
Exogenous super antigens 190

F

Fas ligand 245
Fatty acids 47
Finishing replication 288
First cell 6
Forces controlling protein structure 58
Foreignness 106
Formation of
43s pre-initiation complex 353
80s initiation complex 355
Functional significance of amino acid R-groups 54
Functions of chemokines 209

G

Gene
cards 432
cloning in biotechnology 387
expression data 431
General
features of cell migration and homing 210
immunosuppressive treatment 231
protocol 403
Genetic code 337
Genetics of MHC 124
Genomic
data 430
organization 115
Glycogen 46
GneBank database 431
Good laboratory practice 393
Google 425
Graft rejection is mediated by cell mediated immunity 225
"Green" plastics 420
Group of cytokines 196
Guanosine derivatives 70

H

Hapten and carrier 109
carrier effect 110
Helicase 278
Helix-turn-helix motif 373
Henderson-hasselbalch equation 18
High molecular weight 105
High performance liquid chromatography (HPLC) 63
Histogram 167
Holoenzyme 313
Hot
blot 425
room 395
Human
applications 408
genome project 421
Hybridoma and monoclonal antibody 141
Hydrogen bonding 58
Hydrophobic forces 58
Hyper acute rejection 230

I

ICAM family 208
Identification of
a clone from a gene library 386
the recombinant clone 385
Immune
diversity 90
evasion by parasites 261
system: an overview 85
Immunity to parasites 259
Immunoelectrophoresis 157
Immunofluorescence 163
Immunogenicity, biological factors 111
Immunoglobulin
classes and sub classes 130
domains and variable regions 132
genes and their rearrangement 133
super family 140
synthesis and class switching 177
Immunological tolerance 216
Immunomodulators 249
Immunopathology 91
Immunoregulation 91
Immunoserveillance and anti-tumour effector mechanism 241
Immunosuppressions by anti-tumour antibody 243
Immunosuppressive
cytokines 245
factors 243
Immunotolerance 232
Incubators 398
Inducible and repressible systems of negative regulation 364

Industrial biotechnology 419
Infection immunity 252
and strategies for vaccine development 251
Inflammation and leukocyte extravasation 212
Inflammatory cells 98
Ingestion 101
Inhibition of enzyme catalyzed reactions 34
Initiation of DNA replication in
E coli 287
eukaryotes 288
the replicon model 286
Innate and adaptive immunity 86
Integrin family 208
Intellectual property 436, 437
rights 436
Ion-exchange chromatography 63
Ionic equilibrium 17
Isograft 224
Isometric plots 168

K

Kinetics of
antibody response after immunization 137
graft rejection 224
tumor growth 242
Kozak sequence 340

L

Lac operon 366
Lactose metabolism and operon system of gene regul 366
Lak cells 247
Laminar flow 395
Laminar flow hood 397
Latching 211
Laws of thermodynamics 20, 22
Leucine zipper motif 374
Linkage disequillibrium 117
Lipids 47
Low immunogenicity 244
Lymphocyte migration and homing 212

M

Macrophages 242
Maintaining confidentiality 443
Major histocompatibility complex 115
Malignant transformation and features of malignant 235
Manipulation of DNA 383
Mechanism of graft rejection 227
Messenger RNA 338
MHC I downregulation 244
MHC
gene polymorphism 125
restriction 121
structure, genomic organisation, antigen proce 114
Michaelis-menton kinetics 32
Micropipettes 401
Mismatch repair (MMR) 301
Molecular composition of cells 42
Monoclonal antibodies 141, 240
Mucin family 207
Multivalent subunit vaccines 267
Mutated gene product 237

N

N gonorrhoeae 252
N Maningitidis 252
National legislation 455
NER and human genetic diseases 300
NER in E coli 297
Neutralization 156
Nk-cells 241
Nk-cells and macrophages 254
Nomenclature and classification 236
Nucleic acid 67
Nucleoside and nucleotide structure and nomenclature 67
Nucleus: locus of DNA, RNA and prot synthesis 10

O

Oncofoetal antigen 240
Ontogeny of
B lymphocytes 172
T lymphocytes 181
Opsonization 155
Optical properties of the amino acids 54
Organelles involved in energy transformations 12
Origin and evolution of cells 5

P

Passive immunotherapy 247
Patenting process in India 450
PCR thermal cycler 400
Peptide bond 55, 358
pH meter 402
Phagocytic cells 96
Phospholipase Cγ (PLC) pathway 188
Phospholipids 48
Physicochemical nature 152
Physiological barrier 94
Placement of charged tRNA in the A-site 355
Planning 393
Plasmalogens 49
Pleiotropy 195
Polarization of cytokine response 200
Polya-tail 340
Polynucleotides 71
Polysaccharide vaccine 265
Positive regulation of lac operon (camp receptor) 368

Precipitation reactions (gel techniques) 156
Primase 281
Prior public use 447
Prokaryotic promoters 316
Prokaryotic RNA polymerase 312
Promoters 316
Protease digestion 61
Proteases 258
Protection of intellectual property 439
Protein
  kinase C (PKC) pathway 189
  structure 56
  synthesis 350
  synthesis inhibitors 361
Proteins 50
  of DNA replication 278
Publishing research work 445
Pubmed 427
Purified antigen (Subunit) vaccine 265

Q

Quality of successful vaccine 263

R

Radial immunodiffusion 157
Radioimmunoassay 160
Ras, MAP kinase and SAP kinase pathway 189
Receptors 98
Recombinant
  antigen vaccine 265
  DNA technology 379
  vector vaccine 266
Redundancy 196
Refrigerators and freezers 399
Regulation
  by alternative σ factors 371
  of B cell development 178
  of enzyme activity 37
  of gene expression 363
  MHC gene expression 117
  of transcription in bacteriophage (λ) 372
  of transcription in eukaryotes 373
Related websites 456
Release-factors 360
Restriction fragment length polymorphism (RFLP) 79
Reversible and irreversible reactions 21
RHO dependent termination 323
Ribosome 347
  cycle 349
  recycling factor 360
Risk assessment 404
RNA
  editing 333
  interference (RNAi) 376
  polymerases 312, 316
Rocket electrophoresis 159
Role of coenzymes 27
Rolling (tethering) 211

S

Safety 404
Safety regulations 405
Search engine algorithms 425
Searching
  biological databases 431
  scientific journals and articles 426
Secretion of cytokines 189
Selectin family 206
Sensitization of the host immune cells 227
Sigma subunit 313
Single
  sequence length polymorphisms (SSLPs) 79
  stranded DNA binding protein 280
  nucleotide polymorphisms (SNPs) 78
Size exclusion chromatography 63
Solvation and hydration shells 19
SOS response 304
Specific immunotherapy 232
Sphingolipids 49
Standard opearting procedures 405
Standard state conditions in biological reactions 22
Starch 47
Sterilization and drying equipments 399
Stop codon 358
Storage 396
Strategies for vaccine development 263
Strategy of gene cloning 379
Structure of DNA 72
Super antigen and T cell activation 190
Surface structures 258
Synergism 196
Synthesis of RNA 310
Synthetic nucleotide analogues 71

T

γ, δ TCR 185
T cell
  activation and signal transduction 187
  co-receptor molecules 184
  receptor and co-receptors 183
  subsets 186
T cytotoxic (Tc) cells 186
T-dependent and T-independent antigens 109
Termination 321, 330
Termination of translation 358
Terminology used in genetic analysis 78
Tertiary structure of tRNA 342
T-helper (Th) cells 186
Thermal properties of DNA 73
Thymic selection 182

Tissue typing 226
Topoisomerases 279
Toxoid vaccine 265
Training 403
Transcription 308
Transcription
coupled repair (TCR) in
eukaryotes 297
enhancers and silencers
375
in eukaryotes 323
in prokaryotes 320
Transcriptional
activator proteins 373
regulation in prokaryotes
364
Translation 335
initiation in
eurkaryotes 352
prokaryotes 350
Translocation 358
Transplantation 233
antigens 226
immunity 223
Treatment of autoimmune
diseases 220
Triacylglycerides 48
Triggering 211
tRNA binding sites of ribosome 349
Tryptophan operon 368
Tumor 235
antigens 239 (f)
associated antigens 238
immunity 235
specific antigens 237

U

Useful equipments 403
Using search engines 424
UV transilluminator 402

V

Van der waals forces 58
Variant chain protein 117
Viral gene products 238
Viral infection 253

W

Wash up 396
Western blot (immunoblots)
159
Whole organism vaccine
264
Wobble hypothesis 344

X

Xenograft 224

Z

Zinc finger motif 374